FLUID DYNAMICS

FLUID DYNAMICS

By

A.K. Sharma

&

A.S. Sharma

DISCOVERY PUBLISHING HOUSE PVT. LTD.

NEW DELHI-110 002

First Published - 2007

Reprinted - 2018

ISBN: 978-81-8356-319-2

Fluid Dynamics

Published by:

DISCOVERY PUBLISHING HOUSE PVT. LTD.
4383/4B, Ansari Road, Darya Ganj
New Delhi-110 002 (India)
Phone: +91-11-23279245, 43596064-65
Fax: +91-11-23253475
E-mail: discoverypublishinghouse@gmail.com
sales@discoverypublishinggroup.com
web: www.discoverypublishinggroup.com

Printed at:
Infinity Imaging Systems
Delhi

Preface

This book "Fluid Dynamics" has been design to cover the syllabus of Mathematics required for the B.Sc. (Hon.), M.Sc. and Engineering students of the Indian Universities. The Subject of Fluid Dynamics deals with the laws Governing the pressure and flow of fluids and the application of these laws to engineering practice. The term fluid includes liquid and gases. The term fluid is applied to all the substance which offer no resistance to change of shape. The subject matter has been arranged as to provide a clear and integrated approach to the subject-care has been taken to make the treatment of the subject simple and accessible to the average students.

Suggestion for the improvement of the book will always be most welcome.

Authors

CONTENTS

1

Basic Relationships of Thermodynamics

ISOTHERMAL AND ADIABATIC PROCESSES

When the physical properties of a gas such as pressure, density and temperature are changed due to compression or expansion of the of the gas, it is said to undergo a process. A gas may be compressed or expanded either by isothermal process or by adiabatic process. An isothermal process is that in which the temperature is held constant and it is governed by Boyle's Law according to which

$$PV = P_1V_1 = P_2V_2 = \text{Constant}$$

Or $(P/P) = (P_1/P_1) = (P_2/P_2) = \text{Constant}$

Or $(P/W) = (P_1/W_1) = (P_2/W_2) = \text{Constant}$ ...(1)

On the other hand if during the course of a process the gas neither absorbs heat from, nor given heat to, its surroundings then the process is said to be adiabatic. An adiabatic process is governed by the following law.

$$PV^k = P_1V_1{}^k = P_2V_2{}^k = \text{Constant}$$

Or $(P/P^k) = (P_1/P_1k) = (P_2/P_2{}^k) = \text{Constant}$

Or $(p/w^k) = (p_1/w_1) = (p_2/w_2{}^k) = \text{Constant}$ ...(.2)

In the above equation P is pressure, v is specific volume, P is mass density, w is weight density, and k is adiabatic index. The adiabatic index is given constant by $k = (C_p/C_k)$, The ratio of the specific heat at constant pressure Cp and the specific heat at constant pressure Cp and the specific heat constant volume Cv, for air and other diatomatic gasses in the usual ranges of temperature and pressure K = 1.4.

The process is said to be reversible if the gas and its surroundings could subsequently be completely restored to their initial conditions by adding to (or extracting from) the gas exactly the same amount of heat and work taken from process and it is often termed as isentropic process because as indicated later in this process the entropy does not change. How ever, a reversible process is ideal, *i.e.* it is never achieved in practice.

There are some actual processes which are not isentropic (so that PV^k is not constant) but in which the relation between P and v (or p) is given to a reasonable approximation by Pv^n = con(or p/p^n = constant) where n is a positive constant. Such process are termed as pleiotropic processes

Equation of State

The density p of a particular gas is related to its absolute pressure p and absolute (or thermodynamic) temperature T by the equation of state, which for a perfect gas taken the form.

$$k = p^{RT};$$

or $$pV = m^{RT}$$

in which R is gas constant and V is the volume occupied by the mass m of the gas. The absolute (or Thermodynamic) temperature is expressed in 'Kelvin' (K) when the temperature is measured in C and it is given by

$$T^{o}\ (abs) = TK = 273.\ 15 + t^{o}\ C$$

Future details of the equation of state including the dimensional formula of the gas constant R and its value for air are given in chapter.

Internal Energy

The energy possessed by the molecules of a fluid due to molecular activity is termed as internal energy or microscopic energy. It is to be distinguished from the external or macroscopic energies viz., kinetic and potential energies associated with the fluid mass. The internal energy is a terms of temperature, that is high-or low-temperature implies high or low internal energy respectively. If some quantity of heat is supplied to a certain mass of gas, a part of it may be stored in the gas as internal energy thus producing the rise in temperature of the gas and the remaining part of the heat supplied may be utilized in increasing the volume of the gas. Thus doing some external work.

First Law of Thermodynamics

The first law of the Thermodynamics states that H supplied to a gas must be equal to the increase in the energy $(I_2 - I_1)$ of the gas plus the external work W done by the gas is expanding. Thus

$$H = (I_2 - I_1) + W \quad ...(1.3)$$

However, if instead of heating, the gas is called. Then the heat will be rejected from the gas and hence H will be negative. The value of $(I_2 - I_1)$ may also be negative if the internal energy of the gas decreases, further W is negative, if the gas is compressed and consequently contracts is volume, which means that work is done on the gas from external sources.

For isothermal process, since the temperature remains constant $(I_2 - I_1) = O$ and hence

$$H = W \quad ...(1.3a)$$

on the other hand, for adiabatic process there being neither addition nor removal of heat H = O, and hence.

$$W + (I_2 - I_1) = 0$$

$$W = (I_1 - I_2) \quad ...(1.3b)$$

The work done by a gas in expanding (or on a gas in contracting) is given by

$$W = \int_{V_1}^{V_2} Pdv \quad ...(1.4)$$

For isothermal process introducing the value of p from equation 1.1 in equation 1.4 and integrating it we obtain

$$W = P_1V_1 \log_e (v_2/v_1) = P_1 v_1 \log_e (p_1/p_2)$$

$$\text{or } W = k \log_e (v_2/v_1) = k \log_e (p_1/ p_2) \quad ...(1.4a)$$

in which K is the constant of eduation 1.1. Equation 1.4 (a) gives the work done per unit mass or per unit weight of the gas depending on the specific volume mass or per unit weight of the gas respectively. Similarly for adiabatic process introducing. The value of p from equation 1.2 in equation 1.4 and integrating it we obtain.

$$W = \frac{1}{k-1}(p_1v_1 - p_2v_2) \quad ...(1.4b)$$

Equation 1.4 (b) gives the work done per unit mass of the gas however, if the specific volume is considered as the volume per unit weight of the gas then equation 1.4 (b) gives the work done per unit weight of the gas and it may then equation 1.4 (b) gives the work done per unit weight of the and it may then be expressed as

$$W = \frac{1}{k-1}\left(\frac{p_1}{w_1} - \frac{p_2}{w_2}\right)$$

Entropy

Entropy of a gas may be defined as the measure of the maximum heat energy available for conversion into work. It is a property of the gas and it varies with its absolute temperature and its state if ΔH is the heat transferred per unit weight of gas in a small interval of time and if T is the absolute temperature of the gas at the instant, then the change in the entropy $\Delta \phi$ is defined by the relation

$$\Delta \phi = \frac{\Delta H}{T}$$

$$\text{or} \quad (\phi_2 - \phi_1) = \int_{T_1}^{T_2} \frac{\Delta H}{T} \qquad \text{...(1.5)}$$

For isothermal process since the temperature remains constant

$$(\phi_2 - \phi_1) = \frac{H}{T} \qquad \text{...(1.5a)}$$

But in adiabatic process since there is no transfer of heat, DH = 0, hence,

$$(\phi_2 - \phi_1) = 0; \ \phi = \text{constant} \qquad \text{...(1.5b)}$$

A Process in which the entropy does not change is termed as isentropic process However, the entropy does not change in an adiabatic process, only if it is frictionless. A frictionless adiabatic process is therefore isentropic, which is also reversible.

Enthalpy

The sum of the internal energy and pressure – specific volume product (or pressure divided by mass density, or pressure divided by weight density) (Converted into heat units) is termed as enthalpy. It is a purely mathematical quantity, the consideration of which simplifies certain calculations, since in several problems of Thermodynamics this sum occurs.

Momentum Equation

This is because is momentum equation the change in momentum flux is equated to the force required to cause. The momentum flux is the product of the mass flux (PAV) and the velocity V. By the continuity equation. The mass flux (pav) is constant from section to section. As such the moment equation is completely independent of the compressibility effects and hence for compressible fluids too the momentum equation (for say x direction) may be expressed as

$$\Sigma F_x = (PA\ VV_x)_2 - (PaVV_x)_1$$

Propagation of Elastic ways Due to compression of fluid, velocity of sound If the pressure is rapidly transmitted throughout. The rest of fluid. It so happen because any change in pressure causes and elastic or pressure wave rejecting this change to travel through the fluid at a certain velocity. In a completely incompressible fluid the transmission of new pressure wave is propagated with finite velocity But even in a compressible fluid the changes of pressure are transmitted quite rapidly and as indicated below the velocity of propagation of the elastic wave in a compressible fluid is equal is equal to the velocity of the sound in that fluid medium.

Consider a long rigid tube of uniform cross sectional area A fitted with a past on at on end. The tube is filled with a compressible fluid initially at rest. If the piston is moved suddenly to the right with a velocity V_P, a pressure wave would be propagated Through a distance dx (= V_P dt) while the pressure wave travels a head of the piston through a distance

$$dL = (Cdt).$$

2

Compressible Fluid Flow

INTRODUCTION

It is a established fact that liquids do not change appreciably their volume when subjected to changes in pressure and temperature and that they can be treated incompressible (or constant density fluids) for all practical purposes. On the other hand, the density of gases and vapours varies with both pressure and temperature and hence they are compressible. However, there are many example of flow of gases in which density does not change appreciably, and the theory relating to constant density fluids adequately describes the flow phenomena for such cases.

When the pressure changes produced by flow are large, the assumption of fluid being incompressible is no longer valid and the corresponding changes in density must be considered. It is known through experience that the compression and expansion of a gas involve work done on and by the gas respectively, and this results in changes in the temperature of the gas. This introduces thermodynamics effects in fluid flow problems, and hence the laws of fluid mechanics alone are insufficient for complete solution of compressible flow problems.

In addition, the principles of thermodynamic are necessary to completely determine the quantities involved in a flow problem. The study of the so-called 'compressible fluid flow' is thus a lot more complex than that of constant density fluid flow. In this chapter an attempt has been made to introduce certain basic elements of compressible flow.

There is no sharp line between flows in which the density changes are important and those in which they are not so important. Appreciable changes in density of a gas may be expected :

(1) If the velocity (either of the gas itself or a body moving through it) approaches or exceeds the velocity of sound through the gas,

(2) If the gas is subjected to sudden acceleration or

(3) If there are large changes in elevation. The last conditions is rarely encountered except in meteorology and will not be considered in this chapter.

THERMODYNAMIC PROCESSES OF PERFECT GASES

(1) **Isothermal Process :** The compression and expansion of a gas may take place according to various laws of thermodynamics. If the temperature is held constant, the process is known as 'isothermal' and the pressure-density relationship is given by Boyle's law, Eq. (1)

$$\frac{p}{\rho} = \text{constant} \qquad \text{...(1)}$$

(2) **Adiabatic process :** If the process is such that no heat is added to or with drawn from the gas (heat transfer is zero), it is said to be an 'adiabatic' process.

If the process is reversible (or frictionless) adiabatic, it is called *isentropic* since it is accompanied by no change in entropy, and the following relationship between pressure and density, given by Eq. (2), exists

$$\frac{p}{\rho^k} = \text{constant} \qquad \text{...(2)}$$

where $k = C_v/C_v$

is the ratio of specific heats.

(3) **Polytropic Process :** In general we can express a relation between the pressure and the density for each of the preceding processes by

$$\frac{p}{\rho^n} = \text{constant} \qquad \text{...(3)}$$

where n is a given for each process. Thus,

(i) if n = 0, p = constant, the process is isobaric.

(ii) if n = 1, T = constant, the process is isothermal .

(iii) if n = k, Entropy = constant, the process is isentropic.

THERMODYNAMIC CONCEPTS

Equation of State

A perfect gas is defined as the fluid that has constant specific heats and follows the law given by,

$$p = \rho RT$$

in which p and T are absolute pressure and the absolute temperature respectively, ρ is the density and R the gas constant is known as the *equation of slate* for a perfect gas, and the gas which obeys it is called a perfect gas.

INTERNAL ENERGY AND ENTHALPY

The molecular energy of a compressible fluid is caused by the activity of the molecules, which increases as the temperature of the fluid increases. In a gas, this molecular activity also creates the pressure which is that part of molecular energy usually converted into mechanical work. In thermodynamics, the molecular energy is known as the enthalpy

$$h = u + pv = u + p/\rho \quad ...(1)$$

in which h is the enthalpy or molecular energy per unit mass, v is the specific volume, u is the internal energy per unit mass which is that part of the molecular energy other than the pressure energy per unit mass p/ρ. The internal energy is the energy associated with the kinetic energy of molecules and the forces between the m and depends upon the temperature; high and low temperatures implying high or low internal energies respectively.

SPECIFIC HEATS

The specific heat C is defined as the quantity of heat needed to raise a unit mass of fluid by a unit temperature,

$$C = \frac{dq}{dT}$$

where dq is the heat added to a unit mass of the system and dT is the rise in temperature associated with it.

The value of the specific heat depends upon the process by which the heat is added. We are particularly interested in the specific heats determined at constant volume C_v and at constant pressure Cv. Thus,

$$Cv = \left(\frac{dq}{dT}\right)_{\text{constant volume}} \quad ...(2a)$$

$$Cv = \left(\frac{dq}{dT}\right)_{\text{constant pressure}} \quad ...(2b)$$

REVERSIBLE AND IRREVERSIBLE PROCESS

When the physical properties of a gas (*e.g.* pressure, density ad temperature) are changed, it is said to undergo a process. The process is said to be reversible if the gas and its surroundings could subsequently be completely restored to their initial conditions by adding to (or extracting from) the gas exactly the same amount of heat and work taken from (or added to) it during the process.

A reversible process is an ideal one which is never achieved in practice. Viscous effects and friction dissipate mechanical energy as heat which cannot be converted back to mechanical energy without further changes occurring. Also heat passes by conduction from hotter to cooler parts of the system considered, and heat flow in the reverse direction is not possible. In practice, therefore, all processes are irreversible.

ENERGY

Entropy of a gas may be defined as the measure of availability of heat energy for transformation into mechanical work. If dq is the heat given to the fluid per unit mass due to heat absorbed by the fluid is the given by

$$ds = \frac{dq}{T} \quad ...(3)$$

where T is the absolute temperature of fluid and ds is the change in entropy per unit mass of fluid.

If dq_e is the heat added externally to the fluid per unit mass and dq_t is the heat developed internally per unit mass because of dissipation of mechanical energy into heat due to friction, then the total heat received by the fluid is

$$dq = dq_s + dq_t \quad ...(4)$$

From Eqs. (3) and (4), we obtain

$$ds = \frac{dq_e}{T} + \frac{dq_t}{T} \quad ...(5)$$

Constant entropy (ds = 0) requires dq = 0, a condition which is achieved if no heat passes between the fluid and its surroundings, *i.e.* $dq_e = 0$, and no mechanical energy is converted into heat energy by friction, *i.e.* $dq_t = 0$. In a frictionless process, which is also called a reversible process, $dq_t = 0$.

The process, in which the fluid does not absorb heat from, or give heat to its surroundings, is said to be *adiabatic*. Hence for an adiabatic process $dq_e = 0$. A process in which the entropy does not change (*i.e.* ds = 0) is known as *isentropic*. A frictionless adiabatic process is, therefore, is entropic. In practice such a process is closely approximated if there is little friction and the changes occur rapidly enough for little heat transfer to take place across the boundaries.

If the heat added from external sources is equal to zero (dqe = 0), so the process is adiabatic, from Eq. (5) we obtain

$$ds = \frac{dq_t}{T} > 0 \qquad ...(6)$$

An isentropic process is not equivalent to an adiabatic process.

Since the loss of mechanical energy into heat always corresponds to a positive heat of dissipation ($dq_t > 0$), it is evident that entropy always increases in an adiabatic process ($dq_e = 0$).

FIRST LAW OF THERMODYNAMICS

The principle of conservation of energy, which is generally referred to as the First Law of Thermodynamics, is the energy which can neither be created nor destroyed but can only be changed in form. Experiments have demonstrated that heat is a form of energy which can be expressed in terms of mechanical units through the mechanical equivalent of heat and *vice versa*.

If a small quantity of heat is added to a simple homogeneous system (a system composed of a single gas or liquid whose thermodynamic properties are uniform throughout in all respects), this energy by the first law of thermodynamics can be converted into other forms of energy, such as an increase in the kinetic energy of molecules (the internal energy), kinetic energy of the system proper, and external mechanical work done by the system.

It is assumed that the system is stationary or nearly stationary, so that change in kinetic energy of the system need not be considered. Under these conditions the First Law of Thermodynamics states that a small quantity of heat added to any simple system is equal to the change and kinetic energy of molecules plus the internal mechanical work done by the system. Thus,

$$dq = du + dw \quad ...(1)$$

where dq is the amount of heat added to a unit mass of the system, du is the increase in the internal energy per unit mass of the system, and dw is the external work done per unit mass by the system. If p is the pressure and v the volume per unit mass, then Eq. (1) may be written in the form

$$dq = du + p\,dv = du + pd\left(\frac{1}{\rho}\right) \quad ...(2)$$

The relations among the specific heats Cv, Cv, their ratio, k, and the gas constant R are given by

$$C_v - C_v = R \quad ...(3)$$

$$C_v = \frac{k}{k-1}R \quad ...(4)$$

$$Cv = \frac{1}{k-1}R \quad ...(5)$$

SECOND LAW OF THERMODYNAMICS

The second law of thermodynamics which, like the first law, is the result of observation and experiment may be stated in the following manner:

1. Heat cannot be transferred from a body of lower temperature to one of higher temperature without other simultaneous changes occurring in the two systems or their environments.
2. Heat from a single sources cannot be converted into mechanical work without simultaneous changes in the system or the environment.
3. The transfer of mechanical work into heat by friction is irreversible.
4. It an isolated system (no heat transfer), the entropy cannot decrease. This can be easily seen . On the other hand, the entropy is always increased if the process is irreversible.

FUNDAMENTAL EQUATIONS GOVERNING COMPRESSIBLE FLUID FLOW

In the study of incompressible fluid flow we need only to find the velocity and pressure at every point in space. In case of flow of a compressible fluid we are interested in determining the velocity, pressure, density and temperature of fluid. In order to determine these quantities, one vector and three scalars, we must have one vector equation and three scalar equations. These equations are provided by :

1. Equation of motion,
2. Equation of continuity,
3. Equation of state for a perfect gas, and
4. Energy equation.

EQUATIONS OF MOTION

The equation of motion along a streamline for compressible fluid flow are the same as developed for incompressible fluid.

EQUATION OF CONTINUITY

The equation of continuity in Cartesian co-ordinates for fluid flow in general is derived for compressible fluids may be expressed as

$$\frac{\partial \rho}{\partial t} + \frac{\partial(\rho u)}{\partial x} + \frac{\partial(\rho v)}{\partial y} + \frac{\partial(\rho w)}{\partial z} = 0$$

The mass rate of flow along a narrow stream tube may be expressed by Eq. (1) as

$$\rho AV = \text{constant} \qquad \text{...(1)}$$

The mass rate of flow along a narrow stream tube may be expressed by Eq. (2) as

$$\rho AV = \text{constant} \qquad \text{...(2)}$$

differentiating Eq. (2) and dividing throughout by pAV, we obtain

$$\frac{dp}{\rho} + \frac{dA}{A} + \frac{dV}{V} = 0 \qquad \text{...(3)}$$

The equation of state for a perfect gas is expressed by Eq. (2).

ENERGY EQUATION FOR COMPRESSIBLE FLUIDS

Although the momentum equation is completely independent of compressibility effects, the energy equation is very much dependent upon variation in density. This equation may be applied to any two points along a streamline. If no heat is added to (or extracted from) the fluid between these two points, and no mechanical work is done, we may put q = 0, and w_s = 0 in Eq. (2) and obtain

$$0\left(\frac{V_2^2}{2}+\frac{p_2}{\rho}+gZ_2\right)-\left(\frac{V_2^2}{2}+\frac{p_2}{\rho}+gZ_2\right)+u_2=u_1 \qquad ...(1)$$

Since enthalpy per unit mass is given by Eq. (2.1) as $h = u + p/\rho$ and as points t and 2 were arbitrarily chosen, Eq. (1) may be expressed as

$$h+\frac{V^2}{2}+gZ=\text{constant} \qquad ...(2)$$

Eq. (2) expresses the general form of the energy equation for steady, adiabatic flow in which the fluid (gas, liquid or vapour) neither does work on its surroundings nor has work done on itself. If, however, the fluid is a perfect gas, then the $h = C_vT$ and we obtain

$$C_vT+\frac{V^2}{2}+gZ=\text{constant} \qquad ...(3)$$

From the equation of state for a perfect gas,

$$p=\rho RT \text{ or } T=\frac{p}{\rho R}=\frac{p}{\rho(C_v-C_v)}$$

Substituting for T in Eq. (3)

$$\frac{C_v}{C_v-C_v}\frac{p}{\rho}+\frac{V^2}{2}+gZ=\text{constant} \qquad ...(4)$$

Using $C_v/C_v = k$, the above equation takes the form

$$\frac{k}{k-1}\frac{p}{\rho}+\frac{V^2}{2}+gZ=\text{constant} \qquad ...(5)$$

The potential energy in a compressible fluid exists because of the elevation of the fluid. If the fluid is a gas, the potential energy term will be negligible as compared to other forms of energy because of very low specific weight of the gas. The potential energy per unit mass gZ may, therefore, be omitted from Eq. (5), and the resulting energy equation is,

$$\frac{k}{k-1}\frac{p}{\rho}+\frac{V^2}{2}=\text{constant} \qquad ...(6)$$

and Eq. (2) becomes, $$h+\frac{V^2}{2}=\text{constant} \qquad ...(7)$$

Eq. (6) is more useful when expressed in terms of temperature; using $p = \rho RT$, one gets

$$\frac{k}{k-1}RT + \frac{V^2}{2} = \text{constant} \qquad ...(8)$$

It may be noted that the above equations are applicable to steady adiabatic flow in which the gas neither does work on its surrounding nor has work done on itself.

CONTROLLING PARAMETERS IN COMPRESSIBLE FLOW

There are four para meters which control the flow phenomena of viscous compressible fluids. They are :

(1) The Mach number,

(2) The ratio of specific heats,

(3) the Reynolds number, and

(4) The Prandtl number.

The Mach number is a measure of the effect of fluid compressibility and is defined as the ratio of the free-stream velocity (or the velocity of the body moving through the fluid) and the corresponding velocity of sound in the fluid medium and is expressed as

$$M = \frac{V}{C} \qquad ...(1)$$

in which V is the velocity of fluid (or of the moving body) and C is the velocity of sound in the fluid medium. For isentropic flow from Eqs. (1) and (2)

$$C = \sqrt{kp/\rho} = \sqrt{kRT} \qquad ...(2)$$

The second parameter is the ratio of specific heats of the fluid at constant pressure to that at constant volume, and is expressed by

$$k = \frac{C_v}{C_v} \qquad ...(3)$$

which is a measure of the relative internal comlexity of the fluid molecules.

The third parameter is the familiar Reynolds number which is measure of the viscous effects of the fluid. The fourth parameter is the ratio of the kinematic viscosity to the thermal diffusivity of the fluid and is known as the Prandtl number

$$P = \frac{\mu / \rho}{K / \rho C_v} \quad ...(4)$$

in which P is Prandtl number K is the thermal conductivity. The Prandtl number is a measure of the relative importance of heat conduction and viscosity of a fluid.

In order to have compressible flows around two bodies completely similar not only must the bodies be geometrically similar, but also the four non-dimensional parameters listed above namely the Mach number, ratio of specific heats, Reynolds number and the Prandtl number in the flow systems be equal.

For flow of an inviscid (non-viscous) compressible fluid, the Reynolds number and the Prandtl number are out of consideration and we are thus left with the remaining two controlling parameters, namely the Mach number and the ratio of the specific heats.

COMPRESSIBLE FLOW REGIMES

As already stated one of the most important non-dimensional flow parameters occurring in the analysis of compressible flow is the Mach number M. Based on the value of the Mach number, five flow regimes are generally defined as follows:

1. **Incompressible flow regime:** The Mach number is small as compared to unity ($0< M < 0.3$) for a perfect gas. In this range the effects of compressibility are generally considered to be negligible.
2. **Subsonic flow regime :** The Mach number is less than unity but of sufficient magnitude to be outside the incompressible flow regime ($0.3 < M < 1.0$).
3. **Transonic flow regime:** The Mach number is in the immediate neighbourhood of unity, that is, it ranges from values slightly less than one to slightly greater than one ($0.9 < M < 1.1$).
4. **Supersonic flow regime :** The Mach number is larger than one ($M > 1$).
5. **Hypersonic flow regime :** The Mach number is very much larger than one ($M > 1$). The dividing line between the supersonic and hypersonic regimes maybe taken as approximately 5.

MACH CONE, MACH LINE AND SHOCK WAVE

A small pressure disturbance (or pressure impulse) in a stationary compressible fluid propagates with the speed of sound uniformly in all directions. Hence it may be stated that the pressure disturbances are propagated as spherical wave surfaces.

Consider a tiny object (such as a small projectile) moving from right to left in a stationary fluid at a velocity smaller than the velocity of sound $(0 < V < 0)$. The motion of the object creates pressure disturbances which are propagated spherically outward from the object with the speed of sound C. If the object were not in motion, the wave fronts would spread out spherically and would have the positions shown in Fig. 2.1(a) for successive time intervals $dt = (t_2 - t_1) = (t_2 - t_2)$.

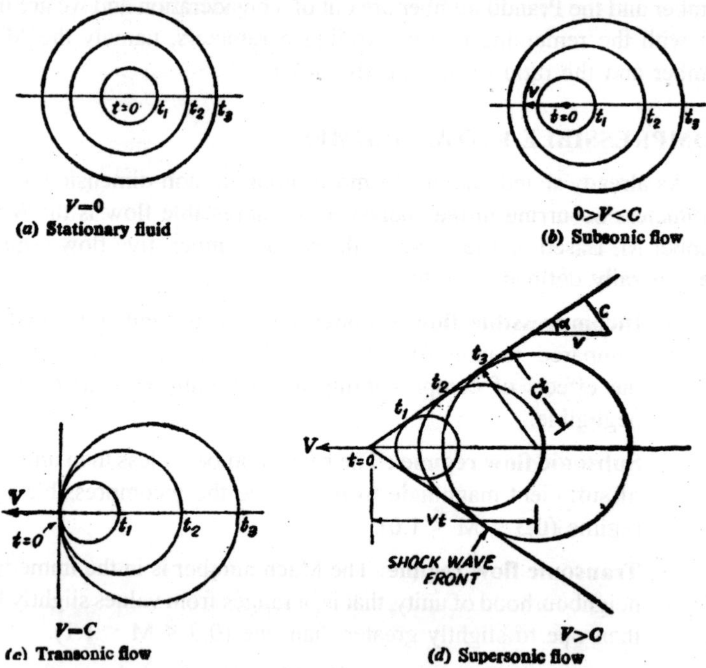

Fig. 2.1 : Wave patterns for a tiny object moving with different velocities.

The position of wave fronts for 0 < V < C is shown in shown in Fig. 1(b). The portion of the wave of front of the object progresses more slowly than for Fig. 2.1 (a), the velocity now being (C – V). If the velocity of objects is increased to the speed of sound C, so that V = C, (e), the waves do not move ahead of the object but remain stationary. The wave fronts merge to form a plane wave front which is tangential to the upstreams side shown, and the downstream wave moves at a speed of (V + C).

In this case, the pressure wave is unable to move upstream against the on coming flow and the fluid a head of this plane wave front is not influenced by the motion of body. Finally, when the velocity of the object is increased beyond the speed of sound so that V > C and M < 1, the individual pressure waves now combine to form a conical wave front.

This conical pattern of wave front is known as the Mach cone. Fluid ahead of the cone is undisturbed but suddenly undergoes changes in pressure, temperature and density, as it passes through the *Mach cone.* An actual wave through which the flow experiences sudden changes in density, pressure and temperature is called a ***shock wave.***

We have thus seen that the effect of pressure disturbance is restricted to the interior of the envelope of the spherical pressure waves which is a cone whose vertex coincides with the sources of disturbance (the assumed small projectile) and whose semi-vertex angle known as the Mach angle, is given by Eq. (1).

$$\sin \alpha = \frac{C}{V} \qquad \text{....(1)}$$

In two dimensional flow, the Mach cone reduces to a pair of intersecting lines each of which is called *Mach line* or *Mach wave,* From Eq. (1), it is evident that the Mach number,

$$M = \frac{V}{C} = \frac{1}{\sin \alpha} = \text{cosec a} \qquad \text{...(2)}$$

The Mach cone divides the fluid into two regions:

(1) the region inside the cone which is disturbed by the motion of the projectile and

(2) the region outside the cone which remains undisturbed.

The former region was referred to by von Karman as the *zone of action* and the latter as the *zone of silence.* This phenomenon of shock wave characterizes the nature of supersonic flow.

If an aeroplane is flying at a subsonic speed, the noise created by its engines can be heard long before the plane reaches directly overhead;

however, the situation created by a supersonic aeroplane is entirely different as the observer cannot hear the noise until the plane has crossed him.

SOLVED EXAMPLES

Example 1:

Find the sonic velocity for the following fluids :

(i) Crude oil of specific gravity 0.8 and bulk modulus of 15,600 kgf/cm²

(ii) Mercury having a bulk modulus of 270,00 kgf/cm².

(iii) A schlieren photograph showing the wave formed by a bullet moving in air gave a Mach angle of 40°. Find the speed of the bullet if the pressure and temperature of the atmosphere are 0.9 kgf/cm² and – 2° C respectively.

Solution:

The sonic velocity in a liquid is given by

$$C = \sqrt{\frac{k}{\rho}}$$

(i) For crude oil,

$$C = \sqrt{\frac{15,600 \times 10^4}{\left(\frac{0.8 \times 1000}{9.81}\right)}} = 1383 \text{ m/s}$$

(ii) For mercury,

$$C = \sqrt{\frac{27,000 \times 10^4}{\left(\frac{136 \times 1000}{981}\right)}} = 1395.5 \text{ m/s.}$$

(iii) The sonic velocity in air under the given conditions is :

$$C = \sqrt{kRT} = \sqrt{1.4 \times 287 \times (273 - 2)} = 329.98 \text{m/s}$$

Mach angle represents the half angle of the Mach cone, which is related to the Mach number

$$M = \frac{V}{C} = \frac{1}{\sin \alpha}$$

$$\therefore \qquad V = \frac{C}{\sin\alpha} = \frac{329.98}{\sin 40°} = 513.35 \text{ m/s}$$

The speed of the bullet is 513.35 m/s.

Example 2:

A supersonic aircraft flies at an altitude of 2000 m where the air temperature is 3°C. Determine the speed of the aircraft if its sound is heard 4.5 seconds after its passage over the head of an observer.

Solution:

With reference to the figure, let O, represent the observer and P the position of aircraft just vertically over the observer. After 4.5 seconds, the aircraft reaches a position represented by O. Line OQ represents the wave front and a the Mach angle.

The velocity of sound in air

$$C = \sqrt{kRT} = \sqrt{1.4 \times 287 \times (273+3)} = 333 \text{ m/s}$$

From the figure,

$$\tan\alpha = \frac{2000}{4.5V}$$

$$\text{but } M = \frac{V}{C} = \frac{1}{\sin\alpha}$$

$$\text{or} \quad V = \frac{C}{\sin\alpha}$$

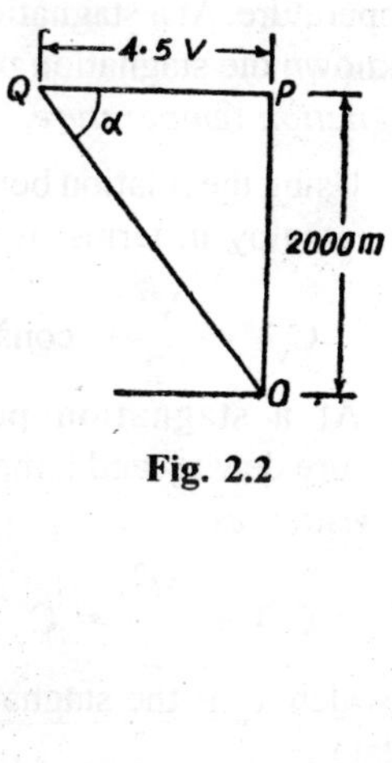

Fig. 2.2

Inserting the value of V in (1),

$$\tan\alpha = \frac{2000\sin\alpha}{4.5 \times 333} \quad \text{or } \cos a = 0.7492$$

$$\sin\alpha = \sqrt{1-\cos^2\alpha} = 0.662^3$$

$$\therefore V = \frac{333}{0.6623} = 502.79 \text{ m/s}$$

The speed of aircraft is 502.79 m/s.

PRESSURE EQUATION OR BERNOULLI'S EQUATION

If p_0 and ρ_0 are the pressure and density respectively at a point where the fluid is at rest,

$$\frac{k}{k-1}+\frac{p}{\rho}+\frac{V^2}{2}=\frac{k}{k-1}\frac{p_0}{\rho_0} \quad ...(1)$$

Eq. (1) is generally referred to as the pressure equation or the Bernoulli's equation for compressible adiabatic flow.

STAGNATION POINT AND ITS PROPERTIES

A stagnation point is characterized by existence of zero velocity at that point; we know that a stagnation point is caused at the tip of Pitot-tube. When a Pitot-state tube is used to measure the velocity of a constant density fluid, the stagnation pressure and the static pressure need not be separately measured, it is sufficient to measure their difference. A high velocity gas stream, however may undergo an appreciable change of density in being brought to rest at the front of the Pitot-static tube, and under these circumstances stagnation and static pressure must be separately measured. In a compressible fluid flow, since the pressure, density and temperature are interrelated, a change in pressure also affects the temperature. At a stagnation point where the velocity is zero, the pressure in known the stagnation pressure and the temperature there is called the *stagnation temperature.*

Using the relation between the enthalpy and velocity and substituting for enthalpy in terms of temperature ($h = C_v T$),

$$C_vT + \frac{V^2}{2} = \text{constant} \quad ...(2)$$

At a stagnation point where the velocity is zero, and the pressure density and temperature are p_0, ρ_0, T_0 respectively. Eq. (2) may be written as

$$C_vT + \frac{V^2}{2} = C_v\,T0 \quad ...(3)$$

in which T_0 is the stagnation temperature. Dividing Eq (3) by CvT, we obtain

$$1 + \frac{V^2}{2C_v^T} = \frac{T_0}{T}$$

we know from that for isentropic flow of a perfect gas $C^2 = kRT$ and from Eq. (2) $C_v = kR\,(k-1)$. Eq. (3) now takes the form

$$1 + \frac{k-1}{2}\frac{V^2}{C^2} = \frac{T_0}{T}$$

and from the definition of Mach number we may write

$$\frac{T_0}{T} = 1 + \frac{k-1}{2}M^2 \quad ...(4)$$

Thus the ratio of the stagnation temperature to the temperature of the undisturbed stream (which is called the static temperature) is direct function of the Mach number and k.

For an isentropic flow of a perfect gas we have the following relations:

$$\frac{p}{\rho^k} = \text{constant, and } p = \rho RT.$$

eliminating r between the two equations, we may write

$$\frac{T^2}{T_1} = \left\{\frac{p^2}{p_1}\right\}\frac{k-1}{k} \quad ...(5)$$

Using Eq. (5), the ratio of stagnation pressure p0 to the undisturbed stream pressure p (known as the static pressure) can be related directly to the temperature ratio T_0/T of Eq. (4) :

$$\frac{p_0}{p} = \left(\frac{T_0}{T}\right)^{\frac{k}{k-1}} = \left(1 + \frac{k-1}{k}M^2\right)^{\frac{k}{k-1}} \quad ...(6)$$

Eq. (6) shown that the pressure ratio p_0/p is also a function of M and k.

Finally, we may find the ratio of the stagnation density p_0 and the static density ρ for isentropic flow

$$\frac{\rho_0}{p} = \left(\frac{p_0}{p}\right)^{\frac{1}{k}} = \left(1 + \frac{k-1}{2}M^2\right)^{\frac{1}{k-1}} \quad ...(7)$$

Like temperature ratio T_0/T and pressure ratio p_0/p, the density ratio ρ_0/ρ is also a function of the Mach number M and the ratio of specific heats k. Eqs. (4), (6) and (7) are not valid for supersonic flow because of the shock wave forming a head of the Pitot tube and so the fluid is not brought to rest isentropically.

If the right hand side of Eq. (6) is expanded by the binomial theorem, we obtain

$$\frac{p_0}{p} = 1 + \frac{k}{2}M^2 + \frac{k}{8}M^4 + ...$$

$$= 1 + \frac{kM^2}{2}\left[1 + \frac{M^2}{4} + ...\right]$$

$$p_0 = p + \rho\frac{p^k M^2}{2}\left[1 + \frac{M^2}{4} + ...\right] \qquad ...(8)$$

Comparison of Eq. (8) with the stagnation pressure $p_0 = p + \frac{pV^2}{22}$ for a constant-density fluid shows that the effects of compressibility have been isolated in the bracketed quantity and that these effects depend only upon the Mach number. The bracketed quantity may thus be considered a *compressibility correction factor*. It is seen that for M< 0.2, the compressibility affects the pressure difference (p_0 –p) by less than 1% and the simple formula for flow at constant density is then sufficiently accurate. For larger value of M as the terms of the binomial expansion become significant, the compressibility effect must be taken into account. The Pitot-static tube used for measuring aircraft speed needs calibration to take into account the effects of compressibility when the much number exceeds a value of about 0.3.

Example 3:

Air has a velocity of 100 km/hr at a pressure of 0.1kgf/cm² vacuum and a temperature of 47°C. Compute its stagnation properties, and the local Much number. Take atmospheric pressure = 1 kgf/cm², R = 29.25 m kgf/ kg° K and k = 1.4.

What would be the compressibility correction factor for a pitot-static tube to measure the velocity at a M number of 0.8?

Solution:

Absolute temperature of air = 273 + 47 = 320°K

Sonic velocity $C = \sqrt{kRT} = \sqrt{1.4 \times 9.81 \times 29.25 \times 320}$

$= 358.53$ m/s

Velocity of air $V = \frac{1000 \times 1000}{60 \times 60} = 277.78$ m/s

Mach number $M = \frac{V}{C} = \frac{277.78}{358.53} = 0.7747$

The stagnation pressure is given by Eq. (6),

$$\frac{p_0}{p} = \left(1 + \frac{k-1}{2} M^2\right)^{k/k-1}$$

$$= \left(1 + \frac{1.4-1}{2} \times 0.7746^2\right)^{1.4/0.4} = 1.486$$

$\therefore$ $p_0 = 1.486\ p = 1.486 \times (1 - 0.1) = 1.337 \text{kgf/cm}^2$.

The stagnation temperature

$$\frac{T_0}{T} = 1 + \frac{k-1}{2} M^2 = 1 + \frac{1.4-1}{2} \times 0.7747^2 = 1.1198$$

$\therefore$ $T_0 = 1.1198 \times 320 = 353.33°K$

From the equation of state,

$$\rho_0 = \frac{p_0}{RT_0} = \frac{1.337 \times 10^4}{29.25 \times 358.33} = 1.275 \text{ kg/m}^3 \text{ (SI units)}$$

Compressibility correction factor

$$= 1 + \frac{M^2}{4} + \frac{2-k}{24} M^4 + \ldots$$

$$= 1 + \frac{0.7747^2}{4} + \frac{2-1.4}{24} \times 0.7747^4 = 1.159.$$

Example 4:

Determine the speed which would be indicated on a pitot-static tube of normal calibration (without taking compressibility effect into account) and calculate the percentage error when the tube indicating the speed of an aircraft which is actually flying at 280 m/s. The velocity of sound in air at the aircraft altitude may be take as 340 m/s.

Solution:

The neglecting high power terms of M

$$\frac{p_0 - p}{\frac{1}{2}\rho V^2} = 1 + \frac{M^2}{4}$$

The actual speed of aircracft can be expressed as

$$V^2_{\text{actual}} = \frac{2(p_0 - p)}{\rho(1 + M^2/4)} \quad \ldots(1)$$

The speed recorded by the pitot-static tube with normal calibration

$$V^2 = \frac{2(p_0 - p)}{2} \qquad ...(2)$$

From Eqs. (1) and (2),

$$\frac{V}{V_{actual}} = \sqrt{1 + \frac{M^2}{4}}$$

When the aircraft flies at 280 m/s,

$$M = \frac{280}{340} = 0.8235, \text{ and}$$

$$\frac{V}{V_{actual}} = 1.0814$$

$$\therefore \quad V = 302.8 \text{ m/s.}$$

The pitot-static tube records a speed of 302.8 m/s as against the actual speed of 280 m/s. The percentage error

$$= \frac{V - V_{actual}}{V_{actual}} \times 100 = \frac{302.8 - 280}{280} \times 100 = 8.14\%.$$

Example 5:

At a certain section of a duct in which air is flowing at a temperature of 32°C and pressure of 0.815 kg/cm² with a velocity of 365 m/s. Assuming isentropic flow (reversible adiabatic) determine:

(i) The velocity and temperature at a section where the pressure is 1.25 kg/cm² and

(ii) The Mach number at both the sections.

(iii) The Mach number at both the sections.

Take R = 29.25 m kgf/kg° K and k for air = 1.4.

Solution:

We have

$$\frac{V^2}{2} + \frac{k}{k-1}\frac{p}{\rho} = \text{constant}$$

which may be written as

$$\frac{V_1^2 - V_2^2}{2} = \frac{k}{k-1}\left(\frac{p_2}{\rho_2} - \frac{p_1}{\rho_1}\right) = \frac{1.4}{0.4}\left(\frac{p^2}{\rho_2} - \frac{p_1}{\rho_1}\right) = 3.5\left(\frac{p^2}{\rho_2} - \frac{p_1}{\rho_1}\right)$$

For isentropic flow,

$$\frac{p}{\rho^k} = \text{constant, or } \frac{p_2}{p_1} = \left(\frac{\rho_2}{\rho_1}\right)^k$$

$$\therefore \quad \frac{V_2^2 - V_2^2}{2} = 3.5\frac{p_1}{\rho_1}\left(1 - \frac{p_2}{p_1}\frac{\rho_1}{\rho_2}\right) = 3.5\frac{p_1}{\rho_1}\left\{1 - \left(\frac{p_2}{p_1}\right)^{(k-1)/k}\right\}$$

$$\text{or } V_2^2 - V_1^2 = 7RT_1\left\{\left(1 - \frac{p_2}{p_1}\right)^{(k-1)/k}\right\}$$

Inserting the numerical values :

V_1 =365 m/s,

R = 9.81 × 29.25 = 286.7 J/kg ° K

$$\frac{p_2}{p_1} = \frac{1.25}{0.815} = 1.534,$$

T_1 = 273 + 32 = 305° K

$V_2^2 = (365)^2 + 7 \times 286.7 \times 305\ \{1-1.534)^{1/3.5}\}$

or V_2 = **231.46 m/s**

$$\text{We have , } \frac{T_2}{T_1} = \left(\frac{p_2}{p_1}\right)^{(k-1)/k} = 1.13$$

$\therefore$ T_2 = 1.13 × 305 = **344.65° K** or 71.65°C

The Mach numbers

$$M_1 = \frac{V_1}{C_1} = \frac{365}{\sqrt{kRT_1}} = \frac{365}{\sqrt{1.4 \times 286.9 \times 305}} = \mathbf{1.043}$$

$$M_2 = \frac{V_2}{C_2} = \frac{231.46}{\sqrt{1.4 \times 286.9 \times 344.65}} = 0.622$$

Example 6:

The pressure leads from a Pitot-static tube mounted on an aircraft are connected to a pressure gauge in the cockpit. The deal of the pressure gauge is calibrated to read the aircraft speed in metres/sec. The calibration is done on the ground by applying a known pressure across the gauge and calculating the equivalent velocity using incompressible Fernoulli equation and assuming that the density is 1.224 kg/m³.

The gauge having been calibrated in this way the aircraft is flown at 9200 m, where the density is 0.454 kg/m³ and the ambient pressure is 30,000 N/m². The gauge indicates a velocity of 152 m/s. What is the true speed of the aircraft?

Solution:

The incompressible Bernoulli equation may be written as,

$$\rho \frac{V^2}{2} + p = \text{constant}$$

and the stagnation pressure created at the Pitot-static tube:

$$p_0 = p + \rho \frac{V^2}{2}$$

or $$V = \sqrt{\frac{2(p_0 - p)}{\rho}}$$

inserting the numerical values

$$152 = \sqrt{\frac{2(p_0 - 30,000)}{1.224}}$$

$$\therefore \quad p_0 = 44139.6 \text{ N/m}^2$$

Neglecting compressibility effect, the speed of aircraft when related to r = 0.454 kg/m³ is

$$V = \sqrt{\frac{2(p_0 - p)}{\rho}} = \sqrt{\frac{2(44139.6 - 30,000)}{0.454}}$$

$$= 249.57 \text{ m/s.}$$

$$\text{Sonic velocity } C = \sqrt{kRT} = \sqrt{k\frac{p}{\rho}} = \sqrt{1.4 \times \frac{30,000}{0.454}}$$

$$= 304.15 \text{ m/s}$$

$$\text{Mach number } M = \frac{V}{C} = \frac{249.57}{304.15} = 0.818$$

Compressibility correction factor

$$= \left(1 + \frac{M^2}{4}\right) = \left(1 + \frac{0.818^2}{4}\right) = 1.167$$

∴ True speed of aircraft

$$= \frac{249.57}{\sqrt{1.167}} = \mathbf{231.08\ m/s.}$$

Example 7:

A jet airplane is propelled at a velocity of 1000 km/hr through air in which the atmospheric pressure is 0.9 kg/cm². The temperature is 5°C and the wind velocity is negligible. Calculate the pressure intensity, the temperature, the density of air at the stagnation point on the nose of the jet, and determine its Mach number. Take R = 290 kg-m/metric slug degree C absolute and k = 1.4.

Solution:

Velocity of sound in the atmospheric air

$$C = \sqrt{kRT} = \sqrt{1.4 \times 290 \times (273-5)} = 330.5 \text{ m/s.}$$

From the equation of state

$$\rho = \frac{p}{RT} = \frac{0.9 \times 10^4}{220 \times (273-5)} = 0.116 \text{ metric slug/m}^3.$$

The stagnation pressure

$$p_0 = p + \rho\,\frac{V^2}{2}\left(1 + \frac{M^2}{4} + \ldots\right)$$

$$M = \frac{V}{C} = \frac{1000 \times 1000/60 \times 60}{330.5} = 0.84$$

$$\therefore\ p_0 = 0.9 \times 10^4 + \frac{0.116 \times (278)^2}{2}\left[1 + \frac{(0.84)^2}{3} + \ldots\right]$$

$$= 1.428 \times 10^4 \text{ kg/m}^2$$

$\therefore$ Stagnation pressure = 1.428 kg/cm²

We have

$$\frac{1}{k-1}RT + \frac{V^2}{2} = \frac{k}{k-1}RT_0$$

From which $\frac{V^2}{2} = \frac{k}{k-1}R(T_0 - T)$

$$T_0 = \frac{(k-1)V^2}{2kR} + T$$

$$= \frac{(14.-1)(278)^2}{2 \times 1.4 \times 290} + (273-5) = 307.4°\text{K absolute.}$$

$T_0 = 34.4°C$

Density at the stagnation point,

$$\rho_0 = \frac{p_0}{RT_0} = \frac{1.428 \times 10^4}{290 \times 307.4} = 0.16 \text{ metric slug/m}^3.$$

Example 8:

Measuring the Mach angles α is an accurate means of measuring supersonic velocities for sharp nosed objects such as projectiles. Estimate the velocity and the Mach number of a projectile which reflects a shock wave of angle α = 30° in air at a pressure of 0.95 kg/cm² and at a temperature of 3°C.

Solution:

The Mach angle is defined as

$$\sin \alpha = \frac{C}{V} = \frac{1}{M}$$

Assuming a frictionless process

$$C = \sqrt{kRT} = \sqrt{1.4 \times 290 \times (273+3)} = 344 \text{ m/s}$$

$$\sin 30° = \frac{344}{V} \quad \therefore V = 344 \times 2 = 688 \text{ m/s.}$$

$$\text{Mach number } M = \frac{V}{C} = 2.$$

Rise in Skin Temperature

At a stagnation point where the velocity is reduced to zero, the stagnation temperature is given by,

$$C_v T_0 = C_p T + \frac{V^2}{2}$$

The stagnation temperature T_0 remains constant along the flow. At a stagnation point the dynamic temperature rise ΔT above the static temperature T is

$$\Delta T = T_0 - T = \frac{V^2}{2C_p}$$

For air the formula may be reduced to a simple expression if V is in km per hour and ΔT in degree C_v then

$$\Delta T = \left[\frac{V}{161}\right]^2$$

If an attempt is made to measure a flowing gas by placing a thermometer in a stream, the temperature recorded will be greater than T. Although the stagnation temperature would be reached at the stagnation point on a thermometer bulb, the temperature would rise less at other points on it and so the mean temperature recorded would be somewhat less than the stagnation temperature. The nose cone of a rocket travelling through air at 5100 km/hour must be able to withstand a temperature rise of about 1000 degree C.

Example 9:

A jet propelled aircraft is flying at 1000 km/hr at sea level, air temperature T = 25°C. Calculate the Mach number and rise in skin temperature.

Solution: Velocity of sound in air at 25°C

$$C = \sqrt{kRT} = \sqrt{1.4 \times 290 \times (273 + 25)} = 3.48 \text{m/s}$$

$$\text{Mach number } M = \frac{V}{C} \quad \frac{1000 \times 1000/(60 \times 60)}{348} = 0.798$$

Rise in skin temperature

$$\Delta T = \left(\frac{V}{161}\right)^2 = \left(\frac{1000}{161}\right)^2 = 385°\text{C}.$$

ONE DIMENSIONAL FLOW WITH NEGLIGIBLE FRICTION

The concept of one-dimensional flow all relevant quantities, such as the velocity, pressure, density and temperature, are considered constant over any cross section of the conduit. Thus the flow can be described in terms of only one coordinate, that is the distance along the conduit axis, and time. The pressure is assumed constant over the flow section and the absence of friction (isentropic flow) permits no variation of velocity. The flow of a real fluid is never strictly one-dimensional because of the presence of boundary layer. But in may problems in which the boundary layer is not very thick, the assumption of one-dimensional flow provides satisfactory solutions. With pressure and velocity assumed constant throughout the flow cross-section, constancy of density and temperature follow.

For a cross-section of area A over which velocity V and density ρ are constant, the continuity equation for steady flow is

$$\rho AV = \text{constant} \qquad ...(1)$$

differentiating and dividing throughout by ρAV, we obtain Eq. (2)

$$\frac{d\rho}{\rho}+\frac{dA}{A}+\frac{dV}{V}=0 \qquad ...(2)$$

If significant changes of cross-section area and, therefore, of V and ρ occur over only a short length of conduit, for example in a nozzel, frictional effects may be neglected. Euler's equation of motion for steady, frictionless flow neglecting the gravity and body forces may be written as

$$VdV + \frac{dp}{\rho} = 0$$

Multiplying the second term by dρ/dr, and nothing that the velocity of sound as C_2 = dp/dρ,

$$\frac{d\rho}{\rho} + \frac{VdV}{C^2} = 0.$$

Substituting for $\frac{dp}{r}$,

$$\frac{dA}{A}+\frac{dV}{V}=\frac{VdV}{C^2}$$

or $$\frac{dA}{A}=\frac{dV}{V}\left(\frac{V^2}{C^2}-1\right)$$

which may be expressed in terms of Mach number

$$\frac{dA}{V}=\frac{dV}{V}(M^2-1) \qquad ...(3)$$

An equation similar to Eq. (3) for the pressure change dp may be determined using

$$\frac{dV}{V}=\frac{dp}{V^2}$$

Substituting this value of $\frac{dV}{V}$

$$\frac{dV}{V} - \frac{dp}{\rho V^2} + \frac{dA}{A} = 0$$

$$\frac{d\rho}{\rho} - \frac{dp}{\rho V^2} + \frac{dA}{A} = 0$$

from which $\frac{dA}{A} = \frac{dp}{\rho V^2} - \frac{d\rho}{\rho} = \frac{dp}{\rho V^2}\left(1 - V^2\frac{d\rho}{dp}\right)$

and since $\frac{dp}{d\rho} = C^2$

we obtain, $\frac{dA}{A} = \frac{dp}{\rho V^2}(1 - M^2)$

Solving for dp, $dp = \frac{\rho V^2}{A}\frac{1}{1 - M^2}dA$ ...(4)

Using Eqs. (3) and (4), we can determine as to how the velocity and pressure vary with the change in flow area under subsonic flow conditions. From the equations some far reaching.

Flow Through Ducts of Varying Area

1. **Subsonic Flow (M < 1) :** Eq. (3) shows that for subsonic flow since $(M^2–1)$ is negative, dA must be of opposite sign, that is an increase of cross-sectional area causes a reduction of velocity and vice versa. This result is similar to that obtained for a constant-density fluid.
   Because the term rv^2/A of Eq. (4) is positive, for subsonic flow, dp and dA are of the same sign. That is increase in cross-sectional area causes in pressure in accordance with Eq. (4). We may, therefore, conclude that in subsonic isentropic flow, an increase in pressure and a decrease in velocity.
2. **Supersonic flow (M>1) :** For supersonic flow since M >1, Eq. (3) shows that dA and dV are of the same sign. An increase in the flow area is accompanied by an increase in the flow velocity and *vice versa*. Eq. (4) for supersonic flow that $(1– M^2)$ being negative, dp and dA are of opposite sign. Thus if the cross-sectional area increases, Eq. (4) shows that the pressure must decrease and *vice-versa.*

In conclusion, for a supersonic isentropic flow, an increase in flow area is accompanied by an increase in flow velocity and a decrease in pressure, whereas a decrease in flow area is accompanied by a decrease in velocity and increase in pressure.

3. **Sonic flow (M =1) :** When V = C so that M =1, dA must be zero and (since the second derivative is positive) A must be minimum. If the velocity of flow equals the sonic velocity anywhere, it must, therefore, do so where the cross-section is of minimum area.

Referring to Eq. (4), the value of $(1–M^2)$ is now zero and either dp becomes infinite as M tends to unity or that dA is also zero simultaneously. The expression for dp would then be indeterminate. It can be shown that M = 1 only when dA = 0. This condition could occur if a duct changes from a converging duct to a diverging duct of *vice-versa.*

VARIATION OF DENSITY AND TEMPERATURE WITH AREA

From the expression for velocity of sound, we know

from which, $$C^2 = \frac{dp}{d\rho}$$

$$d^2 = \frac{1}{C^2}dp \qquad ...(1)$$

Eq. (1) shows that the density varies in the same manner as the pressure.

In the above analysis we have not made use of the equation of state. Therefore, the result from Eq. (3) on wards are applicable to the isentropic flow of any compressible fluid. Confining our attention now to the isentropic flow of an ideal gas, using the isentropic flaw of an ideal gas, using the isentropic relation p = (constant) ρ^k and the equation of state p = ρ RT, we can obtain

$$T = (\text{constant})\ p^{(k-1)/k}$$

Differentiating it and eliminating the constant, we get

$$\frac{dT}{T} = \frac{k-1}{k}\frac{dp}{p} \qquad ...(2)$$

From Eq. (2) it is evident that the temperature varies in the same manner as the pressure. Combining Eqs. (1) and (2), we must conclude that the density and temperature very in the same manner as does the pressure.

From the foregoing analysis, we can study the variation of velocity, pressure, density and temperature with changes in flow area for subsonic and supersonic flow conditions. The variation of flow variables with area for isentropic flow of an ideal gas.

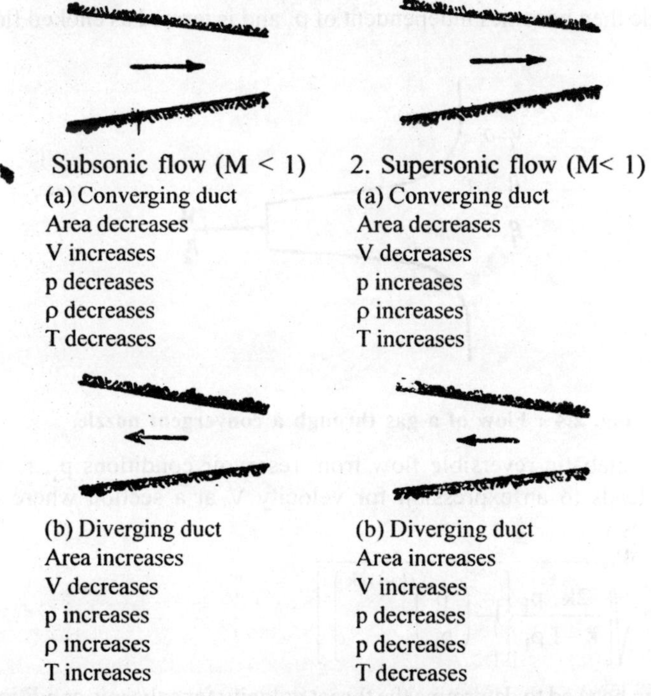

Fig. 2.3 : Variation of Flow variables with Area under Subsonic and Supersonic Flow Conditions.

In a subsonic flow, a diffuser is a diverging duct whereas for a supersonic flow the diffuser has a converging passage.

FLOW THROUGH A CONVERGENT NOZZLE

Consider a convergent nozzle through which a gas is flowing, see Fig. 2.4. The fluid jet is set into motion by the difference of pressure acting on it. It is obvious that when p_2 is close to p_1, the flow will be subsonic throughout. The gas stream would then emerge in a similar way as would a liquid jet except that it would possess considerable compressibility as well as inertia. As the down stream pressure p_2 is reduced, the jet velocity increases.

So long as the stream remains subsonic, the jet velocity will increase due to reduction of downstream pressure p_2. Once, however, the fluid reaches sonic velocity at the throat (M = 1), any further reduction in downstream pressure cannot be propagated upstream. The flow through the nozzle then becomes independent of p_2 and is termed as choked flow.

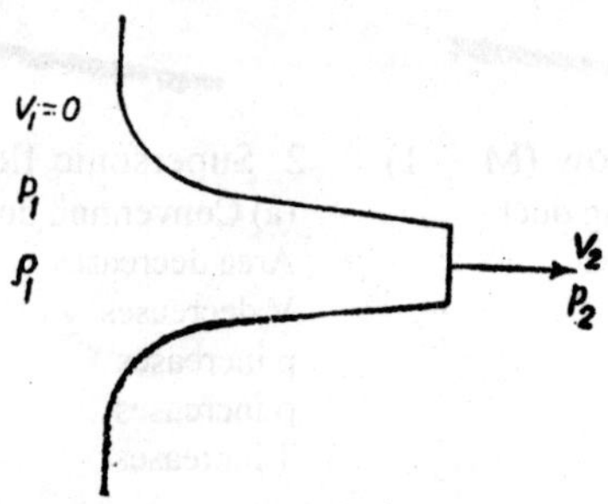

Fig. 2.4 : Flow of a gas through a convergent nozzle.

For adiabatic reversible flow from reservoir conditions p_1, r_1 and $v_1 = 0$, leads to an expression for velocity V at a section where the pressure is p:

$$V = \sqrt{\left[\frac{2k}{k-1}\frac{p_1}{\rho_1}\left\{1-\left(\frac{p}{p_1}\right)^{(k-1)/k}\right\}\right]} \quad \text{...(1)}$$

It can be used to determine the throat velocity for subsonic conditions. The corresponding mass flow rate

$\dot{m} = \rho AV$

For reversible adiabatic flow,

$$\rho = \rho_1\left(\frac{p}{p_1}\right)^{1/k}$$

and hence, $\dot{m} = \rho_1\left(\dfrac{p}{p_1}\right)^{1/k} AV$

$$= A\sqrt{\left[\frac{2k}{k-1}p_1\rho_1\left(\frac{p}{p_1}\right)^{2/k}\left\{1-\left(\frac{p}{p_1}\right)^{(k-1/k)}\right\}\right]} \quad ...(2)$$

In the above expressions, the pressure at the throat is $p = p_2$ so long as the conditions are subcritical. As soon as the Mach number at the throat reaches unity, the throat velocity, pressure and density assume their 'critical' values, V_e, p_e, ρ_e for choked flow. Thus

$$V_0 = \sqrt{\left[\frac{2k}{k-1}\frac{p}{p_1}\left\{1-\left(\frac{p_c}{p_1}\right)^{k-1/k}\right\}\right]}$$

$$m_0 = \sqrt{\left[\frac{2k}{k-1}p_1\rho_1\left(\frac{p_c}{p_1}\right)^{2/k}\left\{1-\left(\frac{p_c}{p_1}\right)^{k-1/k}\right\}\right]} \quad ...(3)$$

Since for the choking flow, the local Mach number reaches unity, and p_2 = pe, when used for these conditions results in

$$\frac{p_c}{p_1} = \left(\frac{2}{k+1}\right)^{k/k-1} \quad ...(4)$$

For air with k = 1.4, the critical pressure ratio

$$\frac{p_c}{p_1} = \left(\frac{2}{1.4+1}\right)^{3.5} = 0.528$$

Thus the mass flowrate through a convergent nozzle is given by Eq. (2) so long as $p_2 > 0.528\ p_1$, and by Eq. (3) when $p_2 < 0.528\ p_1$. Fig. 2.5 shows variation of mass flow are with pressure ratio p_2/p_1.

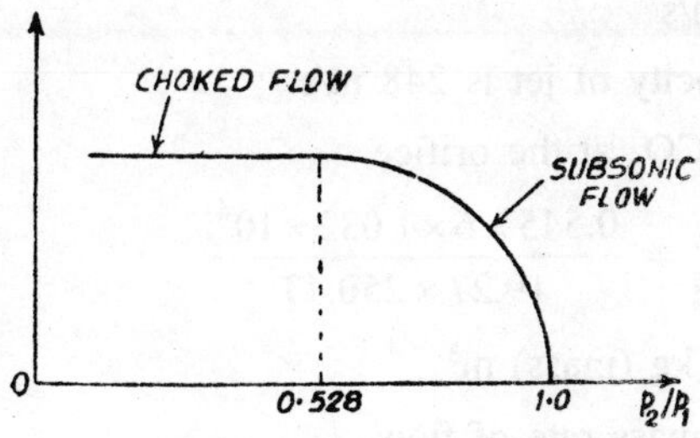

Fig 2.5 : Mass flowrate through a convergent nozzle.

Example 10:

A larger container holds carbon dioxide at a temperature of 15°C and pressure of 6 atmosphere. The gas is discharged through an orifice 10 mm in diameter. Determine the velocity of jet and the mass flowrate. Take k = 1.30 for carbon dioxide and atmospheric pressure = 1.032 kgf/cm², R = 19.27m kgf/kg° K (189 J/kg°K).

Solution:

Density of CO_2 in container

$$\rho_1 = \frac{p_1}{RT_1} = \frac{6 \times 1.032 \times 10^4}{19.27 \times 288} = 11.16 \text{ kg/m}^3$$

Critical pressure ratio for chocked flow

$$\frac{p_c}{p_1} = \left(\frac{2}{k+1}\right)^{k(k-1)} = \left(\frac{2}{1.3+1}\right)^{1.3/0.3} = 0.545$$

$$\frac{p_2}{p_1} = \frac{1 \times 1.032}{6 \times 1.032} = \frac{1}{6} = 0.167$$

Since p_2/p_1, the flow is a chocked one, and the velocity of jet is equal to the sonic velocity in CO_2 at temperature T_e.

$$\frac{T_c}{T_1} = \left(\frac{p_c}{p_1}\right)^{(k-1)/k} = \frac{2}{2.3}$$

or $$T_c = \frac{2}{2.3} \times 288 = 250.43°K$$

and the sonic velocity

$$Ve = C = \sqrt{kRT_e} = \sqrt{1.3 \times 189 \times 250.43}$$

$$= 248 \text{ m/s}$$

∴ The velocity of jet is 248 m/s.

Density of CO_2 at the orifice,

$$\rho_2 = \frac{p_c}{RT_c} = \frac{0.545 \times 6 \times 1.032 \times 10^4}{19.27 \times 250.43}$$

$$= 6.99 \text{ kg (mass) m}^3$$

Hence, the mass rate of flow

$$m = \rho_c AV_c$$

$$= 6.99 \times \frac{\pi}{4} \times \left(\frac{10}{1000}\right)^2 \times 248 = 0.1361 \text{ kg (mass)/sec}$$

$$= \mathbf{8.17 \text{ kg (mass)/ min.}}$$

FLOW VARIABLE IN TERMS OF MACH NUMBER

It is desirable to express changes in velocity, pressure, temperature and density in terms of the Mach number. To obtain these relationships use in made of the continuity, perfect gas, isentropic flow and the energy equations.

Continuity Equation

$$\rho AV = \text{constant}$$

Differentiating and dividing throughout by ρAV

$$\frac{d\rho}{\rho} + \frac{dA}{A} + \frac{dV}{V} = 0.$$

For isentropic flow

$$\frac{p}{\rho^k} = \text{constant}$$

or $$\frac{dp}{p} = k\frac{d\rho}{\rho}$$

For perfect gas

$$p = \rho RT$$

or $$\frac{dp}{p} = \frac{d\rho}{\rho} + \frac{dT}{T}$$

Energy equation

$$C_pT + \frac{V^2}{2} = \text{constant};$$

$$C_p = \frac{k}{k-1}R$$

$$C_pdT + VdV = 0$$

or $$\frac{kR}{k-1}dT + VdV = 0$$

or $$\frac{kR}{k-1}\frac{dT}{V^2}+\frac{dV}{V}=0$$

$$C = \sqrt{kRT} \quad \therefore \; kR = \frac{C^2}{T}$$

or $$\frac{1}{(k-1)M^2}\frac{dT}{T}+\frac{dV}{V}=0 \qquad ...(1)$$

From the Mach number relationship

$$M = V/\sqrt{kRT}$$

$$\frac{dM}{M}=\frac{dV}{V}-\frac{1}{2}\frac{dT}{T} \qquad ...(2)$$

Substituting for dT/T From Eq. (1)

$$\frac{dM}{M}=\frac{dV}{V}\left[1+\frac{(k-1)}{2}M^2\right]$$

or $$\frac{dV}{V} = \frac{1}{\left[1+\frac{(k-1)}{2}M^2\right]}\frac{dM}{M} \qquad ...(3)$$

since the quantity within the brackets is always positive, the trend of variation of velocity and Mach number is similar. For temperature variation, one can write

$$\frac{dT}{T}=\left[\frac{-(k-1)M^2}{1+\frac{k-1}{2}M^2}\right]\frac{dM}{M} \qquad ...(4)$$

Since the right hand side is negative, the temperature changes follow an opposite trend to that of Mach number. Similar trends are obtained for pressure and density,

$$\frac{dp}{p}=\left[\frac{-k.M^2}{1+\frac{k-1}{2}M^2}\right]\frac{dM}{M} \qquad ...(5)$$

and $$\frac{d\rho}{\rho}=\left[\frac{-M^2}{1+\frac{k-1}{2}M^2}\right]\frac{dM}{M} \qquad ...(6)$$

In case of area changes,

$$\frac{dA}{A} = \left[\frac{-(1-M^2)}{1+\frac{k-1}{2}M^2}\right]\frac{dM}{M} \qquad ...(7)$$

The quantity within the brackets may be positive or negative depending upon the magnitude of Mach number. Equation (7) can be integrated to obtain a relationship between the critical throat area, A_c, where the Mach number is unity and the area A at any section where $M \leq 1$

$$\frac{A}{A_c} = \frac{1}{M}\left[\frac{2+(k-1)M^2}{k+1}\right]^{\frac{k+1}{2(k-1)}} \qquad ...(8)$$

Equation (8) is unique in the sense that the Mach number is determined by the area ratio and k only.

Example 11:

(a) *In case of isentropic flow of a compressible fluid through a variable area duct, show that*

$$\frac{A}{A^*} = \frac{1}{M}\left[\frac{1+\frac{1}{2}(k-1)M^2}{\frac{1}{2}(k-1)}\right]^{\frac{(k+1)}{2(k-1)}}$$

where k is the ratio of specific heats, M is the Mach number at a section whose area is A and A is the critical area of flow.*

(b) *A supersonic nozzle is to be designed for air flow with Mach number 3 at the exit section which is 20 cm in diameter. The pressure and temperature, of air at the nozzle exit are to be 0.08 kagf/cm² and 200° K respectively. Determine the reservoir pressure and temperature and the throat area. Take k = 1.4.*

Solution:

Since the velocity in the reservoir is zero, the temperature and pressure there correspond to the stagnation condition.

We have

$$p_0 = p\left[1+\frac{k-1}{2}M^2\right]^{\frac{k}{(k-1)}}$$

$$= 0.08\left[1+\frac{1.4-1}{2}3^2\right]^{3.5} = 2.93 \text{ kgf/cm}^2$$

$$T_0 = T\left(1+\frac{k-1}{2}M^2\right)$$

$$= 200\left[1+\frac{0.4}{2}(3)^2\right] = 560° K$$

From Eq. (8),

$$\frac{A}{A_c} = \frac{1}{3}\left[\frac{2+(1.4-1)(3)^2}{1.4+1}\right]^{\frac{2.4}{2\times 0.4}} = 4.23$$

$$\text{or } A_c = \frac{A}{4.23} = \frac{\frac{\pi}{4}(0.2)^2}{4.23} = 0.00742 \text{ m}^2.$$

NORMAL SHOCK

An abrupt change from supersonic conditions to the subsonic ones takes place through a shock wave (similar to hydraulic jump in open channel flow). Whereas the Mach number and velocity drop across a shock, there occurs a sudden rise in pressure, density temperature and entropy. There may be two types of shocks:

(i) Normal shocks which are almost perpendicular to the flow.

(ii) Oblique shocks which are inclined to the flow direction.

Normal and oblique shocks may mutually interact to form a shock pattern. A shock may be attached or detached with respect to the body over which it may occur. A normal shock may occur in pipes, diverging section of a nozzle, diffuser of a supersonic wind tunnel or in front of bodies having blunt noses. For analysing a normal shock wave, use will be made of the continuity, momentum, perfect gas and energy equations.

The upstream flow is supersonic ($M_1 > 1$) with velocity V_1, pressure p_1, density ρ1 αvδ temperature T_1. On passing through the shock the flow becomes subsonic ($M_1 < 1$) with velocity V_2, pressure p_2, density ρ_2 and temperature T_2. A shock wave involves dissipation of energy and as such cannot be considered isentropic. For a perfect gas undergoing adiabatic flow in a constant area duct $A_1 = A_2 = A$. Considering Fig. 2.6 for a normal shock, the following equations can be written :

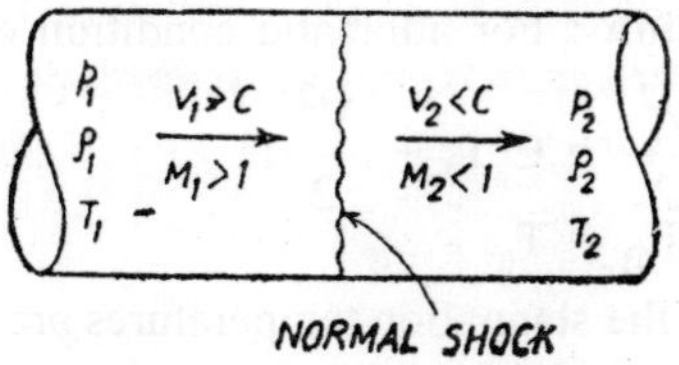

Fig. 2.6 : Normal Shock Wave.

(i) **Continuity equation :** $\rho A_1 V_1 = \rho_2 A_2 V_2$

since $A_1 = A_2$, one gets

$$\rho V_1 = \rho_2 V_2 \qquad ...(1)$$

using the equation of state for a perfect gas and substituting the velocities in terms of the Mach number

$V = CM = M\sqrt{kRT}$, one may write Eq. (1) as

$$\frac{p_1 M_1 \sqrt{kRT_1}}{RT_1} = \frac{p_2 M_2 \sqrt{kRT_2}}{RT_2}$$

or
$$\frac{p_1 M_1}{\sqrt{T_1}} = \frac{p_2 M_2}{\sqrt{T_2}} \qquad ...(2)$$

(ii) **Momentum equation :** Neglecting the effect of boundary friction, the momentum equation gives.

Pressure force = Mass flowrate × Change of velocity

$$(p_1 - p_2)\,A = \rho_1 A V_1 (V_2 - V_1)$$

or
$$(p_1 - p_2) = \rho_2 V_2^2 - \rho_1 V_1^2$$

or
$$p_1 + \rho_1 V_1^2 = p_2 + \rho_2 V_2^2$$

or
$$p_1 + \frac{p_1 V_1^2}{RT_1} = p_2 + \frac{p_2 V_2^2}{RT_2}$$

Substituting $V = MC$, where $C = \sqrt{kRT}$, one gets

$$p_1 + kM_1^2 p_1 = p_2 + kM_2^2 p_2$$

from which
$$\frac{p_2}{p_1} = \frac{1 + kM_1^2}{1 + kM_2^2} \qquad ...(3)$$

Equation (3) represents static pressure ratio across a normal shock. Since M1 > 1 and $M_2 < 1$, it is evident that $p_2 > p_1$, meaning thereby that the static pressure increases across a shock wave.

(iii) **Energy equation :** For adiabatic condition one may write,

$$C_pT_1 + \frac{V_1^2}{2} = C_pT_2 + \frac{V_2^2}{2};$$

and $T_{01} = T_{02}$

indicating that the stagnation temperatures are the same on both sides of a normal shock.

We have

$$\frac{T_0}{T} = 1 + \frac{k-1}{2}M^2$$

Equating the stagnation temperatures,

$$T_1\left(1+\frac{k-1}{2}M_1^2\right) = T_2\left(1+\frac{k-1}{2}M_2^2\right)$$

or

$$\frac{T_2}{T_1} = \frac{1+\frac{k-1}{2}M_1^2}{1+\frac{k-1}{2}M_2^2} \qquad \text{...(4)}$$

since $M_1 > 1$ and $M_2 < 1$, Eq. (4) indicates that $T_2 > T_1$.

From Eq. (2)

$$\frac{p_2}{p_1} = \sqrt{\frac{T_2}{T_1}}\frac{M_1}{M_2}$$

$$= \frac{M_1}{M_2}\left[\frac{1+\frac{k-1}{2}M_1^2}{1+\frac{k-1}{2}M_2^2}\right]^{0.5} \qquad \text{...(5)}$$

Equating Eqs. (5) and (5), one obtains a quadratic relation between M_1 and M_2. Neglecting the obvious trivial solution $M_1 = M_2$ for no-shock, one can write

$$M_2^2 = \frac{2+(k-1)M_1^2}{2kM_1^2-(k-1)} \qquad \text{...(6)}$$

Eq. (6) indicates that with M_1 increasing, the denominator becomes larger and larger resulting in a progressively decreasing value of M_2. By substituting the value of M_2, the following equations are obtained :

$$\frac{p_2}{p_1} = \frac{2kM_1^2-(k-1)}{k+1} \qquad \text{...(7)}$$

$$\frac{T_2}{T_1} = \frac{[(k-1)M_1^2 + 2][2kM_1^2 - (k-1)]}{(k+1)^2 M_1^2} \quad ...(8)$$

$$\frac{\rho_2}{\rho_1} = \frac{p_2/p_1}{T_2/T_1} = \frac{(k-1)M_1^2}{(k-1)M_1^2 + 2} \quad ...(9)$$

Shock Strength

The strength of shock is defined as the ratio of pressure rise across the shock to the upstream pressure, *i.e.*

$$\text{Strength of shock} = \frac{p_2 - p_1}{p_1} = \frac{p_2}{p_1} - 1 = \frac{2k}{k+1}(M_1^2 - 1) \quad ...(2.61)$$

Example 12:

A normal shock wave occur in a duct in which air is flowing at a Mach number of 1.50. The static pressure and temperature upstream of the shock wave are 1.5 bar and 27°C. Determine the pressure, temperature and the Mach number downstream of the shock. Also calculate strength of shock. Take k =14.

Solution:

$$\frac{p_2}{p_1} = \frac{2kM_1^2 - (k-1)}{k+1} = \frac{2 \times 1.4 \times (1.5)^2 - (14.-1)}{1.4+1}$$

$$\therefore \quad p_2 = 2.458 \times 1.5 = 3.687 \text{ bar}$$

$$\frac{T_2}{T_1} = \frac{[(k-1)M_1^2 + 2][2kM_1^2 - (k-1)]}{(k+1)^2 M_1^2}$$

$$= \frac{(0.4 \times 1.5^2 + 2)(2 \times 1.4 \times 1.5^2 - 0.4)}{2.4^2 \times 1.5^2} = 1.32$$

$$\therefore \quad T_2 = 1.32\,(273 + 27) = 396.06° \text{ K or } 123°\text{C}$$

$$M_2^2 = \frac{2 + (k-1)M_1^2}{2kM_1^2 - (k-1)} = \frac{2 + 0.4 \times 1.5^2}{2 \times 1.4 \times 1.5^2 - 0.4} = 0.49 \backslash M^2 = 070$$

$$\text{Strength of shock} = \frac{p_2}{p_1} - 1 = 2.458 - 1 = 1.458.$$

Example 13:

Establish an analogy between shock waves and open-channel waves. Show that the critical velocity in open-channel is equivalent to the sonic velocity in compressible flow and the liquid depth in the channel to the fluid density.

Solution:

The speed of an elementary surface wave in an open channel is

$$V = \sqrt{gy} \quad ...(1)$$

where y is the depth of liquid in the channel. From the continuity equation, for a channel of constant width

$$V.y = \text{constant} \quad ...(2)$$

and the continuity equation for compressible flow in a tube of constant cross-section is

$$V.\ \rho = \text{constant} \quad ...(3)$$

Comparison of the two continuity Eqs. (2) and (3) reveals an analogy between the open channel depth y and the compressible fluid density ρ.

The analogy between the critical velocity in open channel and sonic velocity of compressible flow can be established from the consideration of energy equations. For open channel flow, the energy equation gives

$$\frac{V^2}{V_g} + y = \text{constant}$$

differentiation of which yields

$$VdV + gdy = 0$$

When the velocity in the channel is such that $V = V_0 = \sqrt{gy}$, where V_0 = critical velocity, elimination of g yields

$$VdV + V^2{}_2 \frac{dy}{y} = 0 \quad ...(4)$$

For a compressible fluid flow, we may write the energy equation, from Eq. (4), as

$$VdV + \frac{dp}{\rho} = 0$$

but the sonic velocity $C^2 = \dfrac{dp}{d\rho}$, eliminating dp_1 we obtain

$$VdV + C^2 \frac{dp}{\rho} = 0$$

Comparing the two energy equations, Eqs. (4) and (5), we find that

1. The flow depth y and the density r are analogous and
2. The critical velocity V0 in the open-channel is analogous to the sonic velocity C.

FLOW THROUGH A CONVERGENT DIVERGENT NOZZLE

Consider a convergent-divergent nozzle as shown in Fig. 2.6 with T_0, p_0 and ρ_0 as the stagnation properties and T, p, ρ the static properties at the exit. When the back pressure equals the inlet pressure there will be no flow and the pressure will remain constant throughout the nozzle. The flow starts as the back pressure is reduced. Depending upon the value of the back pressure, the following flow conditions may be obtained :

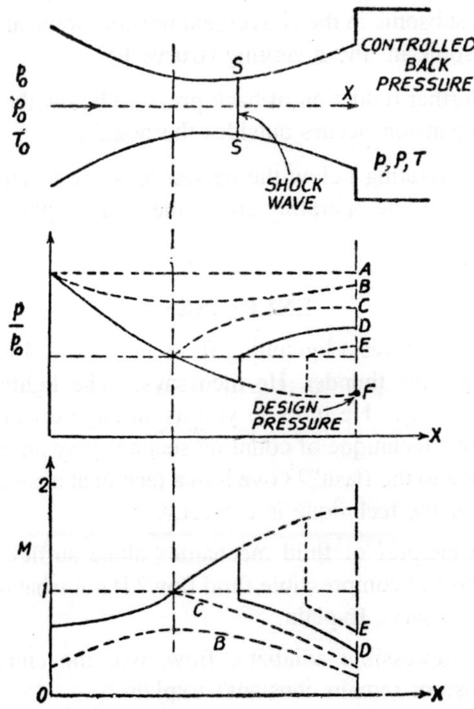

Fig. 2.7 : Flow conditions in a convergent nozzle.

1. When the back pressure is reduced to a value such that the pressure at the throat is above the critical pressure the flow is

subsonic throughout the nozzle. Flow in the nozzle is similar to that in a venturimeter (Curve B).

2. Further lowering of back pressure may cause the sonic velocity at the throat, and the flow in the diverging portion returns to subsonic conditions. (Curve C).
3. With the sonic velocity attained at the throat, further reduction of pressure does not affect the flow in the convergent portion. Low exit pressure such as that represented by D causes the flow to be supersonic upto a certain distance from the throat and then a shock wave occurring within the diverging portion causes subsonic condition to exist in the remaining portion.
4. When the exit pressure equals the nozzle design outlet pressure, there is a steady acceleration in the entire passage. The flow being subsonic in the convergent portion, sonic at the throat and supersonic in 'the diverging (Curve F).
5. Still further reduction of back pressure below the design value, the expansion occurs outsides the nozzles.

A nozzle operating below the design pressure is said to be under expanding and the one operating above the design value is called over expanding.

EXERCISES

1. A man observes a lightening flash and counts 4 seconds before he hears the thunder. He then says, "The lighting was about 65 km away." He turns to you as an engineer and asks "How does this technique of counting seconds give an estimate of the distance to the flash"? Give him a technical answer and tell him whether the technique is correct.
2. Are principles of fluid mechanics alone sufficient to analyse problems of compressible fluid flow? If not what other concepts are necessary? Explain.
3. For compressible adiabatic flow, will the entropy increase, decrease or remain constant? Explain.
4. State the fundamental equations which govern the compressible fluid flow and what are the controlling non-dimensional parameters governing such a flow?

5. Show that the Reynolds number for isothermal compressible gas flow in a uniform conduit remains constant.

6. A model of an aeroplane is put in high-speed wind tunnel, which parts of the aeroplane you expect to become the hottest? Explain with reasons.

7. Comment on the correctness of the statement : "In a compressible flow of an ideal gas the sum of he thermal and kinetic energies is constant."

8. How does the velocity and pressure vary with area for (i) Subsonic flow, and (ii) Supersonic flow.

9. Show that the velocity of sound in an adiabatic gas flow is given by

$$C = \sqrt{kRT} .$$

10. Show that for an isothermal atmosphere the pressure variation in the vertical direction is given by

$$p = p_0 e^{-z}/p_0/\rho_e$$

where z is the distance vertically upward and p_0, ρ_0 are the pressure and density at z = 0. Assuming gravitational acceleration to be constant, compare the pressure given by this equation to the hydrostatic pressure distribution $p = p_0 - \gamma z$. Show this comparison graphically.

11. Show that the velocity of sound in the atmosphere under adiabatic conditions varies with the temperature and hence find out whether the sonic velocity will decrease or it crease in magnitude with increasing elevation above the earth.

12. Show that for subsonic flow in a pipe the velocity must increase in the downstream direction. Take into account the losses in the pipe.

13. Determine the velocity of sound in air at 20°C, 50°C and 100°C. Take gas constant it = 290 kg m/ metric slug degree C absolute, k = 14.

14. Air is flowing in a duct a atmospheric pressure with a velocity of 500 m/s. If the air temperature is 70°C, what is the Mach number?

15. An aeroplane climbs from a low altitude over a hot desert to a high altitude where the temperature is significantly lower. If the

airspeed is constant, what change, if any will occur in the flight Mach number ?

16. How is the first law of thermodynamics different from the Bernoulli's theorem? Why is the Bernoulli's theorem inadequate for compressible flow problems?
17. Explain what do you mean by

    (i) Transonic flow (ii) Hypersonic flow

    (iii) Shock wave (iv) Mach angle.
18. Prove that the temperature at the stagnation point is greater than the temperature in the free-stream.
19. Upto which limit of Mach number can we neglect the compressibility effects and treat the fluid as incompressible?

3

Impulse Momentum Equation and Its Applications

INTRODUCTION

The impulse momentum principle is another very useful principle, in addition to the continuity and the energy principles, the application of which leads to the solution of several fluid flow problems. On the basis of the impulse momentum principle impulse momentum equations are developed. These equations are often used in conjuction with the energy and the continuity equations in order to obtain the solutions of the problems of fluid flow which cannot be solved simply by applying the continuity and the energy equations. In the following paragraphs, the impulse momentum equations applicable to the problems of fluid flow have been derived and some of their applications are discussed.

IMPULSE-MOMENTUM EQUATIONS

The impulse-momentum equations are derived from the impulse-momentum principle (or simply momentum principle) which states that the impulse exerted on any body is equal to the resulting change in momentum of the body. In other words, this principle is a modified form of Newton's second law of motion. Newton's second law of motion states that the resultant external force acting on any body in any direction is equal to the rate of change of momentum of the body in that direction. Thus for any arbitrarily chosen direction x, it may be expressed as,

$$F_x = \frac{d(M_x)}{dt} \qquad ...(1)$$

in which F_x represents the resultant external force in the x-direction and M_x represents the momentum in the x-direction. Equation 1 may also be written as

$$F_x(dt) = d(M_x) \qquad ...(2)$$

In equation 2 the term $F_x(dt)$ is impulse and the term $d(M_x)$ is the resulting change of momentum. Equation 2 is thus known as impulse momentum equation. The impulse-momentum relationship in the form as indicated by equation 2 is, however, applicable to finite or discrete bodies, for which the action of any force may take place and be completed in a finite period of time.

On the other hand, steady flow of fluid involves a motion which is continuous and it is not completed in a finite period of time. Therefore the momentum equation has to be expressed in a form particularly suited to the solution of fluid flow problems as indicated below.

Consider as a free body the fluid mass included between sections 1– 1 and 2–2 within a certain flow passage as shown in Fig. 3.1.

The fluid mass of the free body 1–1 and 2–2 at time t moves to a new position 1′ –1′ and 2′ –2′ at time (t + dt). Section 1′–1′ and 2′ –2′ are curved because the velocities of flow at these two sections are non-uniform.

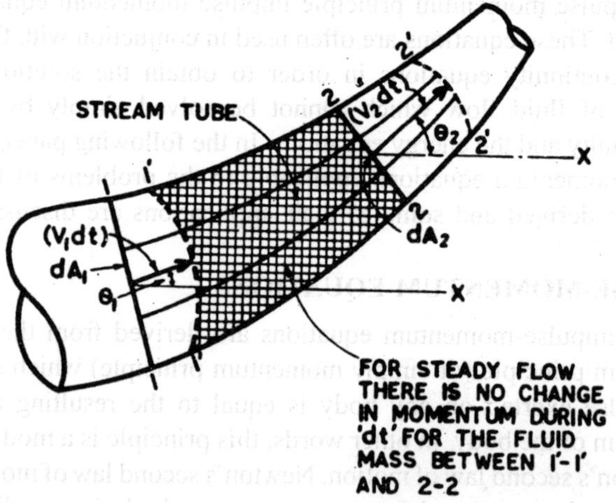

Fig. 3.1: Change of momentum of fluid mass in a flow passage.

For steady flow the following continuity equation holds:

$$\begin{bmatrix}\text{Fluid mass within} \\ \text{section 1-1and1}'-1'\end{bmatrix} = \begin{bmatrix}\text{Fluid mass within} \\ \text{section 2-2and }2'-2'\end{bmatrix}$$

Further for an arbitrary direction x the change in momentum of this mass of fluid during a time interval dt is considered, which may be represented as follows:

$$\begin{bmatrix}\text{Change in momentum}\\ \text{of fluid mass}\\ \text{during dt}\end{bmatrix}_x = \begin{bmatrix}\text{Momentum at}\\ \text{(t + dt) of fluid}\\ 1'-1' \text{ and } 2'-2'\end{bmatrix}_x + \begin{bmatrix}\text{Momentum at}\\ \text{t of fluid}\\ 1-1 \text{ and } 2-2\end{bmatrix}_x$$

But

$$\begin{bmatrix}\text{Momentum at}\\ \text{(t + dt) of fluid}\\ 1'-1' \text{ and } 2'-2'\end{bmatrix}_x = \begin{bmatrix}\text{Momentum at}\\ \text{(t + dt) of fluid}\\ 1'-1' \text{ and } 2-2\end{bmatrix}_x + \begin{bmatrix}\text{Momentum at}\\ \text{(t + dt) of fluid}\\ 2-2 \text{ and } 2'-2'\end{bmatrix}_x$$

and
$$\begin{bmatrix}\text{Momentum at}\\ \text{t of fluid}\\ 1-1 \text{ and } 2-2\end{bmatrix}_x = \begin{bmatrix}\text{Momentum at}\\ \text{t of fluid}\\ 1-1 \text{ and } 1'-1'\end{bmatrix}_x + \begin{bmatrix}\text{Momentum at}\\ \text{t of fluid}\\ 1'-1' \text{ and } 2-2\end{bmatrix}_x$$

Moreover, when the flow is steady, the state of the flowing fluid in the flow passage within sections 1′–1′ and 2–2 remains unchanged at all times.

Therefore

$$\begin{bmatrix}\text{Momentum at}\\ \text{(t + dt) of fluid}\\ 1'-1' \text{ and } 2-2\end{bmatrix}_x + \begin{bmatrix}\text{Momentum at}\\ \text{t of fluid}\\ 1'-1' \text{ and } 2-2\end{bmatrix}_x$$

From which it follows that

$$\begin{bmatrix}\text{Change in momentum}\\ \text{of fluid mass}\\ \text{during dt}\end{bmatrix}_x$$

$$= \begin{bmatrix}\text{Momentum at}\\ \text{(t + dt) of fluid}\\ 2-2 \text{ and } 2'-2'\end{bmatrix}_x - \begin{bmatrix}\text{Momentum at}\\ \text{t of fluid}\\ 1-1 \text{ and } '-1'\end{bmatrix}_x$$

The above relationship when expressed in terms of mathematical symbols, it becomes

$$\Sigma d(mv_x) = \int_{A_2} \rho_2 v_2 dt dA_2 (v_2)_x - \int_{A_1} \rho_1 v_1 dt\, dA_1 (v_1)_x$$

where (ρ_2 v_2 dt dA_2) and (ρ_1 v_1 dt dA_1) represent the mass of flow of fluid during the time interval dt in a stream tube across sections 2–2 and 1–1 respectively as shown in Fig. 3.1.

Further according to Newton's second law of motion (Equation 1), the relationship between the resultant external force and the time rate of change of momentum of the fluid flow in the passage may be written in the following form:

$$\Sigma F_x = \frac{\Sigma d(mv_x)}{dt}$$

$$= \int_{A_2} \rho_2 v_2 dA_2 (v_2 \cos\theta_2) - \int_{A_1} \rho_1 v_1 dA_1 (v_1 \cos\theta_1) \quad \text{...(3)}$$

in which $v_2 \cos\theta_2 = (v_2)_x$ and $v_1 \cos\theta_1 = (v_1)_x$ (see Fig. 3.1). Eqn. 3 may be integrated if the velocity distributions of fluid flow at both sections are known. Since in most of the problems of fluid flow we have to deal with only the mean velocity of flow at each section, it is preferable to express the impulse-momentum equation in terms of the mean velocities. Thus if V_1 and V_2 are the mean velocities at a sections 1–1 and 2–2 respectively, then the impulse-momentum equation 3 may be written as

$$\Sigma Fx = \rho_2 A_2 \cos\theta_2 V_2^2 - \rho_1 A_1 \cos\theta_1 V_1^2 \quad \text{...(4)}$$

For a steady flow of incompressible fluid, the impulse-momentum equation for fluid flow may be simplified to the form noted below. The continuity equation for such a flow may be expressed as $Q = A_1V_1 = A_2V_2$ and $\rho_1 = \rho_2 = \rho$. Thus introducing these expressions in equation 4.4 it becomes

$$\Sigma Fx = \rho\, Q\, (V_2 \cos\theta_2 - V_1 \cos\theta_1)$$

or

$$\Sigma Fx = \rho\, Q\, [(V_2)_x - (V_1)x]$$

in which suffix x is introduced to represent the components of the velocities in the x-direction. The term ΣF_x should include all the external forces acting on the free body of the fluid under consideration.

If D' Alembert's principle is applied to the flow system, the system is brought into relative static equilibrium with the inclusion of inertia forces. The resulting impulse-momentum equation takes following form:

$$\Sigma F_x = \rho Q\, (V_2)_x + \rho Q (V_1)_x = 0 \quad \text{...(6)}$$

The intertia force (ρqV) in fluid flow is usually called the momentum flux.

It may however be noted that the impulse-momentum equation given above has been derived for one direction only, but the same method may be extended to derive the corresponding equations for the other directions of reference as well. Accordingly the impulse-momentum equations for y and z directions may be written as

$$\Sigma Fy = \rho Q\ [V_2)_y - (V_1)_y] \quad ...(7)$$

$$\Sigma Fz = \rho Q\ [V_2)_z - (V_1)_y] \quad ...(8)$$

Further the general impulse-momentum equation for steady flow of fluid may be written in a vector form as

$$\Sigma F \rightarrow \rho QV_2 + \ \rightarrow \rho QV_1 = 0 \quad ...(9)$$

The impulse-momentum equations are often called simply momentum equations. From these equations it may be noted that if the resultant external force that acts on the fluid mass is zero, the momentum of the fluid mass remains constant. This principle is known as the *Law of Conservation of Momentum.*

APPLICATIONS OF THE IMPULSE-MOMENTUM EQUATION

The impulse-momentum equation, together with the energy equation and the continuity equation provides the basic mathematical relationships for solving various engineering problems in fluid mechanics. Since the impulse momentum equation relates the resultant external forces on a chosen free body of fluid or control volume in a flow passage, to the change of momentum flux at the two end sections, it is especially valuable in solving those problems in fluid mechanics in which detailed information about the flow process within the control volume may be either not available or rather difficult to evaluate. Thus in order to apply the impulse-momentum equation, a control volume is first chosen which includes the portion of the flow passage which is to be studied. The boundaries of the control volume are usually extended upto such an extent that its end sections lie in the region of uniform flow. All the external forces acting on this control volume are then considered and the momentum equations in the corresponding directions of reference are applied to evaluate the unknown quantities.

In general the impulse-momentum equation is used to determine the resultant forces exerted on the boundaries of a flow passage by a stream of flowing fluid as the flow changes its direction or the magnitude of velocity or both. The problems of this type include the pipe bend, jet propulsion, propellers and stationary and moving plates or vanes. The application of impulse-momentum equation to the problems of pipe bends, jet propulsion and propellers is illustrated in the following paragraphs. However the problems of stationary and moving vanes are discussed in the chapter of *Impact of Free jets*. Furthermore, the impulse-momentum equation may also be applied to solve the problems of non-uniform flow in which an abrupt change of flow section occurs. The problems of this type include sudden enlargement in pipes, hydraulic jump in open channels etc.

MOMENTUM CORRECTION FACTOR

The above derived impulse-momentum equations in terms of the mean velocities of flow are based on the assumption that the velocity of flow at each section is uniform, that is the velocity is same at different points on the same section. However, in actual practice the velocity is not uniform over a cross section of the flow passage, on account of which the momentum flux computed on the basis of the mean velocity of flow at any section is not equal to the actual momentum flux flowing through the section. The actual momentum flux flowing through any section maybe obtained by integrating the momentum flux flowing through different elementary areas of the cross-section. Thus if v is the velocity of flowing fluid at any point through an elementary area 'dA' of the cross-section, then the mass of fluid flowing per unit time is (ρvdA) and the corresponding momentum flux flowing through this elementary area is ($\pi v^2 dA$). The total momentum flux flowing through the entire cross section Ais equal to

$$\int_A \rho v^2 dA = \frac{w}{g}\int_A v^2 dA$$

which may be evaluated from the known velocity distribution at the cross section.

It is however more convenient to express the momentum flux flowing through any cross-section in terms of the mean velocity of flow. But the actual momentum flux is always greater than that computed by using the mean velocity of flow. Hence in order to account for this difference in the values of the momentum flux due to the non-uniform velocity distribution at any cross-section a factor called *momentum correction* factor represented by β (Greek beta) is introduced, so that the momentum flux computed by using the mean velocity V may be expressed as ($\beta\rho AV^2$) and it is then equal to the actual total momentum flux flowing through the entire cross-section. Thus equating the two, the value of the momentum correction factor may be obtained as

$$\beta\pi AV^2 = \rho \int_A v^2 dA$$

Therefore $$\beta = \frac{1}{AV^2} \int_A v^2 dA \qquad ...(10)$$

Mathematically the square of the average is less than the average of the squares, that is $V < \frac{1}{A} \int_A v^2 dA$, the numerical value of β will always be greater than 1. The actual value of β depends on the velocity distribution at the flow section. If the velocity is uniform over the entire cross-section, β will be equal to 1. For turbulent flow in pipes the value of β lies between 1.02 and 1.05. However, for laminar flow in pipes the value of b is 1.33.

In view of the above discussion, if the velocity distribution is non-uniform, then the momentum correction factor will be required to be introduced in the impulse-momentum equations expressed in terms of mean velocity at each section. Thus equations 5, 7 and 8 are modified as follows:

$$\Sigma F_x = \rho Q\ [\beta_2\ (V_2)_x - \beta_1 (V_1)_x]$$
$$\Sigma F_y = \rho Q\ [\beta_2\ (V_2)_y - \beta_1 (V_1)_y]$$
$$\Sigma F_z = \rho Q\ [\beta_2\ (V_2)_z - \beta_1 (V_1)_z]$$

in which β_1 and β_2 are the momentum correction factors at sections 1–1 and 2–2 respectively. However, in most of the problems of turbulent flow, the value of β is nearly equal to 1 and therefore it may be assumed as one, without any appreciable error being introduced. Accordingly the impulse-momentum equations for the reference directions may be represented by equations 5, 7 and 8.

FORCE ON A PIPE BEND

As state earlier the impulse-momentum equation is applied to determine the resultant force (or thrust) exerted by a flowing fluid on a pipe bend. Fig. 3.2 shows a reducing pipe bend through which a fluid of density ρ flows steadily from section 1 to 2. It is desired to find the force exerted by the flowing fluid on the pipe bend. For this the portion of the bend lying between sections 1 and 2 may be chosen as a control volume and all the external forces acting on this may be considered as indicated below.

1. At sections 1 and 2 the fluid in the control volume will be subjected to pressure forces p_1A_1 and p_2A_2 by the fluid adjacent to these sections as shown in Fig. 3.2, where p_1, p_2 and A_1, A_2 are the mean pressure and cross-sectional areas at sections 1 and 2 respectively.

2. The boundary surface of the bend will exert forces on the fluid in the control volume. These forces will be distributed non-uniformly over the curved surface of the bend. But for the ease of computation it is assumed that these distributed forces are equivalent to a single concentrated force R, which has R_x and R_y as its components along x and y directions respectively as shown in Fig. 3.2. It may however be stated that according to Newton's third law of motion, the force exerted by the flowing fluid on the bend will be equal and opposite to R, which is required to be determined.

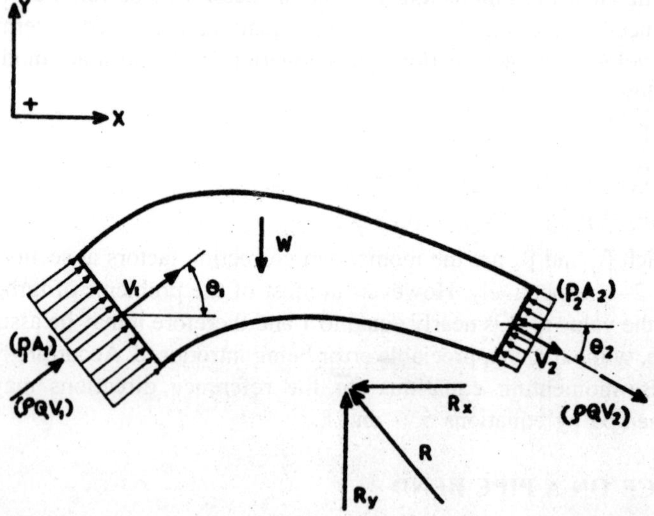

Fig. 3.2 : Change of momentum of flow in a reducing pipe bend.

3. The self-weight W of the fluid in the control volume will be acting in the vertical downward direction.

Thus by applying the impulse-momentum equation in both x and y directions the following expressions are obtained.

$$p_1A_1 \cos\theta_1 - p_2A_2 \cos\theta_2 - R_x = \rho Q (V_2 \cos\theta_2 - V_1 \cos\theta_1) \quad ...(12)$$

For y direction:

$$p_1A_1 \sin\theta_1 + p_2A_2 \sin\theta_2 + R_y - W = \rho Q (-V_2 \sin\theta_2 - V_1 \sin\theta_1) \quad ...(13)$$

From the above equations R_x and R_y can be determined from which the magnitude and direction of the force R exerted by the bend on the fluid can be computed. The force (or thrust) exerted by the fluid on the bend is equal and opposite to R.

Often the continuity equation, the energy equation and the equation of state are required to be used to determine the unknown flow characteristics to be used in the above equations.

For a horizontal bend in equation 13 the term W, representing the weight of the fluid in the bend will be eliminated.

The quantity on the right hand side of equations 12 and 13 is often termed as *dynamic thrust* exerted by the flowing fluid on the bend and vice-versa, in order to distinguish it from the static pressure forces appearing on the left hand side of these equations.

ANGULAR MOMENTUM PRINCIPLE–MOMENT OF MOMENTUM EQUATION

The angular momentum principle states that the torque exerted on any body is equal to the rate of change of angular momentum. The torque is defined as the moment of the force and the angular momentum is defined as the moment of momentum; the moments being taken about the axis of rotation.

Consider a fluid mass δm which is rotating about the z-axis as shown in Fig. 3.3 (a). Let V_x and V_y be its velocity components in x any y directions respectively.

If $a_x = \frac{dV_x}{dt}$ and $a_y = \frac{dV_y}{dt}$ represent the acceleration components of the fluid mass, one obtains

$$\delta F_x = \frac{dV_x}{dt}\delta m, \delta F_y = \frac{dV_y}{dt}\delta m$$

where δF_x and δF_y are the components of external forces causing the acceleration.

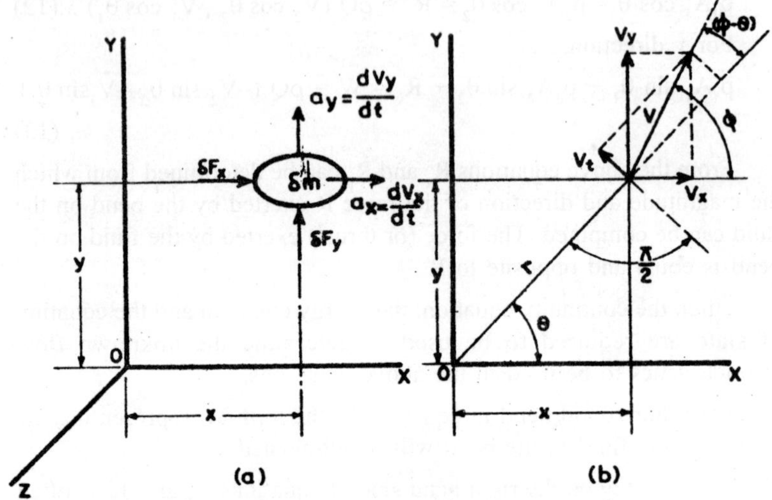

Fig. 3.3 : Fluid mass subjected to torque–sketch.

The moment of the external forces about z-axis (counter-clockwise being considered positive) or the torque δT_z, is then obtained as

$$\delta T_z = (x\delta F_y - y\,\delta F_x)$$

$$= \left(x\frac{dV_y}{dt} - y\frac{dV_x}{dt}\right)\delta m$$

By the rules of differentiation

$$\frac{d}{dt}(xV_y - yV_x) = \frac{dx}{dt}V_y - \frac{dy}{dt}V_x + x\frac{dV_y}{dt} - y\frac{dV_x}{dt}$$

$$= V_x V_y - V_y V_x + x\frac{dV_y}{dt} - y\frac{dV_x}{dt}$$

Therefore, since $(V_x V_y - V_y V_x) = 0$ and δm is constant,

$$\delta T_z = \frac{d}{dt}(xV_y - yV_x)\delta m$$

$$= \frac{d}{dt}[(xV_y - yV_x)\delta m] \qquad ...(27)$$

The quantities $(\delta m V_y)_x$ and $(\delta m V_x)y$ represent the "moments of momentum" or "angular momentum" about the z-axis. Therefore the right hand side of the above expression represents the rate of change of angular momentum about z-axis, and this equal to the torque.

In the above derivation, since z-axis is arbitrarily chosen, a torque equation for the x or y axis may also be similarly obtained.

Hence, it may be stated that the resultant external torque about any axis is equal to the rate of change of angular momentum about that axis. This is the law of moment of moment (or law of angular momentum).

It is usually convenient to express $(xV_y - yV_x)$ in terms of V_t and r, where V_t is the tangential velocity and r is the radial distance as defined in Fig. 3.3 (b).

From Fig 4.6 (b) since,

$$x = r\cos\theta;\ y = r\sin\theta$$

$$V_x = V\cos\phi;\ V_y = V\sin\phi$$

Hence

$$(xV_y - yV_x) = r\cos\theta\,(V\sin\phi) - r\sin\theta\,(V\cos\phi)$$

$$= rV\sin(\phi - \theta)$$

$$= rV_t$$

Thus by substituting in equation 27

$$\delta T_z = \frac{d(rV_t\delta m)}{dt} \quad \text{...(28)}$$

Applying equation 27 or 28 to each of the several small fluid masses of a system and summing all the resulting equations, the resultant external torque T_z for a steady flow system is obtained as

$$\Sigma(\delta T_z) = \frac{\Sigma d(rV_t\delta m)}{dt}$$

or $$T_z = \rho Q(r2Vt2 - r1Vt1) \quad \text{...(29)}$$

in which r_2 and V_{t2} and radial distance and tangential velocity at section 2 and r_1 and V_{t1} are same quantities at section 1 of the control volume.

By rewriting equation 29 in the form

$$T_z - \rho Q r_2 V_{t2} + \rho Q r_1 V_{t1} = 0 \quad \text{...(30)}$$

it can be shown that the moment of the momentum flux across an area about any axis equals the moment of all the external forces applied at the centre of the area about the same axis. Further, it may be seen from equation 30 that it the external forces that act on the fluid mass exert no netm om entabouta fixed axis (*i.e.*, Tz = 0), the moment of momentum of the fluid mass with respect to that axis remains constant. This principle is known as the *law of conservation of moment of momentum or the law of conservation of angular momentum.*

The concept of angular momentum is applied in analyzing the flow problems, such as flow through turbomachinery , where torques are more significant in the analysis than forces. The work done by the flowing fluid on a wheel of a radial flow hydraulic turbine which has radially fixed curved vanes has been evaluated by applying the principle of angular momentum.

JET PROPULSION- REACTION OF JET

When a jet of fluid issues form an opening and strikes an obstruction placed in its path, it exerts a force on the obstruction. This force exerted by the jet is known as the action of the jet. Recalling Newton's third law of motion, since every action is accompanied by an equal and opposite reaction, the jet while coming out of opening exerts a force on the opening in the form of back kick. This force exerted by the jet on the source from which it is issued is known as the reaction of the jet and is therefore equal in magnitude but opposite in direction to the action of the jet. Further if the source issuing the jet is free to move, it will start moving in the direction opposite to that of jet. Thus the reaction of jet can be utilized for the propulsion of various bodies. The principle of jet propulsion, which is applied in the propulsion of surface ships, aircrafts, rockets etc., may be explained by some of the examples described below.

(a) Jet Propulsion of Orifice Tank

Consider a jet of fluid of area a being issued under a constant head H from an orifice provided in the side of a tank, which is large enough so that the velocity within the tank may be neglected [see Fig. 3.4(a)]. Let the velocity of the jet be V assumed to be given by $V = C_v \sqrt{2gH}$; and Q be the discharge of fluid coming out of orifice. Applying the impulse-momentum equation, the force on the fluid to change its velocity from O to V is

$$F = \frac{wQ}{g}(V - O) = \frac{waV^2}{g}$$

$= 2waC_v^2H$ ...(14)

This issuing jet will exert a force on the tank which will be the reaction of the jet and its value will be equal to that of F given by equation 14 but in a direction opposite to V. A physical explanation for the existence of the reaction is that at the vena-contract a the pressure of the fluid is reduced to zero gage pressure and there is also a reduction of pressure of the fluid is reduced to zero gage pressure and there is also a reduction of pressure on the tank walls immediately adjacent to the orifice where the velocity of fluid becomes appreciable. However, on the opposite side of the tank at the same depth the pressure over a corresponding area is wH and the difference of pressure between the two sides of the tank gives rise to the reaction force. Further, it is seen from equation 14 that in the ideal case with no friction since $C_v = 1$, the reaction of the jet is twice the hydrostatic force exerted upon an area of the same size as the jet and at the same depth below the surface.

Now if the tank considered above is mounted on a frictionless trolley as shown in Fig. 3.4 (b), and the orifice is initially kept plugged, then as soon as the orifice is opened the jet will be issued and due to the reaction of the jet the tank will start moving with some velocity say u in the direction opposite to the direction of the issuing jet. Thus, the jet issuing from the orifice exerts a propelling force on the tank.

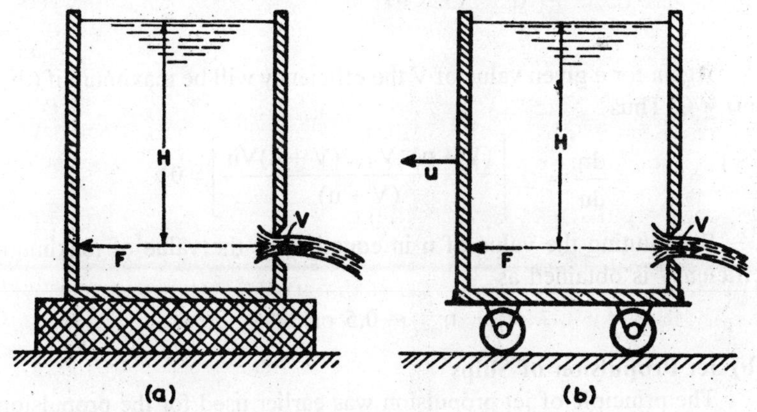

Fig. 3.4 : Jet propulsion of an orifice tank.

As the tank starts moving with a velocity u, the actual velocity of the issue of the jet will be V_r which is the velocity of the issuing jet relative to the moving tank. Thus V_r will be equal to the vectorial difference of the absolute velocity V of the jet and the velocity of propulsion u of the tank, *i.e.*, $V_r = [V-(-u) = (V+u)$. The negative sign for u has been considered because its direction is opposite to that of V. Thus applying impulse-momentum equation :

Propelling force, F = Reaction of jet

$$= \frac{W}{g}[(V+u)-u]$$

where W is the weight of fluid actually coming out per second. Since W = wa (V + u), as the effective velocity of the efflux of jet is V_r = (V + u), the expression for F becomes

$$F = \frac{wa(V+u)}{g}V \quad ...(15)$$

Work done by the jet on the moving tank = (F × u)

Actual kinetic energy of the issuing jet

$$= \frac{WV_r^2}{2g} = \frac{W(V+u)^2}{2g}$$

Hence the this case the efficiency of propulsion

$$\eta = \frac{F \times u}{\frac{WV_r^2}{2g}} = \frac{2Vu}{(V+u)^2} \quad ...(16)$$

Again for a given value of V the efficiency will be maximum if ($d\eta/du$) = 0. Thus

$$\frac{d\eta}{du} = 2\left[\frac{(V+u)^2V - 2(V+u)Vu}{(V+u)}\right] = 0$$

Substituting the value of u in equation 16 the value of maximum efficiency is obtained as

$$\eta_{max} = 0.5 \text{ or } 50\%$$

(b) Jet Propulsion of Ships

The principle of jet propulsion was earlier used for the propulsion of small ships. The ship carries centrifugal pumps which lift water from the surrounding sea and discharge it in the form of a jet by forcing through the orifice provided at the back of the ship, as shown in Fig. 3.5. The reaction produced by the jet entering the sea propels the ship in the direction opposite to that of the jet.

The pump intakes may have two alternative arrangements. In one case the intakes may face in the same direction as that of the issuing jet or the intakes may be on the sides of the ship. In the second arrangement the pump intakes may face in the direction of the motion of the ship. The main difference in the two arrangements is that if the pump intakes face in the direction of the jet then the water has to be sucked by the pumps against the motion of the ship, according more work will be required to be done by the pumps. On the other hand if the pump intakes face in the direction of motion of the ship then water will enter the pipe intakes due to the movement of the ship itself and hence less work will be required to be done by the pumps.

Let V be the absolute velocity of the issuing jet and u be the velocity of the moving ship. Thus, the velocity of the jet relative to the motion of the ship will be $V_r = (V + u)$. Since the effective velocity of the issue of the jet is V_r the kinetic energy available with the water

$$= \frac{WV_r^2}{2g}$$

where W represents the weight of water issuing from the jet per second. If a represents the area of the issuing jet, then $W = wQ = (waV_r)$.

Applying the impulse-momentum equation in the direction of the jet,

$$\text{Propelling force } F = \frac{W}{g}[V + u) - u] = \frac{W}{g}V \qquad ...(17)$$

The above expression for the propelling force may be readily derived by bringing the ship to a stationary state before the impulse-momentum equation is applied. For this a velocity equal in magnitude to that of the ship but in opposite direction, *i.e.*, –u, is applied to the whole system. Thereby bringing the ship to rest, but making the effective velocity of the jet as (V + u) and also developing a velocity equal to u in the same direction as that of the jet for the water in the surrounding sea. The application of the impulse-momentum equation will then provide the expression for the propelling force as given by equation 17.

The work done per second on the ship by the reaction of the jet is equal to

$$(F \times u) = \frac{WV_u}{g} = \frac{waV_r(V_r - u)u}{g} \qquad ...(18)$$

which is the output of the system.

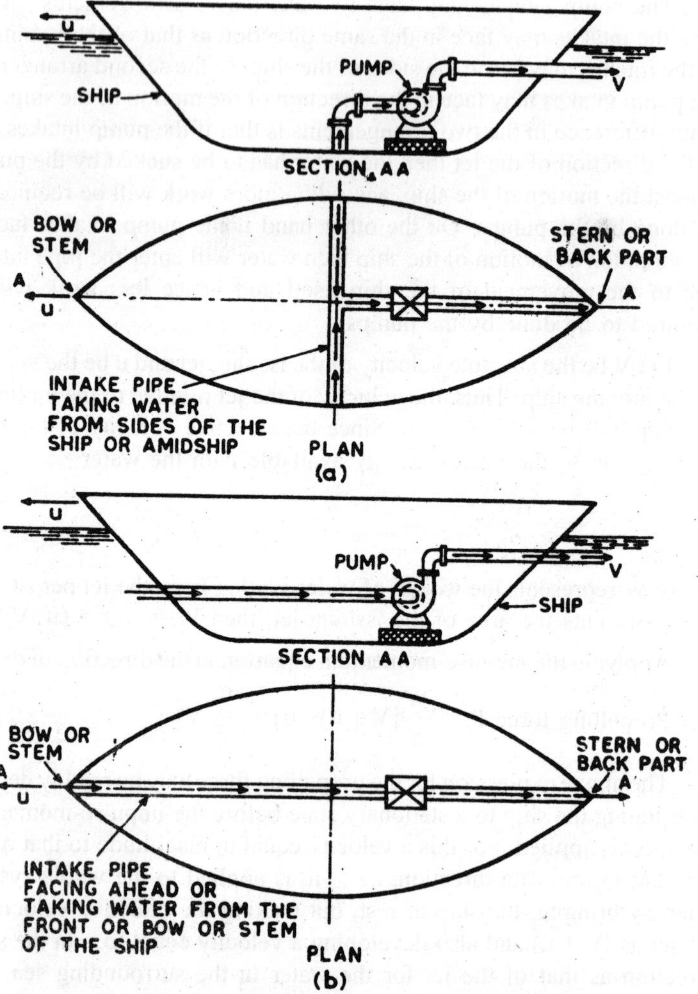

Fig. 3.4 : Jet propulsion of ships.

Now if it is assumed that the pump intakes face in the same direction as that of the issuing jet, then the energy required to be supplied will be equal to the kinetic energy of the jet. Thus energy supplied per second

$$\frac{WV_r^2}{2g} = \frac{waV_r^3}{2g}$$

∴ Efficiency of the propulsion

$$\eta = \frac{\dfrac{waV_r(V_r - u)u}{g}}{\dfrac{waV_r^3}{2g}}$$

$$= \frac{2(V_r - u)u}{V_r^2}$$

$$= \frac{2Vu}{(V+u)^2}$$

For a given jet velocity V, the condition for maximum efficiency of propulsion is given by $\left(\dfrac{d\eta}{du}\right) = 0$

Thus $\dfrac{d\eta}{du} = \dfrac{2V[(V+u)^2 - 2(V+u)u]}{(V+u)^4} = 0$

u = V, (since V ≠ 0 and also V ≠ –u)

Hence for maximum efficiency of propulsion u = V and by substitution

$$\eta_{max} = \frac{2u^2}{(2u)^2} = 0.5 \text{ or } 50\%$$

In the above derivation the loss of head due to friction etc. in the intake and ejecting pipes has been neglected. But if this loss of head is to be considered and is equal to H_L, then the corresponding loss of energy per sec = (WH_L). In which case the work done by the pump or the total energy supplied per second.

$$= \left(\frac{WV_r^2}{2g} + WH_L\right)$$

and then the efficiency of jet propulsion

$$\eta = \frac{\dfrac{WVu}{g}}{\left(\dfrac{WV_r^2}{2g} + WH_L\right)} = \frac{2Vu}{[V_r^2 + 2gH_L]}$$

In the above case it was assumed that the pump intake faces in the direction of the jet or is on one side of the ship. But if the pump intake faces in the direction of the motion of the ship, then since the water is possessing an initial kinetic energy equal to $\left(\frac{Wu^2}{2g}\right)$, corresponding to the velocity of the moving ship, the energy required to be supplied is reduced by this amount. Hence in this case the energy supplied per second

$$= \left[\left(\frac{W}{2g}V_r^2\right) - \left(\frac{W}{2g}u^2\right)\right] = \frac{W}{2g}(V_r^2 - u^2)$$

However in this case also the work done by this jet on the ship will be same as represented by equation 18. Accordingly the efficiency of propulsion will be

$$\frac{\frac{W}{g}(V_r - u)u}{\frac{W}{g}(V_r^2 - u^2)} = \frac{2u}{(V_r + u)} = \frac{2u}{(V + 2u)}$$

In this case, however it is not possible to derive a practical condition for maximum efficiency. But for u = V, which is the condition for the maximum efficiency in the previous case, corresponding value of the efficiency for this case will be

$$\eta = \frac{2u}{u + 2u} = \frac{2}{3} = 0.667 \text{ or } 66.7\%$$

Since in actual practice the velocity of the ship u will normally be less than the velocity of the jet V, and therefore the limiting value of u is equal to V. Accordingly the above obtained value of the of the efficiency may be considered as the maximum possible efficiency for this case.

Again in this case also if the head loss due to friction etc. in the intake and ejecting pipes is equal to H_L, then total energy supplied per second.

$$= \left[\frac{W}{2g}(V_r^2 - u^2) + WH_L\right]$$

According the efficiency of jet propulsion becomes

$$\eta = \frac{\frac{WVu}{g}}{\left[\frac{W}{2g}(V_r^2 - u^2) + WH_L\right]}$$

$$= \frac{2Vu}{[(V_r^2 - u^2) + 2gH_L]}$$

It may however be stated that the jet propulsion in ships is now not commonly adopted because the overall efficiency for such units is much lower than that of screw propeller units.

MOMENTUM THEORY OF PROPELLERS

A propeller is a revolving mechanism which uses the torque of a shaft to produce axial thrust. The conversion of torque into axial thrust is done by propeller by changing the momentum of the fluid in which it is submerged. When a propeller submerged in an undisturbed fluid rotates, it exerts a force on the fluid and pushes the fluid bac' wards. The reaction to this force on the fluid provides a forward force on the propeller itself and this force is the so-called 'propeller thrust' which is used for propulsion. Although the complete design of a propeller cannot be done according to the momentum theory, yet the application of this theory leads to some useful results as indicated by simple analyses of the problem below.

Fig. 3.6 shows a propeller moving to the left with velocity V through still fluid. By applying a velocity equal in magnitude to that of the propeller but in opposite direction (*i.e.*, –V) on the entire system, the propeller may be considered to be a stationary 'actuati g rotor' occupying a fixed position while the fluid on which it directly acts). The fluid enters the slip stream with (that is, the fluid on which it directly acts). The fluid enters the slip stream with velocity V at section 1 upstream from the propeller where the flow is undisturbed, and the fluid velocity increases as it approaches and leaves the propeller. At section 2 some distance behind the propeller, the fluid leaves the slip stream with velocity V_j. For a propeller to operate in a body of still fluid, the pressure at some distance a head of and behind the propeller (*i.e.*, at sections 1 and 2) and the pressure over the slip stream boundary are the same being equal to the pressure of the undisturbed fluid. As shown in the lower portion of Fig. 3.6, the pressure decreases from the value p_0, rises at the propeller, and then drops to p_0 again. It is assumed that all fluid elements passing through the propeller have their pressure increased by exactly the same amount Δp. The rotational effect of the propeller is neglected. Therefore, the thrust on the propeller is equal to

$$T_p = \frac{\pi D^2}{\Delta}(\Delta p) \qquad ....(19)$$

By applying the momentum equation to the free body of fluid between sections 1 and 2 the slip stream boundary, the only force acting on it in the flow direction is propeller thrust T_p, since the outer boundary of the body is everywhere at the same pressure.

Therefore $T_p = \rho Q(V_j - V)$ ...(20)

where Q is the rate of flow through the slip stream.

$$T_p = \frac{\pi D^2}{4}(\Delta p)$$

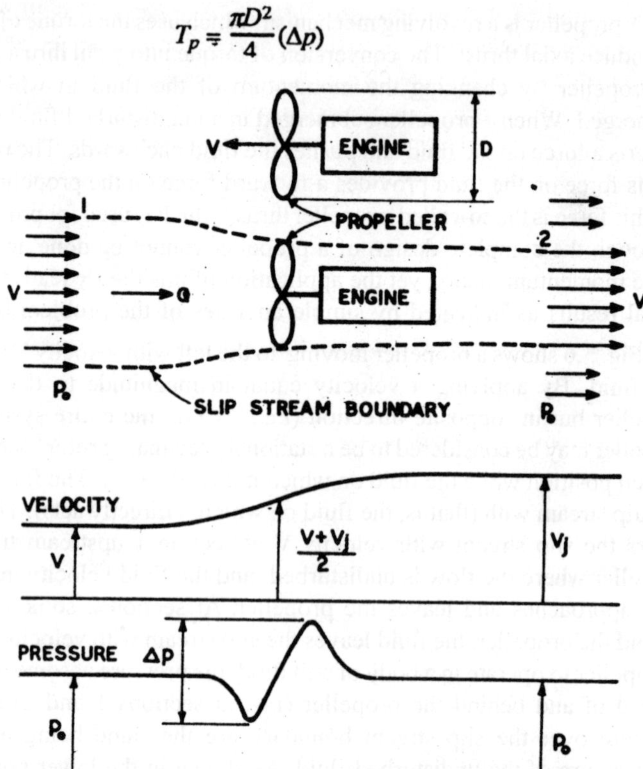

Fig. 3.6 : Propeller in a fluid stream.

Applying the energy equation between sections 1 and 2

$$p_0 + \rho\frac{V^2}{2} + \Delta p = p_0 + \rho\frac{V_j^2}{2} \quad ...(21)$$

in which Δp is the work which the propeller performs on the fluid in the slip stream. Simultaneous solution of the above equations yields

$$Q = \frac{\pi D^2}{4}\frac{V + V_j}{2} \quad ...(22)$$

Equation 22 shows that the velocity of flow at the propeller is the average of the approaching velocity V and the exit velocity V_j. This result known as Froude's theorem after William Froude (1810–79) is one of the principle assumptions in propeller design.

If the undisturbed fluid be considered stationary, the propeller advances through it at velocity V. The rate at which the useful work is done by the propeller is then equal to the product of the propeller thrust T_p and the velocity V. That is

$$\text{Power output} = T_p V = \rho Q (V_j - V)V \qquad ...(23)$$

In addition to the useful work, some power is lost in increasing the kinetic energy of flow in the slip stream, which is given as

$$\text{Power lost} = \rho Q \frac{(V_j - V)^2}{2} \qquad ...(24)$$

since $(V_j - V)$ is the velocity of the downstream fluid relative to earth. The power supplied to the propeller by the engine is the sum of power output and power lost. Thus

$$\text{Power input} = \left[\rho Q (V_j - V)V + \rho Q \frac{(V_j - V)^2}{2}\right] \qquad ...(25)$$

The theoretical propulsive efficiency η_{th}, sometimes known as the Froude efficiency is given by the ratio of the equations 23 and 25:

$$\eta_{th} = \frac{\text{Power output}}{\text{Power input}}$$

$$= \frac{\rho Q (V_j - V)V}{\left[\rho Q (V_j - V)V + \rho Q \frac{(V_j - V)^2}{2}\right]}$$

$$= \frac{2}{1 + (V_j / V)} \qquad ....(26)$$

It may however be noted that this efficiency does not account for friction or the effects of the rotational motion imparted by the propeller to the fluid, and hence it is considered to be theoretical. Further the propulsive efficiency, which is a function of the velocity ratio (V_j/V), increases as the ratio (V_j/V) decreases. As may be seen the efficiency will have a limiting value of 100 per cent if (V_j/V) is equal to one. This condition is however impossible, because such a propeller will produce no thrust due to zero velocity change. In practice, an aircraft propeller may have a maximum efficiency of about 0.85 to 0.9 times the value given by equation 26. However, owing to compressibility effects, the efficiency of an aircraft propeller drops rapidly with speeds above 640 km per hour ship propeller efficiencies are usually less, owing to restrictions in diameter.

SOLVED EXAMPLES

Example 1:

Velocity distribution for laminar flow of real fluid in a pipe is given as $v = V_{max}\ [1-r^2/R^2)]$, where V_{max} is velocity at the centre of the pipe, R is pipe radius, and v is velocity at radius r from the centre of the pipe. Determine the momentum correction factor.

Solution:

From equation 10, Momentum correction factor is given as

$$\beta = \frac{1}{AV^2}\int_A v^2 dA$$

$$\text{Mean velocity } V = \frac{Q}{A} = \frac{\int vdaA}{A}$$

$$= \frac{\int^R 0V_{max}\left(\frac{R_2 - r^2}{R^2}\right)(2\pi rdr)}{\pi R^2} = \frac{V_{max}}{2}$$

$$\text{Thus } \beta = \frac{1}{(\pi R^2)\left(\frac{V_{max}}{2}\right)^2}\int_0^R V_{max}^2\left(\frac{R^2 - r^2}{R^2}\right)^2(2\pi rdr)$$

$$= \frac{8}{R^6}\left(\frac{1}{2}R^6 - \frac{1}{2}R^6 + \frac{1}{6}R^6\right) = \frac{4}{3} = 1.33.$$

Example 2:

A bend in pipeline conveying water gradually reduces from 0.6 m to 0.3 m diameter and deflects the flow through angle of 60°. At the larger end the gage pressure is 171.675 kN/m² [1.75 kg (f) cm²]. Determine the magnitude and direction of the force exerted on the bend, (a) when there is no flow, (b) when the flow is 876 litres/s.

Solution:

(a) When there is no flow the pressure at both the sections of the bend is same, *i.e.*,

$$p_1 = p_2 = 171.675 \text{ kN/m}^2 \qquad \text{(SI units)}$$

or $$p_1 = p_2 = 1.75 \text{ kg/cm}^2 \qquad \text{(metric units)}$$

Let R be the force exerted on the bend and R_x and R_y be its components as shown in the accompanying figure 3.7.

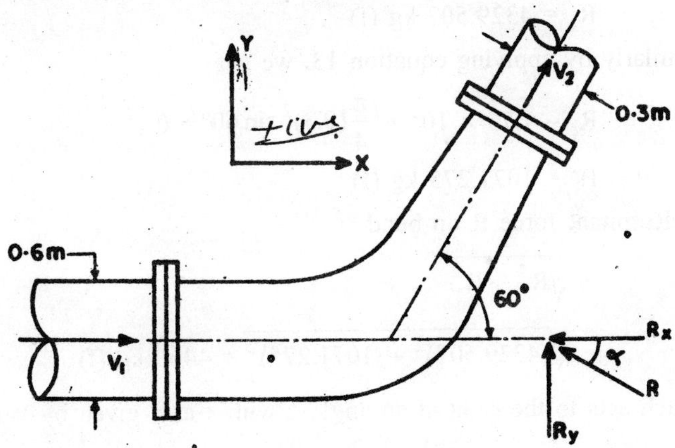

Fig. 3.7

SI units

By applying equation 12, we get

$$171.675 \times \frac{\pi}{4}(0.6)^2 - 171.675 \times \frac{\pi}{4}(0.3)^2 \cos 60^\circ . - R_x = 0$$

or $R_x = 4.2.472$ kN

Similarly by applying equation 8.13, we get

$$R_y - 171.675 \times \frac{\pi}{4}(0.3)^2 \sin 60^\circ = 0$$

or $R_y = 10.509$ kN

∴ Resultant force R on the bend

$$= \sqrt{R_x^2 + R_y^2}$$

$$= \sqrt{(42.472)^2 + (10.509)^2} = 43.753 \text{ kN}$$

which acts to the right at an angle α with x-axis given by

$$\alpha = \tan^{-1}\left(\frac{10.509}{42.472}\right) = 13°54'$$

Metric units

By applying equation 12, we get

$$1.75 \times 10^4 \times \frac{\pi}{4}(0.6)^2 - 1.75 \times 10^4 \times \frac{\pi}{4}(0.3)^2 \cos 60° - R_x = 0$$

or $\quad R_x = 4329.507$ kg (f)

Similarly by applying equation 13, we get

$$R_y - 1.75 \times 10^4 \times \frac{\pi}{4}(0.3)^2 \sin 60° = 0$$

or $\quad R_y = 1071.275$ kg (f)

∴ Resultant force R on bend

$$= \sqrt{R_x^2 + R_y^2}$$

$$= \sqrt{(4329.507)^2 + (1071.275)^2} = 4460 \text{ kg (f)}$$

which acts to the right at an angle a with x-axis given by

$$\alpha = \tan^{-1}\left(\frac{1071.275}{4329.507}\right) = 13.54$$

(b) For continuity of flow

$$Q = A_1V_1 = A_2V_2$$

$$876 \times 10^{-3} = \frac{\pi}{4}(0.6)^2 V_1 = \frac{\pi}{4}(0.3)^2 V_2$$

$$V_1 = 3.1 \text{ m/s; and } V_2 = 12.4 \text{ m/s}$$

SI units

Neglecting frictional losses, and by applying Bernoulli's equation, we get

$$\frac{p_1}{w} + \frac{V_1^2}{2g} = \frac{p_2}{w} + \frac{V_2^2}{2g}$$

or

$$\frac{171.675 \times 10^3}{9810} + \frac{(3.1)^2}{2 \times 9.81} = \frac{p_2}{w} + \frac{(12.4)^2}{2 \times 9.81}$$

or $\qquad \dfrac{p_2}{w} = (17.50 + 0.49\text{-}7.84) = 10.15 \text{ m}$

$\therefore \qquad p_2 = (10.15 \times 9810)$

$= 99572 \text{ N/m}^2 = 99.572 \text{ kN/m}^2$

By applying equation 12, we get

$$171675 \times \frac{\pi}{4}(0.6)^2 - 99572 \times \frac{\pi}{4}(0.3)^2 \cos 60° - R_x$$

$$= (1000 \times 876 \times 10^{-3})\,(12.4 \cos 60° - 3.1)$$

or $\qquad R_x = 42305 \text{ N} = 42.305 \text{ kN}$

Similarly by applying equation 8.13, we get

$$R_y - 99572 \times \frac{\pi}{4}(0.3)^2 \sin 60°$$

$$= (1000 \times 876 \times 10^{-3})\,(12.4 \sin 60°)$$

or $\qquad R_y = 15502 \text{ N} = 15.502 \text{ kN}$

$\therefore$ Resultant force F on the bend

$$= \sqrt{R_x^2 + R_y^2}$$

$$= \sqrt{(42.305)^2 + 15.502)^2} = 45.056 \text{ kN}$$

which acts to the right at an angle a with x-axis given by

$$\alpha = \tan^{-1}\left(\frac{15.502}{42.305}\right) = 20° \, 8'$$

Metric units

Neglecting frictional losses, and by applying Bernoulli's equation, we get

$$\frac{p_1}{w} + \frac{V_1^2}{2g} = \frac{p_2}{w} + \frac{V_1^2}{2g}$$

or $$\frac{1.75 \times 10^4}{1000} + \frac{(3.1)^2}{2 \times 9.81} = \frac{p_2}{w} + \frac{(12.4)^2}{2 \times 9.81}$$

or $\qquad \dfrac{p_2}{w} = (17.5 + 0.49\text{-}7.84) = 10.15 \text{ m}$

$p_2 = (10.15) \times 1000)$

$$= 10150 \text{ kg (f) m}^2 = 1.015 \times 10^4 \text{ kg (f)/m}^2$$

By applying equation 12, we get

$$1.75 \times 104 \times \frac{\pi}{4}(0.3)^2 \cos 60^\circ - R_x$$

$$= \frac{1000 \times 876 \times 10^{-3}}{9.81}(12.4 \cos 60^\circ - 3.1)$$

or $R_x = 4312.46$ kg (f)

Similarly by applying equation 13, we get

$$R_y - 1.015 \times 10^4 \times \frac{\pi}{4}(0.3)^2 \sin 60^\circ$$

$$= \frac{1000 \times 876 \times 10^{-3}}{9.81}(12.4 \sin 60^\circ)$$

or $R_y = 1580.27$ kg (f)

∴ Resultant force F on the bend

$$= \sqrt{R_x^2 + R_y^2}$$

$$= \sqrt{(4312.46)^2 + (1580.27)^2} = 4593 \text{ kg (f)}$$

which acts to the right at an angle α with x-axis given by

$$\alpha = \tan^{-1}\left(\frac{1580.27}{4312.46}\right) = 20^\circ 8'$$

Example 3:

Water flows through a 0.9 m diameter pipe at the end of which there is a reducer connecting to a 0.6 m diameter pipe. If the gage pressure at the entrance to the reducer is 412.02 kN/m²/4.2 kg (f) cm²] and the velocity is 2 m/s, determine the resultant thrust on the reducer, assuming that the frictional loss of head in the reducer is 1.5 m.

Solution:

For continuity of flow

$$\frac{\pi}{4}(0.9)^2 \times 2 = \frac{\pi}{4}(0.6)^2 \times V_2$$

$$V_2 = 4.5 \text{ m/s}$$

SI units

Applying Bernoulli's equation, we have

$$\frac{p_1}{w}+\frac{V_1^2}{2g}=\frac{p_2}{w}+\frac{V_2^2}{2g}+jf$$

$$\frac{412.02\times10^3}{9810}+\frac{(2)^2}{2\times9.81}=\frac{P_2}{w}+\frac{(4.5)^2}{2\times9.81}+1.5$$

or $\frac{p_2}{w}$ = (42.0 + 0.24 – 1.032 – 1.5) = 39.672 m

∴ p_2 = (39.672 × 9810)

= 389182 N/m² = 389.182 kN/m²

Let F_x be the force exerted by the reducer on the fluid, acting opposite to the direction of flow, then applying equation 12, we get

$$412.02\times10^3\times\frac{\pi}{4}(0.9)^2-389.182\times10^3\times\frac{\pi}{4}(0.6)^2-F_x$$

$$=100\times\frac{\pi}{4}(0.9)^2\times2(4.5-2)$$

or F_x = 148896 N =148.896 kN

∴ The resultant thrust exerted by the fluid on the reducer = 148.896 kN, which is acting in the direction of flow.

Metric units

Applying Bernoulli's equation, we have

$$\frac{p_1}{w}+\frac{V_1^2}{2g}=\frac{p_2}{w}+\frac{V_1^2}{2g}+h_f$$

or $$\frac{4.2\times10^4}{1000}+\frac{(2)^2}{2\times9.81}=\frac{P_2}{w}+\frac{(4.5)^2}{2\times9.81}+1.5$$

or $\frac{p_2}{w}$ = (42.0 + 0.24 –1.032 – 1.5) = 39.672 m

∴ p_2 = (39.672× 1000)

= 39.672 × 10³ kg(f) m²

Let F_x be the force exerted by the reducer on the fluid, acting opposite to the direction of flow, then applying equation 12, we get

$$4.2\times104\times\frac{\pi}{4}(0.9)^2-39.672\times10^3\times\frac{\pi}{4}(0.6)^2-F_x$$

$$=\frac{1000}{9.81}\times\frac{\pi}{4}(0.9)^2\times2(4.5-2)$$

or $F_x = 15178\text{kg (f)}$

∴ The resultant thrust exerted by the fluid on the reducer = 15178 kg(f). Which is acting in the direction of flow.

Example 4:

A tank 1.5 m high stands on a trolley and is full of water. It has an orifice of diameter 0.1 m at 0.3 m from the bottom of thank tank. If the orifice is suddenly opened, what will be the propelling force on the trolley? Coefficient of discharge of the orifice is 0.60.

Solution:

Discharge from the orifice

$$= \text{Cda}\sqrt{2gH}$$

$$= 0.60 \times \frac{\pi}{4}(0.1)^2(2\times 9.81\times 1.2)^{1/2}$$

$$= 0.023\text{m}^2/\text{s}$$

Velocity of the jet issuing from the orifice

$$= \frac{Q}{a} = \frac{0.023\times 4}{\pi\times(0.1)^2} = 2.93\text{m/s}$$

$$\text{Propelling force F} = \frac{wQV}{g}$$

$$= \frac{9810\times 0.023\times 2.39}{9.81} = 67.39\text{N}.$$

Example 5:

The resistance to motion of a vessel is 24.525 kN at a velocity of 4.5 m/s. The jet efficiency is to be 80% and the mechanical efficiency of the pumps is 75% Hydraulic losses in the ducts are 5% of the relative kinetic energy at exit. Determine (a) the velocity of the jet; (b) the orifice area at exit; (c) the power required to drive the pumps for the given speed of the vessel, assuming that the water is drawn in through the intakes facing in the direction of the motion of the ship.

Solution:

(a) u = 4.5 m/s and the efficiency = 80%

But efficiency

$$\eta = \frac{2u}{V_r + u}$$

or $0.8 = \dfrac{2 \times 4.5}{V_r + 4.5}$

$V_r = 6.75$ m/s

and $V = 2.25$ m/s

(b) Work done per second

$= \text{resistance} \times \text{velocity of vessel}$

$= (24.525 \times 4.5) = 110.363$ kN m/s

$$= \frac{W}{g}(V_r - u)u$$

$$= \frac{W}{g}(6.75–4.5)\ 4.5$$

$$W = \frac{110.363 \times 9.81}{4.5 \times 2.25}$$

$= 106.929$ kN/s $= 106929$ N/s

$W = waV_r$

$$\therefore a = \frac{W}{wV_r} = \frac{106929}{9810 \times 6.75} = 1.615 \text{ m}^2$$

Area of orifice at exit

$= 1.615$ m^2

(c) Power supplied to jet

$$= \frac{W}{2g}(V_r^2 - u^2)$$

Loss of energy in ducts

$$= 0.05 \frac{V_r^2}{2g} \times W$$

$\therefore$ Total power required from pumps

$$= \frac{W}{2g}[(V_r^2 - u^2) + 0.05V_r^2]$$

$$= \frac{106929}{2 \times 9.81}[(6.75^2 - 4.5^2) + 0.05 \times 6.75^2]W$$

$= 150369 \text{ W} = 150.396 \text{ kW}$

The mechanical efficiency of the pumps is 75%

∴ Power required to drive pumps

$$= \frac{150.369}{0.75} = 200.5 \text{ kW}$$

Example 6:

A ship whose resistance is 24.525 kN is to be driven at 5 m/s by means of a jet of water directed under water. The velocity of the jet is to be 7.5 m/s relative to the ship. The efficiency of the pump operating the jet is estimated to be 80%, the frictional resistance of the pipes being equal to 3 m of water. Calculate :

(a) The power required to drive the pump;

(b) The overall efficiency of the system in the following cases:

(i) the water enters the ship through an inlet facing a head:

(ii) the water enters through an inlet in side of ship.

Solution:

$u = 5 \text{ m/s};$

$V_r = 7.5 \text{ m/s};$

$R = 24.525 \text{ kN}$

$\eta_\rho = 0.8;$

and $H_L = 3 \text{ m}$

The reaction of the jet should be just equal to the resistance to the motion of the ship

$$F = R = 24.525 \text{ kN}$$

But $$F = \frac{W}{g}V = \frac{W}{g}(V_r - u)$$

or $$24.525 = \frac{W}{9.81}(7.5-5)$$

∴ $$W = 96.236 \text{ kN} = 96236 \text{ N}$$

(a) (i) when the water enter the ship through an inlet facing ahead, the output of the pump

$$= \left[\left(\frac{WV_r^2}{2g} - \frac{Wu^2}{2g}\right) + WH_L\right]$$

$$= W\left[\frac{(V_r^2 - u^2)}{2g} + H_L\right]$$

∴ Output of the pump

$$= 96236\left[\frac{(7.5^2 - 5^2)}{2\times 9.81} + 3.0\right]$$

$$= 441989 \text{ W} = 441.989 \text{ kW}$$

∴ Input of the pump

$$= \frac{\text{Output}}{\eta_\rho}$$

$$= \frac{441.989}{0.8} = 552.486 \text{ kW}$$

∴ Power required to drive the pump

$$= 552.486 \text{ kW}$$

(ii) When the water enters through an inlet in the side of the ship, the output of the pump

$$= \left[\frac{WV_r^2}{2g} + WH_L\right]$$

$$= W\left[\frac{V_r^2}{2g} + H_L\right]$$

∴ Output of the pump

$$= 96236\left[\frac{7.5^2}{2\times 9.81} + 3.0\right]$$

$$= 564614 \text{ W} = 564.614 \text{ kW}$$

∴ Input of the pump

$$= \frac{\text{Output}}{\eta} = \frac{564.614}{0.8}$$

$$= 705.768 \text{ kW}$$

(b) (i) The overall efficiency of the system for this case

$$\eta = \frac{F\times u}{\text{Imput of the jump}}$$

$$= \frac{24.525 \times 10^3 \times 5}{552.486 \times 10^3}$$

$$= 0.222 \text{ or } 22.2\%$$

(ii) The overcall efficiency of the system for this case

$$\eta = \frac{F \times u}{\text{Imput of the pump}}$$

$$= \frac{24.525 \times 10^3 \times 5}{705.768 \times 10^3}$$

$$= 0.174 \text{ or } 17.4\%.$$

Example 7:

The diameter of a pipe bend is 0.3 m at inlet and 0.15 m at outlet and the flow is turned through 120° in a vertical plane. The axis at inlet is horizontal and the centre of the outlet section is 1.5 m below the centre of the inlet section. The total volume of fluid contained in the bend is 0.085m³. Neglecting friction, calculate the magnitude and direction of the force exerted on the bend by the water flowing through it at 225 l/s when the inlet pressure is 137.34 kN/m².

Solution:

For continuity of flow

$$Q = A_1V_1 = A_2V_2$$

$$\text{or } 225 \times 10^{-3} = \frac{\pi}{4}(0.3)^2 V_1 = \frac{\pi}{4}(0.15)^2 V_2$$

$$\therefore \quad V_1 = 3.18 \text{ m/s;}$$

$$\text{and} \quad V_2 = 12.73 \text{ m/s}$$

Neglecting friction losses, by applying Bernoulli's equation, we get

$$\frac{p_1}{w} + \frac{V_1^2}{2g} + Z_1 = \frac{p_2}{w} + \frac{V_2^2}{2g} + Z_2$$

$$\text{or} \quad \frac{137.34 \times 10^3}{9810} + \frac{(3.18)^2}{2 \times 9.81} + 1.5 = \frac{p_2}{w} + \frac{(12.73)^2}{2 \times 9.81} + 0$$

$$\text{or} \quad \frac{p_2}{w} = (14.0 + 0.515 + 1.5 - 8.26) = 7.755 \text{ m}$$

$$\therefore \quad p_2 = (7.755 \times 9810)$$

$= 76077 \text{ N/m}^2 = 76.077 \text{ kN/m}^2$

By applying equation 12, we get

$$137.34 \times 10^3 \times \frac{\pi}{4}(0.3)^2 - 76.077 \times 10^3 \times \frac{\pi}{4}(0.15)^2 \cos 120° - F_x$$

$$= 1000 \times 225 \times 10^{-3} (12.73 \cos 120° - 3.18)$$

$$\text{or } F_x = \frac{\pi}{4}(0.5)^2[549.36 \times 10^3 + 38.039 \times 10^3] + 225\ (6.37 + 3.18)$$

$$\text{or } F_x = 12529 \text{ N} = 12.529 \text{ kN}$$

Similarly by applying equation 13, we get

$$F_y + (0.085 \times 9810) - 76.077 \times 10^3 \times \frac{\pi}{4}(0.15)^2 \sin 60°$$

$$= 100 \times 225 \times 10^{-3} (12.73 \sin 60°)$$

$$\text{or} \quad F_y = (2480.51 - 833.85 + 1164.28)$$

$$= 2810.94 \text{ N} = 2.811 \text{kN}$$

$$\text{Thus} \quad F = \sqrt{F_x^2 + F_y^2}$$

$$= \sqrt{(12.529)^2 + (2.811)^2} = 12.84 \text{ kN}$$

$$\tan \alpha = \frac{2.811}{12.529} = 0.2244$$

$$\therefore \alpha = 12°39'$$

∴ Force of 12.84 kN acts on the bend at an angle of 12°39' upwards from inlet axis.

Example 8:

A boat travelling at 12 m/s in fresh water has a 0.6 m diameter propeller which takes 4.25 m³ of water per second between its blades. Assuming that the effects of the propeller hub and the boat hull on flow conditions are negligible, calculate the propeller hub and the boat hull on flow conditions are negligible, calculate the thrust on the boat, the theoretical efficiency of the propulsion, and the power input to the propeller.

Solution:

From equation 22, we have

$$Q = \frac{\pi D^2}{4}\frac{V + V_j}{2}$$

$$\text{or } 4.25 = \frac{\pi}{4}(0.6)^2\frac{12 + V_j}{2}$$

$\therefore V_j = 18.06$ m/s

From equation 8.20 we have

$$T_p = \rho Q\,(V_j - V)$$

or $\quad T_p = 1000 \times 4.25\ (18.06 - 12)$

$$T_p = 25755 \text{ N} = 25.755 \text{ kN}$$

Again from equation 26, we have

$$\eta_{th} = \frac{2}{1 + (V_j/V)}$$

$$= \frac{2}{1 + (18.06/12)} = 0.798 \text{ or } 79.8\%$$

From equation 25

$$\text{Power input} = \left[\rho Q(V_j - V)V + \rho Q\frac{V_j - V)^2}{2}\right]$$

$$= \rho Q\ (V_j - V)\left[V + \frac{V_j - V}{2}\right]$$

$$= 1000 \times 4.25\ (18.06 - 12)\left[12 + \frac{18.06 - 12}{2}\right]$$

$$= 387098 \text{ W} = 387.098 \text{ kW.}$$

Example 9:

A water sprinkler has 10 mm diameter nozzles at either end of a rotating arm, each of which is discharging water in opposite direction at right angle to the rotating arm, at a velocity of 8 m/s. If the axis of rotation is at a distance of 0.15 m from one end and 0.2 m from the other, determine the torque required to hold the arm stationary. If friction is neglected, determine the constant angular speed of the arm.

Solution:

The rate of change of moment of momentum is the required to hold the arm stationary.

Initial moment of momentum is zero. Final moment of momentum

$= \rho Q(V_2 r_2 + V_1 r_1)$

$\therefore$ Torque $T = \rho Q(V_2 r_2 + V_1 r_1)$

$$= 1000 \times \left[\frac{\pi}{4} \times (0.01)^2 \times 8\right][8 \times 0.2 + 8 \times 0.15]$$

$= 1.759$ N-m

If the angular velocity of the sprinkler is ω, then the absolute velocities of flow through the nozzle are

$V_1 = 8 - 0.15\,\omega$

and $V_2 = 8 - 0.2\,\omega$

Since the moment of momentum of flow entering is zero and there is no friction, the moment of momentum leaving the sprinkler must also zero.

Thus $\rho Q[8 - 0.15\omega) \times 0.15 + (8 - 0.2\omega) \times 0.2] = 0$

or $\omega = \dfrac{2.8}{0.0625} = 44.8$rad/s.

EXERCISES

1. To propel a light aircraft at an absolute velocity of 250 km per hour against a head wind of 48 km per hour a thrust of 10.3 kN [1050 kg (f)] is required. Assume a theoretical efficiency of 90% and a constant air density of 1.207 kg/m^3 (0.123msl/m^3). determine the diameter of ideal propeller required and the power needed to drive it. [2.54 m; 947 kW (1288 h.p.)]
2. A pipeline 0.6 m diameter conveying oil (sp. gr. 0.85) at the flow rate of 1800 litres per second has a 90° bend in the horizontal plane. The pressure at the entrance to the bend is 147.15 kN/m^2 [1.5 kg (f)/cm^2] and loss of head in the bend is 2 m of oil. Find the magnitude and direction of the force exerted by the oil on the bend.

   [69.368 kN{7071 kg (f)} to the right at 42°15' with horizontal]
3. A nozzle at the end of a 80 mm hose produces a jet 40 mm in diameter. Determine the longitudinal stress in the joint at the base of the nozzle when it is discharging 1200 litres of water per minute. [358.37N {36.53 kg (f)} tensile]

4. A motor boat is driven at 4 m/s by means of a jet of water issuing from an opening 100 mm square directly behind the boat and having discharge equal to 0.5 m^3/s. Find the driving force. The coefficient of contraction of the jet is 0.62.

   [38.322 kN [3906.5 kg (f)]

5. Oil of sp. gr. 0.8 flows through a horizontal pipe of diameter 100 mm which is provided with a nozzle 25 mm diameter. If the pressure at the base of the nozzle is 784.8 kN/m^2 [8 kg (f)/cm^2), find the force exerted on the nozzle.

   [5.439 kN {554.39 kg (f)}]

6. The velocity distribution in a pipe is given by

$$v = V_{max}\left(1 - \frac{r}{R}\right)^K$$

   where R is the radius of the pipe, r is any radius at which the velocity is v and K is a constant index. Find the momentum correction factor.

$$\left[\frac{(K+1)^2(K+2)^2}{2(2K+1)(2K+2)}\right]$$

7. Two large plates are spaced 50 mm apart. If the velocity profile between the plates is represented by $v = V_{max}\,(1 - 1600\,r^2)$ where r is measured in m from the centre line between the plates, determine the momentum correction factor.

8. Water a pressure of 294.3 kN/m^2 [3kg(f) cm^2] flows through a horizontal pipe of 100 mm diameter with a velocity of 2 m/s.

   (a) If the diameter of the pipe gradually reduces to 50 mm what is the axial force on the pipe assuming no loss of energy.

   (b) If a bend is connected to the pipe which turns through 30° and tapers uniformly from 100 mm to 50 mm. Find the force exerted on the bend other than those due to gravity.

   [(a) 1.698 kN (173.08 Kg (f);
   (b) 1.8.04 kN {183.94 Kg (f)}]

9. A small ship is fitted with jets of total are 0.65 m^2. The velocity through the jets relative to ship is 9 m/s and ship's speed is 18.5 km/hour. The engine efficiency is 85%; the pump efficiency is 65% and pipe losses are equal to 10% of the kinetic energy of

the jets. Determine the propelling force and the overall efficiency. Water is drawn in amid ship (*i.e.*, at the middle of the ship from sides) and sea water weighs 10.055 kN/m^3 [1025 kg (f) m^3].

[23.145 kN {2359.4 kg (f)}: 24.6%]

10. A jet propelled boat has two jets each of 100 mm diameter. The boat travels at 30 km/hour. If the resistance of the motion of the boat is 88.29 u^2 N [9u^2 kg (f) where u is the velocity of the boat in m/s, calculate:

    (a) the velocity of the jet relative to boat;

    (b) the power required to drive the pump if the pump efficiency is 80%; and

    (c) the overall efficiency.

    Assume that the pump intake is facing the direction of motion of the boat.

    (a) 24.36 m/s;

    (b) 125.30 kW (170.33 h.p.);

    (c) 50.96%]

11. A pipe of 1m diameter carrying 2.5 m^3/s of water, is deflected through a 90° bend. The ends of the bend are anchored by the rods at right angles to the bend (one tie rod at each end). Find the tension in each rod. Also determine the resultant dynamic thrust on the bend and the direction of this thrust.

    [7.958 kN {811.2 kg (f); 1.254 kN {1147.2 kg (f)} at 45° with horizontal]

12. A 100 mm diameter orifice at the end of a 150 mm diameter pipe yields a jet of oil 85 mm in diameter. What force will be exerted upon the orifice plate when the pressure intensity of the approaching flow is 13.734 kN/m^2 [0.14 kg (f) cm^2].

13. If a 0.522 kW (0.75 h. p.) motor is required by a ventilating fan to produce a 0.6 m stream of air having a velocity of 12 m/s, what is the efficiency of the fan? Take ρ = 1.207 kg/m^3 (0.123 msl/m^3).

4

Conservation of Momentum

The momentum of a body is defined as the product of the mass of the body and its velocity *i.e.*, $\frac{mq}{g_0}$, and has the dimension of force-time. In the flow of fluids the momentum M per unit volume is given by

$$M = \frac{\sigma q}{g_0} = \rho q$$

Since velocity is a vector quantity so momentum is likewise, a vector quantity, having magnitude and direction both.

EULER'S EQUATION OF MOTION ALONG A STREAMLINE

Consider an elementary section of a stream tube. Let δs be the length the fluid particle moving along a streamline in the positive direction is $\rho\delta A\ \delta s$. The force acting on the element are of two types:

(i) Body forces, and

(ii) Surface forces exerted due to hydrostatic pressure on the end areas of the particle.

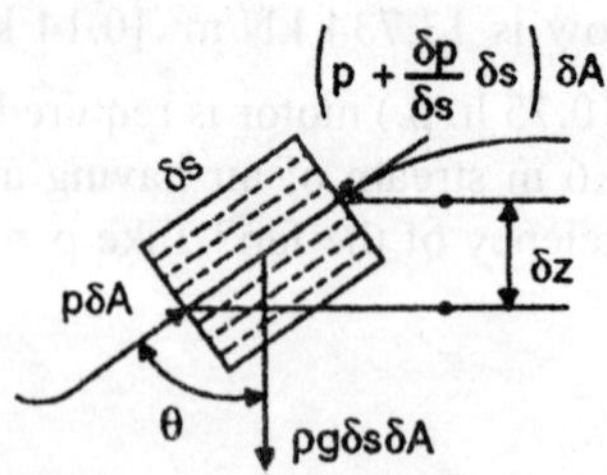

Fig. 4.1

The body force is $\rho Fs\ \delta a\ \delta s$. On the upstream face the pressure force is $p\ \delta A$ in the (+ s) direction and on the downstream face it is $\left(p + \frac{\partial p}{\partial s}\delta s\right) dA$ acting in the (–s) direction. The total along the path δs with tangential unit vector is given by

$$= \rho F_s\ \delta s\ \delta A + \left[p\delta A - \left(P + \frac{\partial p}{\delta s}\delta s\right)\delta A\right]$$

$$= \rho\ F_s\ \delta s\ \delta A - \frac{\partial p}{\partial s}\delta s\ \delta A.$$

The acceleration of the fluid flowing along δs is $\frac{Dq}{Dt}$. By using Newton's second law of motion the equation of momentum along the path is given by

$$\frac{Dq}{Dt}\rho\delta s\ \delta A = \rho F_s \delta s \delta A - \frac{\partial p}{\partial s}\delta s\ \delta A$$

$$\text{or}\quad \frac{Dq}{Dt} = F_s - \frac{1}{\rho}\frac{\partial p}{\partial s}$$

$$\text{or}\quad \frac{Dq}{Dt} + q\frac{\partial q}{\partial s} = F_s - \frac{1}{\rho}\frac{\partial p}{\partial s}, \qquad ...(1)$$

known as the Euler's equation of motion for one-dimensional flow.

Consider the body force due to the pull of gravity. The gravity force is $\rho g\delta s\ \delta A$, its component along the s direction are

$$\rho Fs\ \delta s\ \delta A = -\rho\ \delta s\ \delta A\ g\cos\theta$$

$$\Rightarrow F_s = -g\cos\theta.$$

Since δz is the increase in elevation of the particle for a displacement δs then

$$F_s = -g\ (\partial z/\partial s),\ \text{as}\ \cos\theta = (\partial z/\partial s) \qquad ...(2)$$

From (1) and (2), we obtain

$$\frac{\partial q}{\partial t} + q\frac{\partial q}{\partial s} = -g\ \frac{\partial z}{\partial s} - \frac{1}{\rho}\frac{\partial p}{\partial s}. \qquad ...(3)$$

For steady flow $\partial q/\partial t = 0$, the equation (3) reduces to

$$q\frac{\partial q}{\partial s} = -g\ \frac{\partial z}{\partial s} - \frac{1}{\rho}\frac{\partial p}{\partial s},$$

where **q**, z and p are functions of s only. The partial derivatives may be replaced by the total derivatives

$$\frac{dp}{\rho} + g\ dz + \mathbf{q}\ d\mathbf{q} = 0 \qquad \text{...(4)}$$

$$\int \frac{dp}{\rho} + \mathbf{q}z + \frac{1}{2}\mathbf{q}^2 = \text{constant.}$$

which is an alternative form of *Euler's equation of motion along a streamline* for inviscid and steady flow. It may be integrated if ρ is known as a function of p or is a constant. The pressure along a stream line can be determined without assuming the existence of a velocity potential.

EQUATION OF MOTION OF AN INVISCID FLUID

Consider any arbitrary closed surface 5 drawn in the region occupied by the incompressible fluid at an instant t. We know by Newton's second law of motion that the total force acting on this mass of fluid is equal to the rate of change of linear momentum. The forces are due to (i) the normal pressure thrusts on the boundary, and (ii) the external force (*e.g.*, gravity) F per unit mass.

Let ρ be the density of the fluid particle P with in the closed surface and dτ be the volume enclosing P. The mass of the element ρ dτ will always remain constant. Consider **q** be the velocity of the fluid particle P then the momentum of the volume is

$$M = \int q\rho\ d\tau\,. \qquad \text{...(1)}$$

The time rate of change of momentum is given by differentiating (1) with regard to t, thus we have

$$\frac{dM}{dt} = \int \frac{dq}{dt}(\rho d\tau) + \int \mathbf{q}\frac{d}{dt}(\rho d\tau) = \int \frac{dq}{dt}.\rho d\tau \qquad \text{...(2)}$$

The second integral vanishes as the mass (ρdτ) remains constant for all time.

Let F be the external force per unit mass acting on fluid particle P then the total force on volume is

$$= \int F\rho d\tau\,. \qquad \text{...(3)}$$

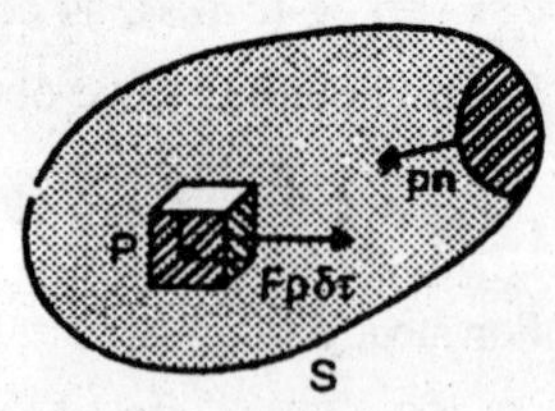

Fig. 4.2

Again, let p be the pressure at a point on the surface along the outward drawn unit normal $\hat{\mathbf{n}}$ then the force on the fluid particle due to the actions of the surrounding fluid is

$$= -\int p\hat{\mathbf{n}}dS = -\int \nabla p d\tau. \qquad ...(4)$$

The equation for the momentum balance is written as

Rate of momentum accumulation = Rate of momentum in – Rate of momentum out + forces acting on system.

$$\int \frac{d\mathbf{q}}{dt}.\rho\, d\tau = -\int F\rho d\tau - \int \nabla \rho d\tau,$$

$$\text{or} \quad \int \left[\rho \frac{d\mathbf{q}}{dt} - \rho F + \nabla p\right] d\tau = 0.$$

Since the volume of integration enclosed in the surface is arbitrary, we can reduce this volume to a point. Therefore

$$\rho \frac{d\mathbf{q}}{dt} - \rho F + \nabla p = 0,$$

$$\text{or} \quad \frac{d\mathbf{q}}{dt} = F - \frac{1}{\rho}\nabla p, \qquad ...(5)$$

known as *Euler's equation of motion at all points of the fluid* which applies only to ideal fluids, the dissipative effects have not been considered.

The equation (5) may be expressed as

$$\frac{\partial \mathbf{q}}{\partial t} + (\mathbf{q}.\nabla)\mathbf{q} = F - \frac{1}{\rho}\nabla p \left(\text{Since} \frac{d}{dt} \equiv \frac{\partial}{\partial t}\mathbf{q}.\nabla\right)$$

$$\text{or} \quad \frac{\partial \mathbf{q}}{\partial t} + \nabla\left(\frac{1}{2}\mathbf{q}^2\right) - \mathbf{q} \times \text{curl}\, \mathbf{q} = F - \frac{1}{\rho}\nabla p, \qquad ...(6)$$

$$\text{or} \quad \frac{\partial \mathbf{q}}{\partial t} + \nabla\left(\frac{1}{2}\mathbf{q}^2\right) + \underline{\omega} \times \mathbf{q} = F - \frac{1}{\rho}\nabla p, \ (\text{Since}\ \underline{\omega} = \nabla \times q) \qquad ...(7)$$

known as *Lamb's Hydrodynamical equations* which is a non-linear equation due to the convective term (q. ∇) q on the L. H. S. in (6).

$$\frac{\partial u}{\partial t} + u\frac{\partial u}{\partial x} + v\frac{\partial u}{\partial y} + w\frac{\partial u}{\partial z} = X - \frac{1}{\rho}\frac{\partial p}{\partial x},$$

$$\frac{\partial v}{\partial t} + u\frac{\partial v}{\partial x} + v\frac{\partial v}{\partial y} + w\frac{\partial v}{\partial z} = Y - \frac{1}{\rho}\frac{\partial p}{\partial y},$$

$$\frac{\partial w}{\partial t}+u\frac{\partial w}{\partial x}+v\frac{\partial w}{\partial y}+w\frac{\partial w}{\partial z}=Z-\frac{1}{\rho}\frac{\partial p}{\partial z}. \quad ...(8)$$

In spherical polar coordinates (ρ, θ, ϕ), the equation (5) becomes

$$\frac{Dq_r}{Dt}-\frac{q_\theta^2+q_\phi^2}{r}=Fr-\frac{1}{\rho}\frac{\partial p}{\partial r},$$

$$\frac{Dq_\theta}{Dt}+\frac{q_r q_\theta - q_\phi^2\cot\theta}{r}=F_\theta-\frac{1}{\rho}\frac{\partial p}{r\partial\theta},$$

$$\frac{Dq_\theta}{Dt}+\frac{q_r q_\theta - q_\theta q_\phi\cot\theta}{r}=F_\phi-\frac{1}{\rho}\frac{1}{r\sin\theta}\frac{\partial p}{\partial\phi},$$

where $$\frac{D}{Dt}\equiv\frac{\partial}{\partial t}+q_r\frac{\partial}{\partial r}+\frac{q_\phi}{r}\frac{\partial}{\partial\theta}+\frac{q_\phi}{r\sin\theta}\frac{\partial q}{\partial f} \quad ...(9)$$

q_r, q_θ and q_ϕ are the velocity components in the r, θ and ϕ directions respectively.

In cylindrical polar coordinates (r, θ, z), the equation (5) becomes

$$\frac{Dq_r}{Dt}-\frac{q_\theta^2}{r}=F_r-\frac{1}{\rho}\frac{\partial p}{\partial r},$$

$$\frac{Dq_\theta}{Dt}+\frac{q_r q_\theta}{r}=F_\theta-\frac{1}{\rho}\frac{\partial p}{r\partial\theta}$$

$$\frac{Dq_z}{Dt}=F_z-\frac{1}{\rho}\frac{\partial p}{\partial z},$$

where $$\frac{Dq_z}{Dt}\equiv\frac{\partial}{\partial t}+q_r\frac{\partial}{\partial r}+\frac{q_\theta}{r}\frac{\partial}{\partial\theta}+q_z\frac{\partial}{\partial z}, \quad ...(10)$$

q_r, q_θ and q_z are the velocity components in the r, θ and z direction respectively.

EQUATION OF MOTION OF AN INVISCID FLUID

(Cartesian Coordinates)

Consider ρ be the density and p be the pressure at a point O_1 (x,y, z) and u, v, w be the velocity components parallel to the coordinate axes. Construct a small parallelopiped with edges of length δx, δy, and δz parallel to their respective axes. Let X, Y, Z be the components of external force per unit mass at time t at the point O_1.

Force on the plane through O1 (x, y, z) parallel to the face PQRS

$= p\delta y\ \delta z$

$= f\ (x,\ y,\ z)$ (say)

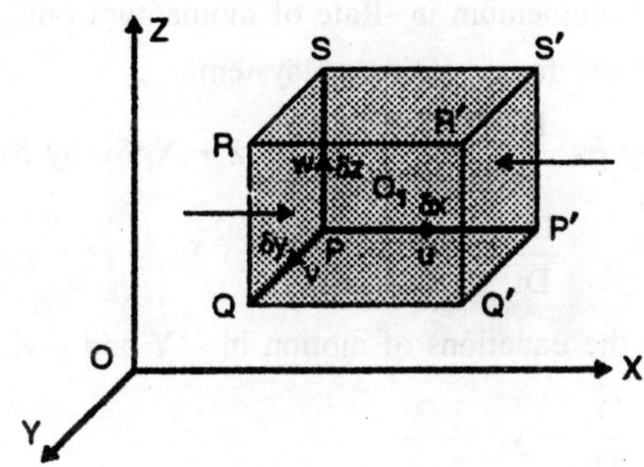

Fig 4.3

Pressure on the face PQRS $= f\left\{\left(x - \frac{1}{2}\delta x\right), y, z\right\}\delta y \delta z$

Expanding by Taylor's theorem, we have

$$= \left[f(x,y,z) - \frac{1}{2}\delta x \frac{\partial}{\partial x} f(x,y,z) + ...\right]\delta y \delta z$$

Force on the opposite face P′Q′R′S′ $= f\left\{\left(x - \frac{1}{2}\delta x\right), y, z\right\}\delta y \delta z$

$$= \left[f(x,y,z) - \frac{1}{2}\delta x \frac{\partial}{\partial x} f(x,y,z) + ...\right]\delta y \delta z$$

Thus the force along X-axis becomes

$$= \left[\left(f - \frac{1}{2}\delta x \frac{\partial f}{\partial x}\right) - \left(f - \frac{1}{2}\delta x \frac{\partial f}{\partial x}\right)\right]\delta y \delta z$$

$$= -\frac{\partial f}{\partial x}\delta x\ \delta y\ \delta z \text{ (to first order approximation)}$$

$$= -\frac{\partial p}{\partial x}\delta x\ \delta y\ \delta z$$

Mass of the fluid element $= \rho \delta x\ \delta y\ \delta z$ ...(1)

External force on the fluid element $= X\rho\ \delta x\ \delta y\ \delta z$ ...(2)

along X-direction

Thus the equation for the momentum balance determines

Rate of momentum accumulation

= Rate of momentum in -Rate of momentum out

+ Sum of the forces acting on system

$$\rho\ \delta x\ \delta y\ \delta z\ \frac{Du}{Dt} = -\frac{\partial p}{\partial x}\delta x\ \delta y\ \delta z + X\rho\delta x\ \delta y\ \delta z$$

$$\Rightarrow \qquad \frac{Du}{Dt} = X - \frac{1}{\rho}\frac{\partial p}{\partial x}.$$

Similarly the equations of motion in – Y and – Z directions are

$$\frac{D_v}{Dt} = y - \frac{1}{p}\frac{\partial p}{\partial y}$$

and $$\frac{Dw}{Dt} = Z - \frac{1}{\rho}\frac{\partial p}{\partial z}.$$

known as *Euler's equation of motion at all points of the fluid*, which applies only to ideal fluid or perfect fluid. The dissipative effects have not been considered.

CAUCHY'S INTEGRAL

Consider $Q = V + \int \frac{dp}{\rho}$.

Let (x, y, z) be the co-ordinates of a particle at any time t, whose initial co-ordinates are (a, b, c). Consider the density r as a function of pressure p. Differentiating partially (1) with regard to a, we have

$$\frac{\partial Q}{\partial a} = \frac{\partial V}{\partial a} + \frac{1}{\rho}\frac{\partial p}{\partial a}. \qquad ...(2)$$

Similarly we can write other two equations.

Equations of motion are

$$\frac{\partial^2 x}{\partial t^2} = -\frac{\partial V}{\partial x} - \frac{1}{\rho}\frac{\partial p}{\partial x},$$

$$\frac{\partial^2 y}{\partial t^2} = -\frac{\partial V}{\partial y} - \frac{1}{\rho}\frac{\partial p}{\partial y},$$

$$\frac{\partial^2 z}{\partial t^2} = -\frac{\partial V}{\partial z} - \frac{1}{\rho}\frac{\partial p}{\partial z} \qquad ...(3)$$

Multiplying (3) by $\frac{\partial x}{\partial a}, \frac{\partial y}{\partial a}$ and $\frac{\partial z}{\partial a}$ respectively and adding, we have

$$\frac{\partial^2 x}{\partial t^2}\frac{\partial x}{\partial a}+\frac{\partial^2 y}{\partial t^2}\frac{\partial y}{\partial a}+\frac{\partial^2 z}{\partial t^2}\frac{\partial z}{\partial a}=-\left(\frac{\partial V}{\partial a}+\frac{1}{\rho}\frac{\partial p}{\partial a}\right)$$

or
$$\frac{\partial^2 x}{\partial t^2}\frac{\partial x}{\partial a}+\frac{\partial^2 y}{\partial t^2}\frac{\partial y}{\partial a}+\frac{\partial^2 z}{\partial t^2}\frac{\partial z}{\partial a}=-\frac{\partial Q}{\partial a},$$

Similarly
$$\frac{\partial^2 x}{\partial t^2}\frac{\partial x}{\partial b}+\frac{\partial^2 y}{\partial t^2}\frac{\partial y}{\partial b}+\frac{\partial^2 z}{\partial t^2}\frac{\partial z}{\partial b}=-\frac{\partial Q}{\partial b},$$

and
$$\frac{\partial^2 x}{\partial t^2}\frac{\partial x}{\partial c}+\frac{\partial^2 y}{\partial t^2}\frac{\partial y}{\partial c}+\frac{\partial^2 z}{\partial t^2}\frac{\partial z}{\partial c}=-\frac{\partial Q}{\partial c}. \qquad ...(4, 5, 6)$$

Since $\frac{\partial}{\partial b}\left(\frac{\partial Q}{\partial c}\right)=\frac{\partial}{\partial c}\left(\frac{\partial Q}{\partial b}\right)$.

Q will be eliminated by differentiating (5) and (6) partially with regard to c and b respectively and subtracting, we have

$$\left(\frac{\partial^2 u}{\partial b\partial t}\frac{\partial x}{\partial c}-\frac{\partial^2 u}{\partial c\partial t}\frac{\partial x}{\partial b}\right)+\left(\frac{\partial^2 v}{\partial b\partial t}\frac{\partial x}{\partial c}-\frac{\partial^2 v}{\partial c\partial t}\frac{\partial y}{\partial b}\right)$$
$$+\left(\frac{\partial^2 w}{\partial b\partial t}\frac{\partial z}{\partial c}-\frac{\partial^2 w}{\partial c\partial t}\frac{\partial z}{\partial b}\right)=0$$

or
$$\left(\frac{\partial}{\partial t}\left(\frac{\partial u}{\partial b}\frac{\partial x}{\partial c}-\frac{\partial u}{\partial c}\frac{\partial x}{\partial b}\right)-\frac{\partial u}{\partial b}\frac{\partial^2 x}{\partial t\partial c}+\frac{\partial u}{\partial c}\frac{\partial^2 x}{\partial t\partial b}\right)$$
$$+ \text{two similarexpression} = 0$$

or
$$\frac{\partial}{\partial t}\left(\frac{\partial u}{\partial b}\frac{\partial x}{\partial c}-\frac{\partial u}{\partial c}\frac{\partial x}{\partial b}\right)+\frac{\partial}{\partial t}\left(\frac{\partial v}{\partial b}\frac{\partial y}{\partial c}-\frac{\partial v}{\partial c}\frac{\partial y}{\partial b}\right)$$
$$+\frac{\partial}{\partial t}\left(\frac{\partial w}{\partial b}\frac{\partial z}{\partial c}-\frac{\partial w}{\partial c}\frac{\partial z}{\partial b}\right)=0.$$

$$\left(\text{Since}\frac{\partial^2 x}{\partial t\partial c}=\frac{\partial u}{\partial c},\frac{\partial^2 x}{\partial t\partial b}=\frac{\partial u}{\partial b}\right)$$

By integrating with regard to t, we have

$$\frac{\partial u}{\partial b}\frac{\partial x}{\partial c}-\frac{\partial u}{\partial c}\frac{\partial x}{\partial b}+\frac{\partial v}{\partial b}\frac{\partial y}{\partial c}-\frac{\partial v}{\partial c}\frac{\partial y}{\partial b}+\frac{\partial w}{\partial b}\frac{\partial z}{\partial c}-\frac{\partial w}{\partial c}\frac{\partial z}{\partial b}$$

$$=\frac{\partial w_0}{\partial b}-\frac{\partial v_0}{\partial c}, \quad ...(4)$$

where u_0, v_0, w_0 are initial values.

Initially $\frac{\partial x}{\partial a}=1$, $\frac{\partial x}{\partial b}=0$, $\frac{\partial x}{\partial c}=0$ etc. as $x = a$, $y = b$, $z = c$

Since $\frac{\partial u}{\partial a}=\frac{\partial u}{\partial x}\frac{\partial x}{\partial a}+\frac{\partial u}{\partial y}\frac{\partial y}{\partial a}+\frac{\partial u}{\partial z}\frac{\partial z}{\partial a}$.

Then relation (2) becomes

$$\left(\frac{\partial w}{\partial y}-\frac{\partial v}{\partial z}\right)\frac{\partial(yz)}{\partial(bc)}+\left(\frac{\partial u}{\partial z}-\frac{\partial w}{\partial x}\right)\frac{\partial(zx)}{\partial(bc)}$$

$$+\left(\frac{\partial v}{\partial x}-\frac{\partial u}{\partial y}\right)\frac{\partial(xy)}{\partial(bc)}=\frac{\partial w_0}{\partial b}-\frac{\partial v_0}{\partial c}$$

or $\xi\frac{\partial(yz)}{\partial(bc)}+\eta\frac{\partial(zx)}{\partial(bc)}+\zeta\frac{\partial(xy)}{\partial(bc)}=\xi_0$,

where ξ, η, ζ are the vorticity components. ...(5)

Similarly other two expressions are

$$\xi\frac{\partial(yz)}{\partial(ca)}+\eta\frac{\partial(zx)}{\partial(ca)}+\zeta\frac{\partial(xy)}{\partial(ca)}=\eta_0, \quad ...(6)$$

and $$\xi\frac{\partial(yz)}{\partial(ab)}+\eta\frac{\partial(zx)}{\partial(ab)}+\zeta\frac{\partial(xy)}{\partial(ab)}=\zeta_0, \quad ...(7)$$

where $\xi=\frac{\partial w}{\partial y}-\frac{\partial v}{\partial z}$, $\eta=\frac{\partial u}{\partial z}-\frac{\partial w}{\partial x}$

and $\zeta=\frac{\partial v}{\partial x}-\frac{\partial u}{\partial y}$

Also, the equation of continuity in Lagrangian form is

$$\rho\frac{\partial(xyz)}{\partial(abc)}=\rho_0.$$

Multiplying (5), (6) and (7) by $\frac{\partial x}{\partial a}$, $\frac{\partial x}{\partial b}$, $\frac{\partial x}{\partial c}$, respectively and adding, we have

$$\frac{\xi}{\rho} = \frac{\xi_0}{\rho_0}\frac{\partial x}{\partial a} + \frac{\eta_0}{\rho_0}\frac{\partial x}{\partial b} + \frac{\zeta_0}{\rho_0}\frac{\partial x}{\partial c},$$

$$\frac{\eta}{\rho} = \frac{\xi_0}{\rho_0}\frac{\partial y}{\partial a} + \frac{\eta_0}{\rho_0}\frac{\partial y}{\partial b} + \frac{\zeta_0}{\rho_0}\frac{\partial y}{\partial c}$$

$$\frac{\zeta}{\rho} = \frac{\xi_0}{\rho_0}\frac{\partial z}{\partial a} + \frac{\eta_0}{\rho_0}\frac{\partial z}{\partial b} + \frac{\zeta_0}{\rho_0}\frac{\partial z}{\partial c}. \qquad ...(9)$$

Thus the components ξ, η, ζ of a particle can be obtained at any instant if its path is known. These are known as Cauchy's integral When velocity potential exists then $\xi = \eta = \zeta = 0$. Thus from (9) we conclude that these quantities are always zero if their initial values are zero.

SOLVED EXAMPLES

Example 1:

Show that the velocity field

$$u\ (x,y) = \frac{A(x^2 - y^2)}{(x^2 + y^2)^2},\ v(x,y) = \frac{2Axy}{(x^2 + y^2)^2}, w = 0,$$

satisfies the equation of motion for inviscid incompressible flow. Determine the pressure associated with this velocity field.

Solution:

$$\text{Here } u\ (x,y) = \frac{A(x^2 - y^2)}{(x^2 + y^2)^2},\ v\ (x, y) = \frac{2Axy}{(x^2 + y^2)^2},\ w = 0. \qquad ...(1)$$

$$\frac{\partial u}{\partial x} = A\frac{-2x(x^2 + y^2)^2 - 4x(x^2 - y^2)(x^2 + y^2)}{(x^2 - y^2)^4} = \frac{2Ax(3y^2 - x^2)}{(x^2 + y^2)^3},$$

$$\frac{\partial u}{\partial y} = A\frac{-2y(x^2 + y^2)^2 - 4y(x^2 - y^2)(x^2 + y^2)}{(x^2 - y^2)^4} = -\frac{2Ay(3x^2 - x^2)}{(x^2 + y^2)^3},$$

$$\frac{\partial v}{\partial x} = 2A\frac{y(x^2 + y^2) - 4x^2y(x^2 + y^2)}{(x^2 + y^2)^4} = \frac{2Ay(y^2 - 3x^2)}{(x^2 - y^2)^3},$$

$$\frac{\partial v}{\partial y} = 2A\frac{x(x^2 + y^2)^2 - 4xy^2(x^2 + y^2)}{(x^2 + y^2)^4} = \frac{2Ax(x^2 - 3y^2)}{(x^2 + y^2)^3}$$

The equations of motion for steady flow are given by

$$u\frac{\partial u}{\partial x}+v\frac{\partial u}{\partial y}+w\frac{\partial u}{\partial z}=-\frac{1}{\rho}\frac{\partial p}{\partial x},$$

$$u\frac{\partial v}{\partial x}+v\frac{\partial v}{\partial y}+w\frac{\partial v}{\partial z}=-\frac{1}{\rho}\frac{\partial p}{\partial y},$$

$$u\frac{\partial w}{\partial x}+v\frac{\partial w}{\partial y}+w\frac{\partial w}{\partial z}=-\frac{1}{\rho}\frac{\partial p}{\partial z}, \quad ...(2,3,4)$$

From (1) and (2, 3, 4), we have

$$\frac{2A^2x}{(x^2+y^2)^3}=\frac{1}{\rho}\frac{\partial p}{\partial x},$$

$$\frac{2A^2y}{(x^2+y^2)^3}=\frac{1}{\rho}\frac{\partial p}{\partial y},$$

$$0=-\frac{1}{\rho}\frac{\partial p}{\partial z}. \quad ...(5, 6,7)$$

Equation (7) shows that pressure p is independent of z, *i.e.*, p = p (x,y). Therefore

$$dp = (\partial p/\partial x)\,dx + (\partial p/\partial y)\,dy$$

$$\text{or } dp = \frac{2A^2\rho x}{(x^2+y^2)^3}dx + \frac{2A^2\rho y}{(x^2+y^2)^3}dy$$

$$\text{or } dp = 2A^2\rho\frac{xdx+ydy}{(x^2+y^2)^3}$$

By integrating, we have

$$p=\frac{A^2\rho}{2(x^2+y^2)^2}.$$

Proved

Example 2:

The particle velocity for a fluid motion referred to rectangular axes is given by the components.

$$u = A\cos\frac{\pi x}{2a}\cos\frac{\pi z}{2a},\quad v=0,$$

$$w = A\sin\frac{\pi x}{2a}\sin\frac{\pi z}{2a},$$

where A is a constant. Show that this is a possible motion of an incompressible fluid under no body forces in an infinite fixed rigid tube, $-a \leq x \leq a$, $0 \leq z \leq 2a$. Also, find the pressure associated with this velocity field.

Solution:

The equations of motion for a two-dimensional steady, inviscid, incompressible flow under no body force, in cartesian coordinates, are given by

$$u\frac{\partial u}{\partial x} + v + \frac{\partial u}{dy} + w\frac{\partial u}{\partial z} = \frac{1}{\rho}\frac{\partial p}{\partial x},\ 0 = -\frac{1}{\rho}\frac{\partial p}{\partial y},$$

$$u\frac{\partial w}{\partial x} + v + \frac{\partial w}{dy} + w\frac{\partial w}{\partial z} = -\frac{1}{\rho}\frac{\partial p}{\partial z}, \quad \text{...(1, 2, 3)}$$

Here $\quad u = A\cos\frac{\pi x}{2a}\cos\frac{\pi z}{2a},$

$$v = 0,\ w = A\sin\frac{\pi x}{2a}\sin\frac{\pi z}{2a}, \quad \text{...(4)}$$

From the equation (2), it follows that the pressure p is independent of y *i.e.*, p = p (x, z).

Using (4) into (1) and (3), we have

$$\left(A\cos\frac{\pi x}{2a}\cos\frac{\pi z}{2a}\right)\left(-\frac{\pi A}{2a}\sin\frac{\pi x}{2a}\cos\frac{\pi z}{2a}\right) + \left(A\sin\frac{\pi x}{2a}\sin\frac{\pi z}{2a}\right) \times \left(-\frac{\pi A}{2a}\cos\frac{\pi x}{2a}\sin\frac{\pi z}{2a}\right) = -\frac{1}{\rho}\frac{\partial p}{\partial x},$$

or $$\frac{\pi A^2}{2a}\left[\cos\frac{\pi x}{2a}\sin\frac{\pi x}{2a}\cos^2\frac{\pi z}{2a} + \cos\frac{\pi x}{2a}\sin\frac{\pi x}{2a}\sin^2\frac{\pi z}{2a}\right] = \frac{1}{\rho}\frac{\partial p}{\partial x}$$

or $$\frac{\pi A^2}{2a}\cos\frac{\pi x}{2a}\sin\frac{\pi x}{2a} = \frac{1}{\rho}\frac{\partial p}{\partial x}, \quad \text{...(5)}$$

and $$\left(A\cos\frac{\pi x}{2a}\cos\frac{\pi z}{2a}\right)\left(\frac{\pi A}{2a}\cos\frac{\pi x}{2a}\sin\frac{\pi z}{2a}\right) + \left(A\sin\frac{\pi x}{2a}\sin\frac{\pi z}{2a}\right) \times \left(\frac{\pi A}{2a}\sin\frac{\pi x}{2a}\cos\frac{\pi z}{2a}\right) = -\frac{1}{\rho}\frac{\partial p}{\partial z},$$

or $$\frac{\pi A^2}{2a}\left[\cos\frac{\pi z}{2a}\sin\frac{\pi z}{2a}\cos^2\frac{\pi x}{2a} + \cos\frac{\pi z}{2a}\sin\frac{\pi z}{2a}\sin^2\frac{\pi x}{2a}\right] = -\frac{1}{\rho}\frac{\partial p}{\partial z},$$

$$\text{or} \quad \frac{\pi A^2}{2a}\cos\frac{\pi z}{2a}\sin\frac{\pi z}{2a} = \frac{1}{\rho}\frac{\partial p}{\partial z}, \quad \text{...(6)}$$

The equations (5) and (6) show that the velocity components satisfy the equations of motion.

Again $dp = (\partial p/\partial x)dx + (\partial p/\partial z)\,dz$

$$\text{or } dp = \frac{\pi\rho A^2}{2a}\left[\cos\frac{\pi x}{2a}\sin\frac{\pi x}{2a}dx - \cos\frac{\pi z}{2a}\sin\frac{\pi z}{2a}dz\right]$$

By integrating, we have

$$p = \frac{1}{2}\rho A^2\left[\cos^2\frac{\pi z}{2a} - \cos^2\frac{\pi x}{2a}\right] + C,$$

where C is an integration constant. This gives the required pressure distribution. **Proved**

Example 3:

Determine the pressure if the velocity field

$q_r = 0, q_\theta = Ar + B,\ q_z = 0,$

satisfies the equation of motion $\rho\frac{q_\theta^2}{r} = \frac{dp}{dr}$, *where A and B are arbitrary constants.*

Solution:

$$\frac{dp}{dr} = \rho\frac{1}{r}\left(Ar + \frac{B}{r}\right)^2$$

$$\text{or } \frac{dp}{dr} = \rho\left(A^2 r + \frac{B^2}{r^2} + 2AB\frac{1}{r}\right)$$

By integrating, we have

$$p = \rho\left(\frac{1}{2}A^2r^2 + \frac{B^2}{2r^2} + 2AB\log r\right) + C,$$

where C is an integration constant.

Example 4:

Prove that the equation of motion is satisfied for an inviscid, incompressible, steady flow with negligible body force whose velocity components are given by

$$q_r = U\left(1-\frac{A^3}{r^3}\right)\cos\theta,\ q_\theta = -U\left(1+\frac{A^3}{2r^3}\right)\sin\theta,\ q_\phi = 0,$$

where A is constant. Find the resultant velocity when $r \to \infty$.

Solution:

The equations of motion for an inviscid, incompressible and steady flow with negligible external force, in spherical polar coordinates, are given as

$$q_r\frac{\partial p_r}{\partial r}+\frac{q\theta}{r}\frac{\partial q_r}{\partial\theta}+\frac{q\phi}{r\sin\theta}-\frac{\partial q_r}{\partial\phi}-\frac{q_\theta^2+q_\phi^2}{r}=-\frac{1}{\rho}\frac{\partial p}{\partial r},$$

$$q_r\frac{\partial q_\theta}{\partial r}+\frac{q_\theta}{r}\frac{\partial q_\theta}{\partial\theta}+\frac{q\phi}{r\sin\theta}\frac{\partial q\theta}{\partial\phi}+\frac{q_r q_\theta}{r}-\frac{q_\phi^2\cot\theta}{r}=-\frac{1}{\rho}\frac{\partial p}{\partial\theta},$$

$$q_r\frac{\partial q_\phi}{\partial r}+\frac{q_\theta}{r}\frac{\partial q\phi}{\partial\theta}+\frac{q\phi}{r\sin\theta}\frac{\partial q\phi}{\partial\phi}+\frac{q_\phi q_r}{r}-\frac{q_\theta q_\phi\cot\theta}{r}$$

$$=-\frac{1}{\rho}\frac{1}{r\sin\theta}\frac{\partial p}{\partial\phi} \qquad \text{...(1, 2, 3)}$$

Here $q_r = U\left(1-\frac{A^3}{r^3}\right)\cos\theta,\ q_\theta = -U\left(1+\frac{A^3}{2r^3}\right)$

$\sin\theta,\ q_\phi = 0,$ ...(4)

From the relation (4), equations (1,2,3) reduce to

$$U\left(1-\frac{A^3}{r^3}\right)\cos\theta\left(\frac{3UA^3}{r^4}\right)\cos\theta+\frac{U}{r}\left(1+\frac{A^3}{2r^3}\right)\sin\theta$$

$$\times U\left(1-\frac{A^3}{r^3}\right)\sin\theta-\frac{U^2}{r}\left(1+\frac{A^3}{2r^3}\right)\sin^2\theta=-\frac{1}{\rho}\frac{\partial p}{\partial r},$$

$$U\left(1-\frac{A^3}{r^3}\right)\cos\theta\left(\frac{3UA^3}{r^4}\right)\sin\theta+\frac{U}{r}\left(1+\frac{A^3}{2r^3}\right)\sin\theta$$

$$\times U\left(1+\frac{A^3}{2r^3}\right)\cos\theta-\frac{U^2}{r}\left(1-\frac{A^3}{r^3}\right)\left(1+\frac{A^3}{2r^3}\right)\sin\theta\cos\theta$$

$$=-\frac{1}{\rho}\frac{\partial p}{r\partial\theta},$$

$$0 = \frac{1}{\rho}\frac{1}{r\sin\theta}\frac{\partial p}{r\partial\phi}. \qquad ...(5, 6, 7)$$

Equation (7) shows that the pressure p is independent of ϕ, therefore $p = p(r, \theta)$. On simplifying the equation (5) and (6), we have

$$\frac{3U^2A^3}{r^4}\left(1-\frac{A^3}{r^3}\right)\cos^2\theta + \frac{U^2}{r}\left(1-\frac{A^3}{2r^3}-\frac{A^6}{2r^6}\right)\sin^2\theta$$

$$-\frac{U^2}{r}\left(1+\frac{A^3}{r^3}+\frac{A^6}{4r^6}\right)\sin^2\theta = -\frac{1}{\rho}\frac{\partial p}{\partial r},$$

or $$\frac{3U^2A^3}{r^4}\left(1-\frac{A^3}{r^3}\right)\cos^2\theta + \frac{U^2}{r}\left(-\frac{3A^3}{2r^3}-\frac{3A^6}{4r^6}\right)\sin^2\theta = \frac{1}{\rho}\frac{\partial p}{\partial r},$$

or $$\frac{3U^2A^3}{r^4}\left(1-\frac{A^3}{r^3}\right)\cos^2\theta - \frac{3U^2A^3}{2r^4}\left(1+\frac{A^3}{2r^3}\right)\sin^2\theta$$

$$= -\frac{1}{\rho}\frac{\partial p}{\partial r}. \qquad ...(8)$$

and $$\frac{3U^2A^3}{2r^4}\left(1-\frac{A^3}{r^3}\right)\sin\theta\cos\theta + \frac{U^2}{r}\left(1+\frac{A^3}{r^3}+\frac{A^6}{4r^6}\right)\sin\theta\cos\theta$$

$$-\frac{U^2}{r}\left(1-\frac{A^3}{2r^3}-\frac{A^6}{2r^6}\right)\sin\theta\cos\theta - \frac{1}{\rho}\frac{\partial p}{\partial\theta}.$$

or $$\frac{3U^2A^3}{2r4}\left(1-\frac{A^3}{r^3}\right)\sin\theta\cos\theta + \frac{3U^2A^3}{2r^4}\left(1+\frac{A^3}{2r^3}\right)\sin\theta\cos\theta =$$

$$-\frac{1}{\rho}\frac{\partial p}{r\partial\theta}. \qquad ...(9)$$

Differentiating equation (8) with regard to θ, we have

$$-\frac{6U^2A^3}{r^4}\left(1-\frac{A^3}{r^3}\right)\cos\theta\sin\theta$$

$$-\frac{3U^2A^3}{r^4}\left(1+\frac{A^3}{2r^3}\right)\sin\theta\cos\theta = -\frac{1}{\rho}\frac{\partial^2 p}{\partial r\partial\theta}$$

or $$\left(-\frac{9U^2A^3}{r^4}+\frac{9U^2A^6}{2r^7}\right)\sin\theta\cos\theta = -\frac{1}{\rho}\frac{\partial^2 p}{\partial r\partial\theta} - \qquad ...(10)$$

Differentiating equation (9) with regard to r, we have

$$\frac{3U^2A^3}{2}\left(-\frac{3}{r^4}+\frac{6A^3}{r^7}\right)\cos\theta\sin\theta$$

$$+\frac{3U^2A^3}{2}\left(-\frac{3}{r^4}+\frac{3A^3}{r^7}\right)\sin\theta\cos\theta = -\frac{1}{\rho}\frac{\partial^2 p}{\partial r\partial\theta}$$

or $$\frac{9}{2}\frac{U^2A^3}{2}\left(1-\frac{2A^3}{r^3}\right)\cos\theta\sin\theta$$

$$-\frac{9}{2}\frac{U^2A^3}{2}\left(1+\frac{A^3}{r^3}\right)\sin\theta\cos\theta = -\frac{1}{\rho}\frac{\partial^2 p}{\partial r\partial\theta}$$

or $$\left(-\frac{9U^2A^3}{r^4}+\frac{9U^2A^6}{2r^7}\right)\sin\theta\cos\theta = -\frac{1}{\rho}\frac{\partial^2 p}{\partial r\partial\theta} \qquad ...(11)$$

Equation (10) and (11) are identical. Hence, the equation of motion is satisfied.

When r → ∞, the resultant velocity is equal to U. **Proved.**

Example 5:

Prove that the velocity components

$$q_r\,(r,\theta) = -U\left(1-\frac{a^2}{r^2}\right)\cos\theta,\ q_\theta(r,\theta) = U\left(1+\frac{a^2}{r^2}\right)\sin\theta,$$

satisfy the equation of motion for a two-dimensional inviscid incompressible flow. Find the pressure associated with this velocity field. U and a are constants.

Solution:

The equations of motion for a two-dimensional steady, inviscid, incompressible flow in the absence of external force, in spherical polar coordinates, are given by

$$q_r\frac{\partial q_r}{\partial r}+\frac{q_\theta}{r}\frac{\partial q_r}{\partial\theta}-\frac{q_\theta^2}{r} = -\frac{1}{\rho}\frac{\partial p}{\partial r},$$

$$q_r\frac{\partial q_\theta}{\partial r}+\frac{q_\theta}{r}\frac{\partial q_\theta}{\partial\theta}+\frac{q_r q_\theta}{r} = -\frac{1}{\rho}\frac{\partial p}{r\partial\theta},$$

$$0 = -\frac{1}{\rho}\frac{1}{r\sin\theta}\frac{\partial p}{\partial\phi}. \qquad ...(1, 2, 3)$$

Here $q_r(r, \theta) = -U\left(1-\frac{a^2}{r^2}\right)\cos\theta,$

$$q_\theta(r, \theta) = U\left(1+\frac{a^2}{r^2}\right)\sin\theta. \qquad ...(4)$$

From (3), it follows that the pressure p is independent of ϕ *i.e.*, p = p (r, θ).

Using the relation (4) into (1) and (2), we have

$$U\left(1-\frac{a^2}{r^2}\right)\cos\theta+\left(\frac{2Ua^2}{r^3}\right)\cos\theta+\frac{U}{r}\left(1+\frac{a^2}{r^2}\right)\sin\theta$$

$$\times U\left(1-\frac{a^2}{r^2}\right)\sin\theta-\frac{U^2}{r}\left(1+\frac{a^4}{r^2}\right)^2\sin^2\theta = -\frac{1}{\rho}\frac{\partial p}{\partial r},$$

or $$\frac{2U^2a^2}{r^3}\left(1-\frac{a^2}{r^2}\right)\cos^2\theta-\frac{2U^2a^2}{r^3}\left(1+\frac{a^2}{r^2}\right)\sin^2\theta$$

$$= -\frac{1}{\rho}\frac{\partial p}{\partial r}. \qquad ...(5)$$

and $$\frac{2U^2a^2}{r^3}\left(1-\frac{a^2}{r^2}\right)\sin\theta\cos\theta+\frac{U^2}{r}\left(1+\frac{a^2}{r^2}\right)\sin\theta\cos\theta$$

$$-\frac{U^2}{r}\left(1-\frac{a^2}{r^4}\right)\sin\theta\cos\theta = -\frac{1}{\rho}\frac{\partial p}{r\partial\theta}$$

or $$\frac{2U^2a^2}{r^3}\left(1-\frac{a^2}{r^2}\right)\sin\theta\cos\theta+\frac{2U^2a^2}{r}\left(1+\frac{a^2}{r^2}\right)\sin\theta\cos\theta$$

$$= -\frac{1}{\rho}\frac{\partial p}{r\partial\theta}$$

$$\frac{4U^2a^2}{r^3}\sin\theta\cos\theta = -\frac{1}{\rho}\frac{\partial p}{r\partial\theta}. \qquad ...(6)$$

Differentiating (5) with regard to θ, we have

$$-\frac{4U^2a^2}{r^3}\left(1-\frac{a^2}{r^2}\right)\cos\theta\sin\theta\frac{4U^2a^2}{r^2}\left(1+\frac{a^2}{r^2}\right)\sin\theta\cos\theta$$

$$= -\frac{1}{\rho}\frac{\partial^2 p}{\partial r \partial \theta}$$

$$\frac{8U^2a^2}{r^3}\cos\theta\sin\theta = \frac{1}{\rho}\frac{\partial^2 p}{\partial r \partial \theta}. \quad ...(7)$$

Differentiating (6) with regard to r, we have

$$\frac{8U^2a^2}{r^3}\cos\theta\sin\theta = \frac{1}{\rho}\frac{\partial^2 p}{\partial r \partial \theta}. \quad ...(8)$$

Equation (7) and (8) are identical. Hence, the equation of motion are satisfied.

Again, p is a function or r and θ, we have

$$dp = \left(\frac{\partial p}{\partial r}\right) dr + \left(\frac{\partial p}{\partial \theta}\right) d\theta$$

$$\text{or } dp = -2r\ U2a2 \left[\left(\frac{1}{r^3} - \frac{a^2}{r^5}\right)\cos^2\theta - \left(\frac{1}{r^3} + \frac{a^2}{r^5}\right)\sin^2\theta\right] dr$$

$$-\frac{4\rho U^2 a^2}{r^2}(\sin\theta\cos\theta)\, d\theta$$

By integrating, we have

$$p = -2\rho U^2a^2\left[\left(-\frac{1}{2r^2} + \frac{a^2}{4r^4}\right)\cos^2\theta - \left(-\frac{1}{2r^2} - \frac{a^2}{4r^4}\right)\sin^2\theta\right] + C$$

$$\text{or } p = \frac{4\rho U^2 a^2}{r}(\cos^2\theta - \sin^2\theta) + C,$$

where C is an integration constant.

$$\text{or } p = 2\rho U^2a^2\left[\left(\frac{1}{2r^r} - \frac{a^2}{4r^4} - \frac{2}{r^2}\right)\cos^2\theta - \left(\frac{1}{2r^2} - \frac{a^2}{4r^4} - \frac{2}{r^2}\right)\sin^2\theta\right] + C$$

$$\text{or } p = 2\rho U^2a^2\left[-\frac{3}{2r^2}\cos 2\theta - \frac{a^2}{4r^4}\right] + C$$

$$\text{or } p = -\rho U^2a^2\left[-\frac{3}{r^2}\cos 2\theta + \frac{a^2}{2r^4}\right] + C,$$

Which gives the required pressure distribution.

BERNOULLI'S EQUATION (Stream Tube Method)

Consider a stream tube with A1 and A_2 as its two cross-sections. Let $\rho_1 p_1$, q_1 and ρ_2, p_2, q_2 be the density, the pressure, and the velocities at its two ends. Since the velocity normal to the curved surface of the stream tube is always zero therefore the flow will take place only through A_1 and A_2. From the law of conservation, we have

$$\rho_1 A_1 = \rho_2 A_2 = k \text{ (const.)} \quad ...(1)$$

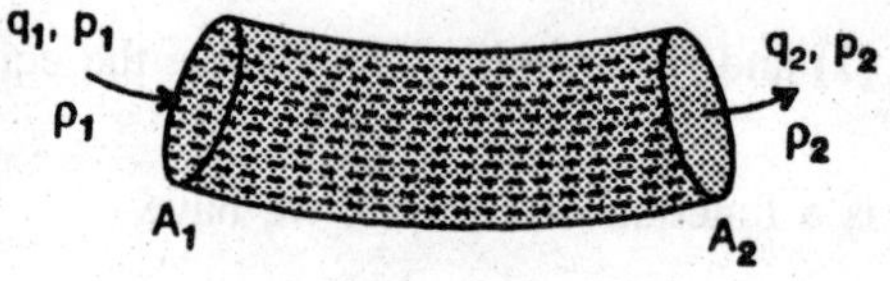

Fig. 4.4

The net force on the tube due to pressure is

$$p_1 A_1 - p_2 A_2 = \frac{p_1}{\rho_1}(\rho_1 A_1) - \frac{p_2}{\rho_2}(\rho_2 A_2)$$

$$\Rightarrow p_1 A_1 - p_2 A_2 = k\left(\frac{p_1}{\rho_1} - \frac{p_2}{\rho_2}\right).$$

The difference of the forces acting on the two cross-sections must be equal to the gain in the energy of the fluid passing through the tube. Thus, we have

$$\frac{p_1}{\rho_1} - \frac{p_2}{\rho_2} = k_2 - k_1$$

$$\text{or } \frac{p_1}{\rho_1} + k_1 = \frac{p_2}{\rho_2} + k_2 = \text{constant}, \quad ...(1)$$

where k_1 and k_2 are the sum of kinetic $\left(\frac{1}{2}q^2\right)$ and potential (Ω) energies per unit mass of the fluid entering and leaving the tube. Thus, we have

$$\frac{p}{\rho} + \frac{1}{2}q^2 + \Omega = \text{const.}$$

Again, let the cross-section A_1 and A_2 of the stream tube tends to zero such that it reduces to a stream line, even then equation (3) holds. Hence the equation holds along a stream line.

CONSERVATIVE FIELD OF FORCE

If the work done by the force F of the field in taking a unit mass from one point A to another point B independent of the path then it is termed as conservative field of force

$$\int_{ACB} F.dr = \int_{ADB} F.dr = -\Omega(\text{say}),$$

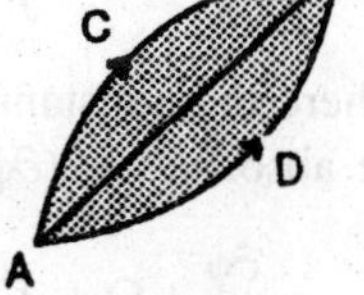

Fig. 4.5

where Ω is a scalar point function whose value depends on the initial and final position A and B. Thus

$$F = -\nabla\Omega.$$

where W is known as force potential which measures the potential energy of the field.

INTEGRATION OF EULER'S EQUATION

The Euler's equation of motion is given as follows :

$$\frac{\partial \mathbf{q}}{\partial t} + \nabla\left(\frac{1}{2}\mathbf{q}^2\right) - \mathbf{q} \times \text{curl}\,\mathbf{q} = F - \frac{1}{\rho}\nabla p. \quad ...(1)$$

Suppose that the body forces are conservative and that the flow is of the potential kind then there exist scalar functions W and f, such that

$$F = \nabla\,\Omega, \qquad \mathbf{q} = -\nabla\phi.$$

From (1) and (2), we have ...(2)

$$\frac{\partial \mathbf{q}}{\partial t} + \nabla\left(\frac{1}{2}\mathbf{q}^2\right) - \mathbf{q} \times \text{curl}\,\mathbf{q} = -\nabla\Omega - \frac{1}{\rho}\nabla p$$

$$\text{or } \frac{\partial \mathbf{q}}{\partial t} - \mathbf{q} \times \text{curl}\,\mathbf{q} = -\nabla\left[\Omega + \int\frac{dp}{p} + \frac{1}{2}\mathbf{q}^2\right]$$

$$\text{or } \frac{\partial}{\partial t}(-\nabla\phi) - \mathbf{q} \times \text{curl}\,\mathbf{q} = -\nabla\left[\Omega + \int\frac{dp}{p} + \frac{1}{2}\mathbf{q}^2\right]$$

$$\text{or } \nabla\left[-\frac{\partial\phi}{\partial t} + \Omega + \int\frac{dp}{\rho} + \frac{1}{2}\mathbf{q}^2\right] = \mathbf{q} \times \text{curl}\,\mathbf{q}. \quad ...(1)$$

IRROTATIONAL FLOW

For an irrotational flows the vorticity vector $\underset{\sim}{\omega}$ = curl **q** = 0 then the equation (3) becomes

$$\nabla\left[-\frac{\partial\phi}{\partial t}+\Omega+\int\frac{dp}{\rho}+\frac{1}{2}\mathbf{q}^2\right]=0.$$

By integrating, we have

$$-\frac{\partial\phi}{\partial t}+\Omega+\int\frac{dp}{\rho}+\frac{1}{2}\mathbf{q}^2=\chi\ (t), \qquad ...(4)$$

where the constant χ (t) is an arbitrary function of time only, c (t) can be absorbed in $(\partial\phi/\partial t)$ then the equation (4) reduces to the form

$$-\frac{\partial\phi}{\partial t}+\Omega+\int\frac{dp}{\rho}+\frac{1}{2}\mathbf{q}^2 = \text{constant}, \qquad ...(5)$$

which is known as Bernoulli's equation for unsteady, irrotational flows.

STEADY AND IRROTATIONAL FLOW

If the motion be steady and irrotational then $\partial\phi/\partial t = 0$, curl $\mathbf{q} = 0$. Since $\mathbf{q} \times$ curl $\mathbf{q} = 0$, it follows that $\mathbf{q}$ and curl $\mathbf{q}$ are parallel *i.e.*, stream lines and vortex lines coincide. For such a motion $\mathbf{q}$ is known a Beltrami vector and the flow is *Beltrami flow.*

Thus the equation (5) reduces to

$$\int\frac{dp}{\rho}+\frac{1}{2}\mathbf{q}^2+\Omega=\text{const.}, \qquad ...(6)$$

where the constant is an absolute constant *i.e.*, independent of the time t.

If the fluid be incompressible and homogeneous then the equation (6) reduces to

$$\frac{p}{\rho}+\frac{1}{2}\mathbf{q}^2+\Omega=\text{const.}=C, \qquad ...(7)$$

where C is a constant along a streamline but varies from one streamline to another. The equation (7) is termed as Bernoulli's *equation for steady and irrotational flows* and holds irrespective of whether the motion is rotational or irrotational.

HELMHOLTZ EQUATIONS

Euler's equations of motion are given by

$$\frac{\partial u}{\partial t}+u\frac{\partial u}{\partial x}+v\frac{\partial u}{\partial y}+w\frac{\partial u}{\partial z}=X-\frac{1}{\rho}\frac{\partial p}{\partial x},$$

$$\frac{\partial v}{\partial t}+u\frac{\partial v}{\partial x}+v\frac{\partial v}{\partial y}+w\frac{\partial v}{\partial z}=y-\frac{1}{\rho}\frac{\partial p}{\partial y},$$

$$\frac{\partial w}{\partial t}+u\frac{\partial w}{\partial x}+v\frac{\partial w}{\partial y}+w\frac{\partial w}{\partial z}=Z-\frac{1}{\rho}\frac{\partial p}{\partial z}, \qquad ...(1, 2, 3)$$

Let V be the potential function of the external forces and the density ρ be the function of the pressure p. Equation (1) may be written as

$$\frac{\partial u}{\partial t}+\left(u\frac{\partial u}{\partial x}+v\frac{\partial u}{\partial x}+w\frac{\partial w}{\partial x}\right)+v\left(\frac{\partial u}{\partial y}-\frac{\partial v}{\partial x}\right)+w\left(\frac{\partial u}{\partial z}-\frac{\partial w}{\partial x}\right)$$

$$=-\frac{\partial V}{\partial x}-\frac{1}{\rho}\frac{\partial p}{\partial x}$$

$$\Rightarrow \frac{\partial u}{\partial t}-\frac{1}{\rho}\frac{\partial}{\partial x}(q^2)-2v\zeta+2w\eta=-\frac{\partial V}{\partial x}-\frac{1}{\rho}\frac{\partial}{\partial x},$$

where $\underline{\Omega}$ (ξ, η, ζ) are the spin components and $q^2 = u^2 + v^2 + w^2$.

$$\Rightarrow \frac{\partial u}{\partial t}-2v\zeta+2w\eta=-\frac{\partial}{\partial x}\left(V+\frac{1}{2}q^2+\int\frac{dp}{\rho}\right)=-\frac{\partial Q}{\partial x}\text{ (let)},$$

where $Q = V + \frac{1}{2}\mathbf{q}^2 + \int\frac{dp}{\rho}$.

Thus $\frac{\partial u}{dt}-2v\zeta+2w\eta=-\frac{\partial Q}{dx}$,

Similarly $\frac{\partial v}{dt}-2w\xi+2u\zeta=-\frac{\partial Q}{dy}$,

and $\frac{\partial w}{dt}-2u\eta+2v\,x=-\frac{\partial Q}{\partial z}$. ...(4, 5,6)

Differentiating (5) and (6) partially with to z and y, we have

$$\frac{\partial^2 v}{\partial z\partial t}-2w\frac{\partial\xi}{\partial z}-2\xi\frac{\partial w}{\partial z}+2u\frac{\partial\zeta}{\partial z}+2\zeta\frac{\partial u}{\partial z}$$

$$=\frac{\partial^2 w}{\partial y\partial t}-2u\frac{\partial\eta}{\partial y}-2\eta\frac{\partial u}{\partial y}+2v\frac{\partial\xi}{\partial y}+2\xi\frac{\partial v}{\partial y}$$

$$\Rightarrow\frac{\partial}{\partial t}\left(\frac{\partial w}{\partial y}-\frac{\partial v}{\partial z}\right)-2u\left(\frac{\partial\eta}{\partial y}+\frac{\partial\zeta}{\partial z}\right)+2v\left(\frac{\partial\xi}{\partial y}\right)+2w\left(\frac{\partial\xi}{\partial z}\right)$$

$$+2\xi\left(\frac{\partial v}{\partial y}+\frac{\partial w}{\partial z}\right)-2\eta\left(\frac{\partial u}{\partial y}\right)-2\zeta\frac{\partial u}{\partial z}=0. \qquad ...(7)$$

But $\dfrac{\partial \xi}{\partial x}+\dfrac{\partial \eta}{\partial y}+\dfrac{\partial \zeta}{\partial z}=0$ ...(8)

From (7) and (8), we have

$$2\frac{\partial \xi}{\partial t}+2u\frac{\partial \xi}{\partial x}+2v\frac{\partial \xi}{\partial y}+2w\frac{\partial \xi}{\partial z}+2\xi\left(\frac{\partial u}{\partial x}+\frac{\partial v}{\partial y}+\frac{\partial w}{\partial z}\right)$$

$$-2\xi\frac{\partial u}{\partial x}-2\eta\frac{\partial y}{\partial y}-2\zeta\frac{\partial u}{\partial z}=0$$

$$\Rightarrow \frac{D\xi}{Dt}+\xi\left(\frac{\partial u}{\partial x}+\frac{\partial v}{\partial y}+\frac{\partial w}{\partial z}\right)=\xi\frac{\partial u}{\partial x}+\eta\frac{\partial u}{\partial y}+\zeta\frac{\partial u}{\partial z} \quad ...(9)$$

The equation of continuity is

$$\frac{D\rho}{Dt}+\rho\left(\frac{\partial u}{\partial x}+\frac{\partial v}{\partial y}+\frac{\partial w}{\partial z}\right)=0. \quad ...(10)$$

From (9) and (10), we have

$$\frac{D\xi}{Dt}+\frac{\xi}{\rho}\frac{D\rho}{Dt}=\xi\frac{\partial u}{\partial x}\eta\frac{\partial u}{\partial y}+\zeta\frac{\partial u}{\partial z}$$

$$\Rightarrow \frac{1}{\rho}\frac{D\xi}{Dt}-\frac{\xi}{\rho^2}\frac{D\rho}{Dt}=\frac{\xi}{\rho}\frac{\partial u}{\partial x}+\frac{\eta}{\rho}\frac{\partial u}{\partial y}+\frac{\zeta}{\rho}\frac{\partial u}{\partial z} \quad ...(11)$$

$$\Rightarrow \frac{D}{Dt}\left(\frac{\xi}{\rho}\right)=\frac{\xi}{\rho}\frac{\partial u}{\partial x}+\frac{\eta}{\rho}\frac{\partial u}{\partial y}+\frac{\zeta}{\rho}\frac{\partial u}{\partial z},$$

Similarly $\dfrac{D}{Dt}\left(\dfrac{\eta}{\rho}\right)=\dfrac{\xi}{\rho}\dfrac{\partial v}{\partial x}+\dfrac{\eta}{\rho}\dfrac{\partial v}{\partial y}+\dfrac{\zeta}{\rho}\dfrac{\partial v}{\partial z},$

and $\dfrac{D}{Dt}\left(\dfrac{\zeta}{\rho}\right)=\dfrac{\xi}{\rho}\dfrac{\partial w}{\partial x}+\dfrac{\eta}{\rho}\dfrac{\partial w}{\partial y}+\dfrac{\zeta}{\rho}\dfrac{\partial w}{\partial z}.$ ...(12, 13)

But $\dfrac{\eta}{\rho}\dfrac{\partial u}{\partial y}+\dfrac{\zeta}{\rho}\dfrac{\partial u}{\partial z}=\dfrac{\eta}{\rho}\left\{\left(\dfrac{\partial u}{\partial y}-\dfrac{\partial v}{\partial x}\right)+\dfrac{\partial v}{\partial x}\right\}+\dfrac{\zeta}{\rho}\left\{\left(\dfrac{\partial u}{\partial z}-\dfrac{\partial w}{\partial x}\right)+\dfrac{\partial w}{\partial x}\right\}$

$$=\frac{\eta}{\rho}(-2\zeta)+\frac{\eta}{\rho}\frac{\partial v}{\partial x}+\frac{\zeta}{\rho}(2\eta)+\frac{\zeta}{\rho}\frac{\partial w}{\partial x},$$

$$=\frac{\eta}{\rho}\frac{\partial v}{\partial x}+\frac{\zeta}{\rho}\frac{\zeta}{\rho}\frac{\partial w}{\partial x} \quad ...(14)$$

Using the relation (14), the equations (11, 12,13) reduce to

$$\frac{D}{Dt}\left(\frac{\xi}{\rho}\right) = \frac{\xi}{\rho}\frac{\partial u}{\partial y} + \frac{\eta}{\rho}\frac{\partial v}{\partial y} + \frac{\zeta}{\rho}\frac{\partial w}{\partial y}$$

$$\frac{D}{Dt}\left(\frac{\eta}{\rho}\right) = \frac{\xi}{\rho}\frac{\partial u}{\partial y} + \frac{\eta}{\rho}\frac{\partial v}{\partial y} + \frac{\zeta}{\rho}\frac{\partial w}{\partial y},$$

$$\frac{D}{Dt}\left(\frac{\eta}{\rho}\right) = \frac{\xi}{\rho}\frac{\partial u}{\partial z} + \frac{\eta}{\rho}\frac{\partial v}{\partial z} + \frac{\zeta}{\rho}\frac{\partial w}{\partial z}, \qquad ...(15, 16, 17)$$

Equations (15), (16), (17) are known as Helmholtz's equation. Let $\xi = \eta = \zeta = 0$ at an instant of time t then

$$\frac{D}{Dt}\left(\frac{\xi}{\rho}\right) = \frac{D}{Dt}\left(\frac{\eta}{\rho}\right) = \frac{D}{Dt}\left(\frac{\zeta}{\rho}\right) = 0$$

$$\frac{D\zeta}{Dt} = \frac{D\eta}{Dt} = \frac{D\zeta}{Dt} = 0, \ \rho = \text{const.}$$

ξ, η, ζ must be constant. Since they are all zero at an instant of time t and have to remain constant.

In general, let $\frac{\partial u}{\partial x}, \frac{\partial v}{\partial x}$,...are all finite and less than a quantity P then

$\frac{\xi}{\rho}, \frac{\eta}{\rho}, \frac{\zeta}{\rho}$ can not increase faster than if they satisfy the equations.

$$\frac{D}{Dt}\left(\frac{\xi}{\rho}\right) =]\frac{D}{Dt}\left(\frac{\eta}{\rho}\right) = \frac{D}{Dt}\left(\frac{\zeta}{\rho}\right) = \frac{P}{\rho}(\xi + \eta + \zeta)$$

Let $\xi + \eta + \zeta = PW$, then

$$\frac{D}{Dt}\left(\frac{\xi}{\rho} + \frac{\eta}{\rho} + \frac{\zeta}{\rho}\right) = \frac{D}{Dt}(W) = 3PW$$

$\Rightarrow W = ke^{3Pt}, W \neq 0$

When $t = 0$,

$W = 0 \Rightarrow k = 0$ *i.e.*,

W be zero at time $t = 0$,

it may be so for all time.

Since W is the sum of three quantities ξ, γ, ζ which cannot be negative. Hence $W = 0$, it follows that each of these three quantities must be zero $\xi = 0 = \eta = \zeta$.

Hence if the motion is irrotational at any instant, it must be so for all time *i.e.*, if once, the velocity potential exists it exists for all time.

This is known as *the principle of Permanance of irrotational motion.*

Example 6:

An elastic fluid, the weight of which is neglected, obeying Boyle's law is in motion in a uniform straight tube; show that on the hypothesis of parallel sections the velocity at any time t at a distance r from a fixed point in the tube is defined by the equation

$$\frac{\partial^2 v}{\partial t^2} + \frac{\partial}{\partial r}\left(2v\frac{\partial v}{\partial t} + v^2\frac{\partial v}{\partial r}\right) = k\,\frac{\partial^2 v}{\partial r^2}.$$

Solution:

Since the fluid obeys Boyle's law then

$$p = k\rho. \qquad \text{...(1)}$$

The equation of continuity and the equation of motion is given by

$$\frac{\partial \rho}{\partial t} + \frac{\partial}{\partial r}(\rho v) = 0, \qquad \text{...(2)}$$

and
$$\frac{\partial v}{\partial t} + v\frac{\partial v}{\partial r} = -\frac{1}{\rho}\frac{\partial p}{\partial r} \qquad \text{...(3)}$$

Using the relation (1), we have

$$\frac{\partial v}{\partial t} + v\frac{\partial v}{\partial r} = -\frac{k}{\rho}\frac{\partial \rho}{\partial r}. \qquad \text{...(4)}$$

Differentiating (4) partially with regard to t, we have

$$\frac{\partial^2 v}{\partial t^2} + \frac{\partial}{\partial t}\left(v\frac{\partial v}{\partial r} + \frac{k}{\rho}\frac{\partial \rho}{\partial r}\right) = 0,$$

$$\Rightarrow \qquad \frac{\partial^2 v}{\partial t^2} + \frac{\partial}{\partial r}\left(v\frac{\partial v}{\partial t} + \frac{k}{\rho}\frac{\partial \rho}{\partial t}\right) = 0$$

$$\Rightarrow \qquad \frac{\partial^2 v}{\partial t^2} + \frac{\partial}{\partial r}\left\{v\frac{\partial v^*}{\partial t} + \frac{k}{\rho}\left(-\frac{\partial}{\partial r}(\rho v)\right)^t\right\} = 0$$

$$\Rightarrow \qquad \frac{\partial^2 v}{\partial t^2} + \frac{\partial}{\partial r}\left\{v\frac{\partial v}{\partial t} + \frac{k}{\rho}\left(\rho\frac{\partial}{\partial r} + v\frac{\partial \rho}{\partial r}\right)^t\right\} = 0$$

$$\Rightarrow \qquad \frac{\partial^2 v}{\partial t^2} + \frac{\partial}{\partial r}\left\{v\frac{\partial v}{\partial t} - k\frac{\partial v}{\partial r} - \frac{k}{\rho}\frac{\partial \rho}{\partial r}v\right\} = 0$$

$$\Rightarrow \frac{\partial^2 v}{\partial t^2} + \frac{\partial}{\partial r}\left\{v\frac{\partial v}{\partial t} - k\frac{\partial v}{\partial r} + \left(\frac{\partial v}{\partial t} + v\frac{\partial v}{\partial r}\right)v\right\} = 0$$

$$\Rightarrow \qquad \frac{\partial^2 v}{\partial t^2} + \frac{\partial}{\partial r}\left(2v\frac{\partial v}{\partial t} + v^2\frac{\partial v}{\partial r} - k\frac{\partial v}{\partial r}\right) = 0$$

$$\Rightarrow \qquad \frac{\partial^2 v}{\partial t^2} + \frac{\partial}{\partial r}\left(2v\frac{\partial v}{\partial t} + v^2\frac{\partial v}{\partial r}\right) = k\frac{\partial^2 v}{\partial r^2}.$$

Proved.

Example 7:

Air obeying Boyles's law is in motion in a uniform tube of small section, Prove that if ρ be the density and v the velocity at a distance x from a fixed point at time t.

$$\frac{\partial^2 \rho}{\partial t^2} = \frac{\partial^2}{\partial x^2}\{\rho(v^2 + k)\}.$$

Solution:

Let p be the pressure and v be the velocity at a distance x from the end of the tube at any time t. The equation of motion and the equation of continuity is given by

$$\frac{\partial v}{\partial t} + v\frac{\partial v}{\partial x} = -\frac{1}{\rho}\frac{\partial p}{\partial x}, \qquad ...(1)$$

and $\dfrac{\partial \rho}{\partial t} + \dfrac{\partial}{\partial x}(\rho v) = 0.$ ...(2)

Since the air obeys Boyle's law, then

$p = k\rho \Rightarrow dp = kd\rho$ ...(3)

From (1) and (3), we have

$$\frac{\partial v}{\partial t} + v\frac{\partial v}{\partial t} = -\frac{k}{\rho}\frac{\partial p}{\partial x}. \qquad ...(4)$$

Differentiating (2) partially with regard to t, we have

$$\frac{\partial^2 \rho}{\partial t^2} + \frac{\partial}{\partial t}\left\{\frac{\partial}{\partial x}(\rho v)\right\} = 0$$

$$\Rightarrow \frac{\partial^2 \rho}{\partial t^2} + \frac{\partial}{\partial x}\left\{\frac{\partial}{\partial t}(\rho v)\right\} = 0$$

$$\Rightarrow \frac{\partial^2 \rho}{\partial t^2} + \frac{\partial}{\partial x}\left\{\rho \frac{\partial}{\partial t} v \frac{\partial \rho}{\partial t}\right\} = 0$$

From (2) and (4), we have

$$\frac{\partial^2 \rho}{\partial t^2} + \frac{\partial}{\partial x}\left\{\rho\left(-v\frac{\partial v}{\partial x} - \frac{k}{\rho}\frac{\partial \rho}{\partial x}\right) - v\frac{a}{\partial x}(\rho v)\right\} = 0$$

$$\Rightarrow \frac{\partial^2 \rho}{\partial t^2} + \frac{\partial}{\partial x}\left\{\rho v \frac{\partial v}{\partial x} + v\frac{\partial}{\partial x}(\rho v) + k\frac{\partial \rho}{\partial x}\right\}$$

$$\Rightarrow \frac{\partial^2 \rho}{\partial t^2} + \frac{\partial}{\partial x}\left\{\frac{\partial}{\partial x}(\rho v.v + k\frac{\partial \rho}{\partial x}\right\} = \frac{\partial^2}{\partial x^2}\{\rho\,(v^2 + k)\}.$$ **Proved.**

Example 8:

Stream is rushing from a boiler through a conical pipe, the diameters of the ends of which are D and d. If V and v be the corresponding velocities of the stream and if the motion be supposed to be that of divergence from the vertex of the cone, prove that

$$\frac{v}{V} = \frac{D^2}{d^2}\exp.\left(\frac{v^2 - V^2}{2k}\right),$$

where k is the pressure divided by the density, and supposed constant.

Solution:

Let ρ_1 and ρ_2 be the densities of steam at the ends of the conical pipe AB and CD. By the principle of conservation of mass the mass of the steam that enters and leaves at the ends AB and CD are of the same. Thus we have

$$\pi\left(\frac{1}{2}d\right)^2 v\rho_1 = \pi\left(\frac{1}{2}D\right)^2 v\rho_2$$

or $$\frac{v}{V} = \frac{D^2}{d^2}\frac{\rho^2}{\rho_1}. \quad ...(1)$$

Let p be the pressure ρ the density and u the velocity at distance r from AB, then the equation of motion is given by

$$u\frac{\partial u}{\partial r} = -\frac{1}{\rho}\frac{\partial p}{\partial r},$$

Since $k = p/\rho \Rightarrow p = k\rho$

$$\Rightarrow u\frac{\partial u}{\partial r} = -\frac{k}{\rho}\frac{\partial \rho}{\partial r}.$$

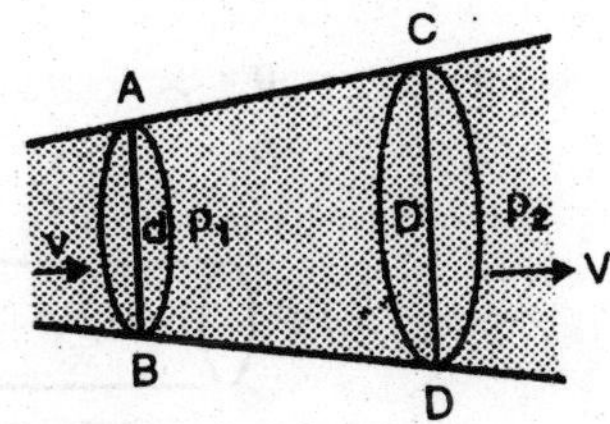

Fig. 4.6

By integrating, we have

$$\frac{1}{2}u^2 = -k \log \rho + k \log A$$

where A is an arbitrary constant.

$$\Rightarrow \log (\rho/A) = -(u^2/2k)$$

$$\Rightarrow \rho = A \exp.(-u^2/2k).$$

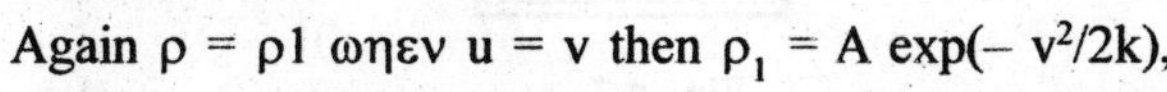

Again $\rho = \rho 1$ ωηεν $u = v$ then $\rho_1 = A \exp(-v^2/2k)$,

and $\rho = \rho^2$ when $u = V$ then $\rho_2 = A \exp(-V^2/2k)$,

$$\Rightarrow \frac{\rho_1}{\rho_2} = \frac{\exp(-v2/2k)}{\exp(-V^2/2k)}.$$

$$\Rightarrow \frac{\rho_2}{\rho_1} = \exp\{(v2 - V^2)/2k\}. \qquad ...(2)$$

From (1) and (2), we have

$$\frac{v}{V} = \frac{D^2}{d^2}\exp\left(\frac{v^2 - V^2}{2k}\right).$$ **Proved.**

Example 9:

A stream in a horizontal pipe, after passing a contraction in the pipe at which its sectional area is A, is delivered at atmospheric pressure at a place where the sectional area is B. Show that if a side is connected with the pipe at the former place, water will be sucked up through it into the pipe from a reservoir at a depth $\frac{S^2}{2g}\left(\frac{1}{A^2} - \frac{1}{B^2}\right)$ *below the pipe; S being the delivery per second.*

Solution:

Let v and p be the velocity and pressure in the tube of sectional area A and V and Π be the velocity and the atmospheric pressure at the sectional area B. The equation of motion is

$$v\frac{\partial v}{\partial r} = -\frac{1}{\rho}\frac{\partial p}{\partial r}. \qquad ...(1)$$

By integrating, we have

$$\frac{1}{2}v^2 = -\frac{p}{\rho} + C, \qquad ...(2)$$

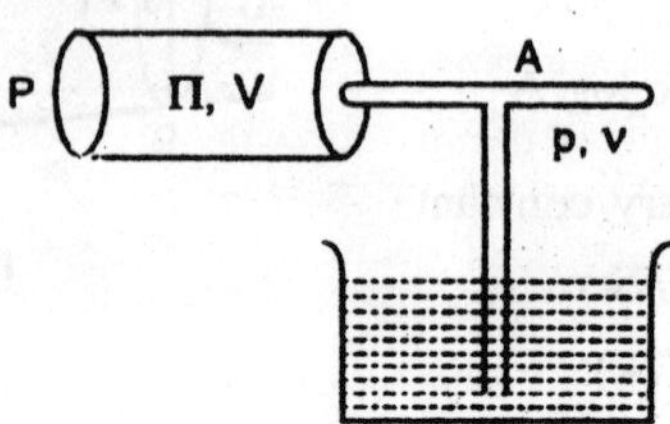

Fig. 4.7

where C is an arbitrary constant.

Since v = V, p = P at the sectional area B then

$$\frac{1}{2}v^2 + \frac{p}{\rho} = \frac{1}{\rho}V^2 + \frac{\Pi}{\rho}; C = \frac{1}{2}V^2 + \frac{\Pi}{\rho} \qquad ...(3)$$

$$\frac{1}{2}(v^2 - V2) = \frac{1}{\rho}(\Pi - p). \qquad ...(4)$$

The delivery of stream per second in the tube of sectional area A and B is given as S. From the equation of continuity, we have

Av = BV = S.

$$\Rightarrow \frac{1}{\rho} = \frac{1}{2}\left(\frac{S^2}{A^2} - \frac{S^2}{B^2}\right). \qquad ...(6)$$

Let h be the height through which the water is sucked up, then gph = dff. of pressure in the tube of section A and section B.

From (6) and (7), we have

$$\frac{1}{\rho}gph = \frac{S^2}{2}\left(\frac{1}{A^2} - \frac{1}{B^2}\right) \Rightarrow h = \frac{S^2}{2g}\left(\frac{1}{A^2} - \frac{1}{B^2}\right)$$ **Proved.**

Example 10:

A pulse travelling along a fine straight uniform, tube filled with gas causes the at time t and distance x from the origin where the velocity is u_0 to become $\rho_0\ \phi\ (vt - x)$. Prove that the velocity u (at time t and distance x from the origin) is given by

$$v + \frac{(u_0 - v)\phi(vt)}{\phi(vt - x)}.$$

Solution:

Let u be the velocity and r be the density of the gas at a distance x then

$$\rho = p_0\phi\ (vt - x). \qquad ...(1)$$

The equation lf continuity is given by

$$\frac{\partial \rho}{\partial t} = \frac{\partial}{\partial x}(\rho u) = 0$$

$$\Rightarrow \frac{\partial \rho}{\partial t} + \rho \frac{\partial u}{\partial x} + u \frac{\partial \rho}{\partial x} = 0. \qquad ...(2)$$

Differentiating (1) partially with regard to t and x respectively, we have

$$\frac{\partial \rho}{\partial t} = \rho_0 v\ \phi,\ (vt - x);\ \frac{\partial \rho}{\partial x} = -\ \rho_0\ \phi'(vt - x). \qquad ...(3)$$

From (2) and (3), we have

$$\frac{du}{v - u} + \frac{\phi'(vt - x)}{\phi(vt - x)}\ dx = 0$$

$$\Rightarrow (v - u)\ \phi\ (vt - x) = A, \qquad ...(4)$$

where A is integration constant to be determined by the given initial condition; $x = 0$, $u = u_0$; $A = (v - u_0)\ \phi\ (vt)$

$$\text{or } u = v + \frac{(u_0 - v)\phi(vt)}{\phi(vt - x)}.$$ **Proved.**

Example 11:

A quantity of liquid occupies a length 2l of a straight tube of uniform small bore under the action of a force to a point in the tube varying as the distance from that point. Determine the motion and the pressure.

Solution:

Consider p be the pressure and u be the velocity at a distance x from the fixed point O. Let x_0 be the distance of the nearer free surface from O. The equation of continuity is $\partial u/\partial x = 0$. The equation of motion becomes

$$\frac{\partial u}{\partial t} = -\ \mu x - \frac{1}{\rho}\frac{\partial p}{\partial x}. \qquad ...(1)$$

Integrating (1) with regard to x, we have

$$x\frac{\partial u}{\partial t} = -\frac{1}{2}\mu x^2 - \frac{p}{\rho} + A.$$

Again $p = 0, x = x_0$

and $p = 0, x = x_0 + 2l.$ ...(3)

From (2) and (3), we have

$$x_0\frac{\partial u}{\partial t} = -\frac{1}{2}\mu x_0^2 + A,$$

and $$(x_0 + 2l)\frac{\partial u}{\partial t} = \frac{1}{2}\mu(x_0 + 2l)^2 + A$$

Eliminating the arbitrary constant A, we have

$$\frac{\partial^2 x_0}{\partial t^2} + \mu(x_0 + l) = 0;\ u = x_0 \quad ...(4)$$

whose solution is given by

$$x_0 + l = B\cos(\sqrt{\mu}t + \alpha), \quad ...(5)$$

where the constants are determined by the initial position and velocity. The pressure at any point is determined as

$$p/\rho = -\frac{1}{2}\mu(x^2 - x_0^2) - (x - x_0)(\partial u/\partial t), \quad \textbf{Ans.}$$

$$\Rightarrow p/\rho = -\frac{1}{2}\mu(x^2 - x_0^2) + \mu(x - x_0)(x_0 + l).$$

SYMMETRICAL FORMS OF THE EQUATION OF CONTINUITY

I. Spherical Symmetry

Consider two concentric spheres whose motion is symmetrical about the centre O. Let q (r, t) be the velocity in the direction OP. MASS gained by the flow through the inner surface is

$4\pi r^2 q_r \rho = F(r, t)$ (say).

Mass lost by the flow through the outer surface is

$= F(r + \delta r, t).$

Mass between the spheres at an instant of time t is

$= 4\pi r^2 \rho\delta r.$

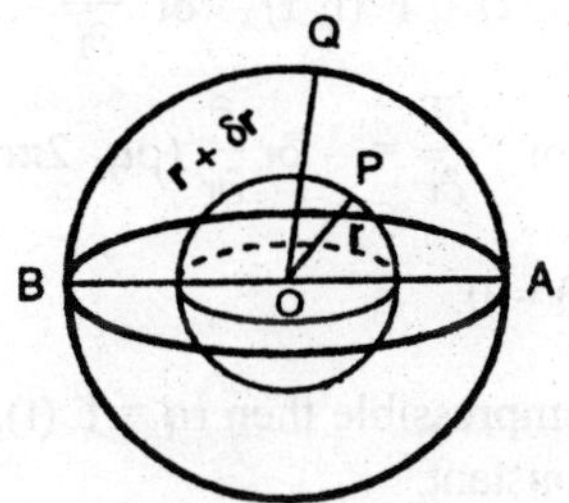

Fig. 4.8

By the principle of conservation of mass, we have

$$\frac{\partial \rho}{\partial t}(4\pi r^2 \delta r) = F\ (r, t) - F\ (r + dr, t)$$

$$\Rightarrow \frac{\partial \rho}{\partial t}(4\pi r^2 \delta r) = F\ (r, t) - F(r, t) - \frac{\partial F}{\partial r}\delta r \ .....$$

$$\Rightarrow \frac{\partial \rho}{\partial t}(4\pi r^2 \delta r) = \delta r \frac{\partial}{\partial r}(4\pi r^2 q\rho)$$

$$\Rightarrow \frac{\partial \rho}{\partial t} + \frac{1}{r^2}\frac{\partial}{\partial r}(\rho q_r r^2) = 0. \qquad ...(1)$$

For an incompressible fluid r is constant throughout the motion *i.e.*, $\partial p/\partial t = 0$, then equation (1) reduces to

$r^2 q_r = F\ (t)$.

In a steady flow F (t) shall be absolute constant.

II. Cylindrical Symmetry

Let q (r, t) be the velocity at any point P which is perpendicular to a fixed axis OZ, r is the perpendicular distance of P from OZ. Consider two cylinders of radii r and r + δr at a unit distance a part with OZ as axis.

Rate of flow across the inner surface = $\rho \mathbf{q}\ (2\pi r)$ = F(r, t).

Rate of flow across the outer surface = F (r + δr, t).

Mass between the two cylinders at any instant of time t

= $2\pi\ r\rho\delta r$.

By the principle of continuity, we have

$$\frac{\partial}{\partial t}(2\pi r\rho\delta r) = F(r, t) - F\ (r + \delta r, t)$$

$$= F(r, t) - F(r, t) - \delta r \frac{\partial F}{\partial r} - ....$$

$$\Rightarrow \frac{\partial \rho}{\partial t} 2\pi r \delta r = -\delta r \frac{\partial F}{\partial r} = -\delta r \frac{\partial}{\partial r}(\rho \mathbf{q}.\ 2\pi r)$$

$$\Rightarrow \frac{\partial \rho}{\partial t} + \frac{1}{r}\frac{\partial}{\partial r}(\rho r \mathbf{q}) = 0.$$

If the fluid be incompressible then **rq** = F (t). In a steady flow F (t) shall be an absolute constant.

Example 12:

A sphere is at rest in an infinite mass of homogeneous liquid of density ρ, the pressure at infinity being Π. Show that if, the radius R of the sphere varies in any manner, the pressure at the surface at any time t is

$$\Pi \ \frac{1}{2}\rho\left\{\frac{d^2}{dt^2}(R^2) + \left(\frac{dR}{dt}\right)^2\right\} +.$$

If R = a (2 + cos nt), show that to prevent cavitation in the fluid, P must not be less that $3\rho a^2 n^2$.

Solution:

The only possible motion which can take place is one in which each element of liquid moves towards the centre, whence the free surface will remain spherical. In an incompressible fluid, the fluid velocity will be radial spherical outside the sphere, so it will be a function of r and the time t only. The equation of continuity reduces to

$$\frac{1}{2}\frac{d}{dr}(r^2 v) = 0$$

$\Rightarrow r^2 v$ = constant = F (t), (let) ...(1)

where v is the velocity of a fluid particle at a distance r from the centre at the time t.

Equation of motion is

$$\frac{\partial v}{\partial t} + v\frac{\partial v}{\partial r} = -\frac{1}{\rho}\frac{\partial p}{\partial r}, \quad \left\{\begin{array}{l}\text{Since } r^2 v = F(t) \\ \dfrac{dv}{dt} = \dfrac{F'(t)}{r^2}\end{array}\right.$$

$$\text{or} \quad \frac{F'(t)}{r^2} + v\frac{\partial v}{\partial r} = -\frac{1}{\rho}\frac{\partial p}{\partial r}.$$

Integrating with regard to r, we have

$$-\frac{F'(t)}{r}+\frac{1}{2}v^2=\frac{p}{\rho}+B, \qquad ...(2)$$

where B is an arbitrary constant.

Initially $r \to \infty$, $v = 0$, $p = \Pi$; $B = (\Pi/\rho)$

$$\Rightarrow -\frac{F'(t)}{r}+\frac{1}{2}v^2=\frac{\Pi}{\rho}-\frac{p}{\rho}$$

$$\Rightarrow p=\Pi+\frac{1}{2}\rho\left(\frac{2F'(t)}{r}-v^2\right)$$

Let P be the pressure on the surface of the sphere of radius R and V be the velocity, then

$$P=\Pi+\frac{1}{2}\rho\left\{\frac{2F'(t)}{R}-V^2\right\} \qquad ...(3)$$

Again from (1) we have

$R^2V = F(t)$, $F(t) = R^2 (dR/dt)$.

Differentiating with regard to t, we have

$F'(t) = R^2(d^2R/dt^2) + 2R(dR/dt)^2$.

From the equation (3), we have

$$P=\Pi+\frac{1}{2}\rho\left\{2R\frac{d^2R}{dt^2}+4\left(\frac{dR}{dt}\right)^2-V^2\right\}$$

$$\text{or } P=\Pi+\frac{1}{2}\rho\left\{2R\frac{d^2R}{dt^2}+4\left(\frac{dR}{dt}\right)^2-\left(\frac{dR}{dt}\right)^2\right\}$$

$$\text{or } P=\Pi+\frac{1}{2}\rho\left\{2\left[R\frac{d^2R}{dt^2}+-\left(\frac{dR}{dt}\right)^2\right]+\left(\frac{dR}{dt}\right)^2\right\}$$

$$\text{or } P=\Pi+\frac{1}{2}\rho\left\{\frac{d^2}{dt^2}(R^2)+\left(\frac{dR}{dt}\right)^2\right\}$$ **Proved.**

Since r'^2v' = constant

Now v' is maximum when r' is minimum *i.e.* $r' = R$

Again $R = a(2 + \cos nt)$, $R^2 = a^2 (2 + \cos nt)^2$ ...(4)

$$\text{or } \frac{dR}{dt}=-an\sin nt$$

$$\frac{d}{dt}(R^2) = -2\,a^2n\,(2 + \cos nt)\sin nt$$

and $$\frac{d^2}{dt^2}(R^2) = 2a^2n^2\sin^2 nt - 2a^2n^2\,(2 + \cos nt)\cos nt$$

Hence

$$P = \Pi + \frac{1}{2}\rho\,\{2a^2n^2\sin^2 nt - a^2n^2r\,(2 + \cos nt)\cos nt + a^2n^2\sin^2 nt\}$$

$$P = \Pi + \frac{3}{2}\,\rho a^2n^2\sin^2 nt - a^2n^2r\,(2\cos nt + \cos^2 nt) \qquad ...(5)$$

From (4), we obtain that R varies form 3a to a. Thus the sphere has greatest radius 3a when nt = 0 or 2mπ. As sphere shrinks from R = 3a, there is a possibility of a cavitation value of pressure on the surface of the sphere *i.e.*, t = 0 or nt = 2mπ then from (5), we have

$$P' = \Pi - 3\rho a^2n^2$$

Hence, to prevent, cavitation in the fluid P′ must be positive *i.e.*, Π must not be less than $3\rho a^2n^2$.

Example 13:

A centre of force attracting inversely as the square of the distance is at the centre of a spherical cavity within an infinite mass of incompressible fluid, the pressure on which at an infinite distance is Π, and is such that the work done by this pressure on a unit of area through a unit of length in one-half the work done by the attractive force on a unit volume of the fluid from infinity to the initial boundary of the cavity, prove that the time of filling up the cavity will be

$$\pi a\sqrt{\left(\frac{\rho}{\Pi}\right)\left\{2-\left(\frac{3}{2}\right)^{3/2}\right\}};$$

a being the initial radius of the cavity, and ρ the density of the fluid.

Solution:

The only possible motion which can take place is one in which each element of liquid moves towards the centre, whence the free surfaces will remain spherical. Let v′ be the velocity at a distance r′ at any time t and p be the pressure there. The equation of continuity reduce to

$$\frac{1}{r^2}\frac{d}{dr'}(r'^2 v') = 0,$$

$$\Rightarrow r'^2 v' = f(t) = \text{Const.} \qquad \text{....(1)}$$

$$\Rightarrow r'^2 v' = f(t) = r^2 v.$$

Equation of motion is

$$\frac{\partial v'}{\partial t} + v'\frac{\partial v'}{\partial r'} = -\frac{\mu}{r'^2}\frac{1}{\rho}\frac{\partial v'}{\partial r'} \qquad \text{...(2)}$$

$$\Rightarrow \frac{f'(t)}{r'^2} + v'\frac{\partial v'}{\partial r'} = -\frac{\mu}{r'^2}\frac{1}{\rho}\frac{\partial v'}{\partial r'}.$$

Integrating with regard to r′, we have

$$-\frac{f'(t)}{r'} + \frac{1}{2}v'^2 = -\frac{\mu}{r'} - \frac{p}{\rho} + A,$$

where A is an arbitrary constant.

Initially $r' = \infty$, $v' = 0$ and $p = \Pi$; $A = \Pi/\rho$.

$$\Rightarrow -\frac{f'(t)}{r'} + \frac{1}{2}v'^2 = \frac{\mu}{r'} + \frac{\Pi - p}{\rho}. \qquad \text{...(3)}$$

Let r be radius of the spherical cavity and v be the velocity at any time t. The pressure p vanishes on the surface of the cavity. Hence equation (3) reduces to

$$-\frac{f'(t)}{r} + \frac{1}{2}v^2 = \frac{\mu}{r} + \frac{\Pi}{\rho}$$

$$\Rightarrow -\frac{1}{r}\left(r^2 v\frac{dv}{dr} + 2rv^3\right) + \frac{1}{2}v^2 = \frac{\mu}{r} + \frac{\Pi}{\rho}$$

$$\left\{\begin{aligned} &\text{Since } f(t) = r^2 v \\ &f'(t) = r^2\frac{dv}{dt} + 2rv\frac{dr}{dt} \\ &f'(t) = r^2\frac{dv}{dr}.\frac{dr}{dt} + 2rv^2 \\ &f'(t) = r^2 v\frac{dv}{dr} + 2rv^2. \end{aligned}\right.$$

$$\Rightarrow 2rv\,dv + 3v^2\,dr = -2\left(\frac{\mu}{r} + \frac{\Pi}{\rho}\right)dr$$

Multiplying both the sides with r^2, we get

$$2r^3 v\,dv + 3v^2 r^2\,dr = -2\left(\mu r + \frac{\Pi}{\rho}r^2\right)dr.$$

By integrating, we have

$$r^3v^2 = -\{\mu r^2 + (2\Pi/3\rho)\, r^3\} + B, \qquad ...(4)$$

where B is an arbitrary constant.

Initially r = a, v = 0; $B = \mu a^2 + (2\Pi/3\rho)\, a^3$.

Equation (4) becomes

$$r^3v^2 = \mu\,(a^2 - r^2) + (2\Pi/3\rho)\,(a^3 - r^3). \qquad ...(5)$$

The work done by the pressure Π on a unit of area through a unit of length $= \frac{1}{2} \times$ work done by the attractive force on a unit volume of the fluid from infinity to the initial boundary of the cavity *i.e.*,

$$\Pi \times 1 \times 1 = \frac{1}{2}\int_{\infty}^{a}\left(\frac{\mu}{r^2}\right)\rho dr = \frac{\mu\rho}{2a} \Rightarrow \mu = \frac{2a\Pi}{\rho}.$$

Substituting the value of m in (5), we get

$$r^3v^2 = \frac{2a\Pi}{\rho}(a^2 - r^2) + \frac{2\Pi}{3\rho}(a^3 - r^3)$$

$$\Rightarrow r^3v^2 = \frac{2\Pi}{3\rho}\{3a\,(a^2 - r^2) + (a^3 - r^3)\}$$

$$\Rightarrow v^2 = \frac{2\Pi\,(3a(a^2 - r^2) + (a^3 - r^3))\}}{3\rho \qquad r^3}$$

$$\Rightarrow \frac{dr}{dt} = \sqrt{\left(\frac{2\Pi}{3\rho}\right)}.\frac{\{3a(a^2 - r^2) + (a^3 - r^2)^{1/2}}{r^{3/2}}$$

$$\Rightarrow dt = -\sqrt{\left(\frac{3\rho}{2\Pi}\right)}\int_{a}^{0}\frac{r^{3/2}dr}{\{3a(a^2 - r^2) + (a^3 - r^3)^{1/2}}$$

Negative sign taken as r decreases when t increases.

Let t be the time of filling up the cavity then

$$t = \sqrt{\left(\frac{3\rho}{2\Pi}\right)}\int_{0}^{a}\frac{r^{3/2}dr}{\sqrt{\{3a(a^2 - r^2) + (a^3 - r^3)\}}}$$

$$\Rightarrow t = \sqrt{\left(\frac{3\rho}{2\Pi}\right)}\int_{0}^{a}\frac{r^{3/2}dr}{(r + 2a)\sqrt{(a - r)}}$$

Let $r = a\sin^2\theta$, $dr = 2a\sin\theta\cos\theta\, d\theta$

$$\Rightarrow t = 2a\sqrt{\left(\frac{3\rho}{2\Pi}\right)}\int_{0}^{\pi/2}\frac{\sin^4\theta}{\sin^2\theta + 2}d\theta$$

$$\Rightarrow t = 2a\sqrt{\left(\frac{3\rho}{2\Pi}\right)}\int_0^{\pi/2}\left(\sin^2\theta - 2 + \frac{4}{2+\sin^2\theta}\right)d\theta$$

$$\Rightarrow t = 2a\sqrt{\left(\frac{3\rho}{2\Pi}\right)}\left[\frac{\pi}{4} - \pi + \int_0^{\pi/2}\frac{4\sec^2\theta}{2\sec^2\theta + \tan^2\theta}d\theta\right]$$

$$\Rightarrow t = 2a\sqrt{\left(\frac{3\rho}{2\Pi}\right)}\left[-\frac{3\pi}{4} + \int_0^{\pi/2}\frac{4\sec^2\theta}{2+3\tan^2\theta}d\theta\right]$$

$$\Rightarrow t = 2a\sqrt{\left(\frac{3\rho}{2\Pi}\right)}\left[-\frac{3\pi}{4} + \frac{4}{3}\sqrt{\left(\frac{3}{2}\right)}.\frac{\pi}{2}\right]$$

$$\Rightarrow t = \pi a\sqrt{\left(\frac{\rho}{\Pi}\right)}\left\{2 - \left(\frac{3}{2}\right)^{3/2}\right\}.$$

Proved.

Example 14:

An infinite mass of homogeneous incompressible fluid is at rest subject to a uniform pressure Π and contains a spherical cavity of radius a, filled with gas at a pressure m Π : prove that, if the inertia of the gas be neglected, and Boyle's law be supposed to hold throughout the ensuing motion, the radius of the sphere will oscillate between the values a and na, where n is determined by the equation

$1 + 3m \log n - n^3 = 0.$

If m be nearly to 1, the time of an oscillation will be

$2\pi\sqrt{(a^2\rho/3\Pi}$

π being the density of the fluid.

Solution:

In an incompressible fluid the velocity v will be radial outside the spherical cavity, so v will be a function of r radial distance from the centre of cavity (*i. e.*, origin) and the time t only. The continuity equation reduces to

$$\frac{1}{r}\frac{d}{dr}(r^2 v) = 0 \Rightarrow r^2 v = f(t).$$

Let v′ be the velocity at a distance r′ from the centre of the cavity at any time t and p be the pressure there, then we have

$r'^2 v' = f(t) = r^2 v.$

Equation of motion is

$$\frac{\partial v'}{\partial t} + v'\frac{\partial v'}{\partial r'} = -\frac{1}{\rho}\frac{\partial p}{\partial r'},$$

or $$\frac{f'(t)}{r'^2} + v'\frac{\partial v}{\partial r'} = -\frac{1}{\rho}\frac{\partial p}{\partial r'},$$

Integrating with regard to r′, we have

$$-\frac{f'(t)}{r'} + \frac{1}{2}v'^2 = -\frac{p}{\rho} + A,$$

where A is an arbitrary constant,

Initially when $r' = \infty$, $v' = 0$,

$$p = \Pi \ ; \ A = \Pi/\rho.$$

Equation (3) becomes

$$-\frac{f'(t)}{r'} + \frac{1}{2}v'^2 = \frac{\Pi - p}{\rho}.$$

Hence motion on the surface of the cavity is

$$-\frac{f'(t)}{r} + \frac{1}{2}v'^2 = \frac{\Pi - p}{\rho}.$$

From Boyle's law, we have

$$\frac{4}{3}\pi r^3 p = \frac{4}{3}\pi a^3 m\Pi \Rightarrow p = (a^3/r^3)m\Pi.$$

Substituting the value of f ′ (t) and p in (5), we get

$$-\left(2rv^2 + r^2 v\frac{dv}{dr}\right) + \frac{1}{2}v^2 = \frac{\Pi}{r} - \frac{a^3}{\rho r^3}m\Pi$$

$$\Rightarrow rv\frac{dv}{dr} + \frac{3}{2}v^2 = -\frac{\Pi}{\rho} - \frac{a^3}{\rho r^3}m\Pi$$

Multiplying with $2r^2$ dr both the sides, we have

$$2r^3 v\, dv + 3r^2 v^2\, dr = \left(-\frac{2\Pi r^2}{\rho} + \frac{2a^2 m\Pi}{\rho r}\right)dr.$$

By integrating, we have

$$r^3v^2 = -\frac{2\Pi}{3\rho}r^3 + \frac{2a^3 m\Pi}{\rho}\log r + B.$$

Initially r = a, v = 0;

$$B = \frac{2\Pi}{3\rho.}a^3 - \frac{2a^2 m\Pi}{\rho}\log a$$

$$r^3v^2 = \frac{2\Pi}{3\rho.}(a^3 - r^3)\frac{2a^2 m\Pi}{\rho}\log\left(\frac{r}{a}\right). \quad ...(6)$$

Since the radius of the sphere oscillates between a andna it follows that at r = a, v = 0 and r = na, v = 0.

Substituting r = na, v = 0 in relation (6), we get

$$\frac{2\Pi}{3\rho.}\left\{a^3 - n^3a^3 + 3ma^3\log\left(\frac{na}{a}\right)\right\} = 0$$

$$1 + 3m\log n - n^3 = 0,\ a \neq 0$$

Which proves the required condition.

Again if m be nearly equal to 1, let r = a + x where x is small then from (5), we have

$$(a + x)^3x^2 = \frac{2\Pi}{3\rho}\{a^3 - (a + x)^3\} + \frac{2a^3\Pi}{\rho}\log\left(\frac{a + x}{a}\right).$$

Since r = a + x, v = dr/dt = x

$$\Rightarrow a^3\left(1 + \frac{x}{a}\right)^3 x^2 = \frac{2\Pi}{3\rho}a^3\left\{1 - \left(1 + \frac{x}{a}\right)^3\right\} + \frac{2a^3\Pi}{\rho}\log\left(1 + \frac{x}{a}\right)$$

$$\Rightarrow \left(1 + \frac{x}{a}\right)^3 x^2 = \frac{2\Pi}{2\rho}\left\{1 - \left(1 + \frac{3x}{a} + \frac{3x^2}{a^2} + ...\right)\right\}$$

$$+ \frac{2a^3\Pi}{\rho}\left\{\frac{x}{a} - \frac{1}{2}\frac{x^2}{a^2} + ...\right\}$$

$$\Rightarrow x^2 = \frac{2\Pi}{3\rho}\left[\left(-\frac{9}{2}\frac{x^2}{a^2} + ...\right)\left(1 + \frac{x}{a}\right)^{-3}\right]$$

$$\Rightarrow x^2 = \frac{2\Pi}{3\rho}\left[\left(-\frac{9}{2^2}\frac{x^2}{a^2} + ...\right)\left(1 + \frac{3x}{a} + \frac{6x^2}{a^2} + ...\right)\right] = \frac{3\Pi}{\rho a^2}x^{2,}$$

neglecting higher orders of x, being very small.

Differentiating with regard to t, we have

$$2x\, x = -\, (3\Pi/\rho a^2).2xx$$

$$\Rightarrow \qquad x = -\, (3\Pi/\rho a^2)\; x,\; x \neq 0,$$

which represents the standard equation of S. H. M. whose time period is

$2p\,\sqrt{(\rho a^2/3\Pi)}$. **Proved.**

Example 15:

An infinite fluid in which as a spherical hollow of radius a is initially at rest under the action of no forces. If a constant pressure Π is applied at infinity, show that the time of filling up the cavity is

$$\pi a^2\, (\rho/\,\Pi)^{1/2}\; 2^{5/6} \left\{\Gamma\left(\frac{1}{3}\right)\right\}^{-3}$$

Solution:

In an incompressible fluid, the fluid velocity v will be radial outside the hollow sphere and will be a function or a (radial distance from the centre of the sphere *i.e.*, origin) and the time t and p be the pressure there. From the continuity equation, we have

$$r'^2 v' = f(t) = r^2 v. \qquad ...(1)$$

The equation of motion is

$$\frac{\partial v'}{\partial r} + v'\frac{\partial v'}{\partial r'} = -\frac{1}{\rho}\frac{\partial p}{\partial r'}$$

$$\text{or} \quad \frac{f'(t)}{r'^2} + v'\frac{\partial v'}{\partial r'} = -\frac{1}{\rho}\frac{\partial p}{\partial r'}$$

Integrating with regard to r′, we get

$$-\frac{f'(t)}{r'} + \frac{1}{2}v'^2 = -\frac{p}{\rho} + A,$$

where A is an arbitrary constant.

initially $r' = \infty$, $v' = 0$ $p = \Pi$; $A = \dfrac{\Pi}{\rho}$.

$$\text{So} \quad -\frac{f'(t)}{r'} + \frac{1}{2}v'^2 = \frac{\Pi - p}{\rho} \qquad ...(2)$$

Let v be the velocity and r be the radius of spherical cavity at any time t $v' = v$, $r' = r$, $p = 0$ (being hollow).

From (2), we get

$$-\frac{f'(t)}{r}+\frac{1}{2}v^2=\frac{\Pi}{\rho} \quad \text{...(3)}$$

$$rv\frac{dv}{dr}+\frac{3}{2}v^2=-\frac{\Pi}{\rho}.$$

⇒ Multiplying both the sides by $2r^2dr$ and integrating, we have

$$2r^3vdv+3r^2v^2\,dr=(2\Pi/\rho)r^2\,dr \quad \text{...(4)}$$

$$\Rightarrow r^3v^2=-(2\Pi/3\rho)\,r^3+B,$$

Initially $r = a$, $v = 0$;

$$B=(2\pi/3\rho)\,a^3.$$

Equation (4) reduces to

$$r^3v^2=\frac{2\Pi}{3\rho}(a^3-r^3)$$

$$\Rightarrow v=\frac{dr}{dt}=-\sqrt{\left(\frac{2\Pi}{3\rho}\right)}\sqrt{\left(\frac{a^3-r^3}{r^3}\right)}$$

(–) ve sign is taken since v increases as r decreases.

Let t be the time of filling up the cavity

$$t=-\sqrt{\left(\frac{3\rho}{2\Pi}\right)}\int_a^0\sqrt{\left(\frac{r^3}{a^3-r^3}\right)}dr$$

Let $r = a\sin^{2/3}\theta$, $dr=-\frac{2a}{3}\sin^{-1/3}\theta\cos\theta d\theta$

$$t=\frac{2a}{3}\sqrt{\left(\frac{3\rho}{2\Pi}\right)}\int_0^{\pi/2}\sin^{2/3}\theta d\theta$$

By applying Gamma function, we have

$$=\frac{2a}{3}\sqrt{\left(\frac{3\rho}{2\Pi}\right)}.\frac{\Gamma\frac{5}{6}\Gamma\frac{1}{2}}{2\Gamma\left(\frac{5}{6}+\frac{1}{2}\right)}$$

$$= \frac{2a}{3}\sqrt{\left(\frac{3\rho}{2\Pi}\right)}.\frac{\Gamma\frac{5}{6}\Gamma\frac{1}{2}}{2\Gamma\frac{4}{3}}$$

$$= \frac{2a}{3}\sqrt{\left(\frac{3\rho}{2\Pi}\right)}.\frac{\Gamma\frac{5}{6}\Gamma\frac{1}{2}}{2\Gamma\frac{1}{3}\Gamma\frac{1}{3}}$$

$$= \frac{2a}{3}\frac{3}{2}\sqrt{\left(\frac{3\rho}{2\Pi}\right)}.\frac{\sqrt{\pi}\Gamma\left(\frac{1}{3}+\frac{1}{2}\right)}{\Gamma\frac{1}{3}}$$

Multiplying N^r and D^r with $\Gamma\frac{1}{3}$, we have

$$= a\sqrt{\left(\frac{3\rho}{2\Pi}\right)}.\sqrt{\pi}\frac{\Gamma\left(\frac{1}{3}+\frac{1}{2}\right)\Gamma\frac{1}{3}}{\left(\Gamma\frac{1}{3}\right)^2} \quad \left[\text{as}\Gamma n\Gamma\left(n+\frac{1}{2}\right)=\sqrt{\pi}\frac{\Gamma n}{2^{2n-1}}\right]$$

$$= a\sqrt{\left(\frac{3\rho}{2\Pi}\right)}.\sqrt{\pi}.\frac{\sqrt{\pi}.\Gamma\frac{2}{3}}{\left(\Gamma\frac{1}{3}\right)^2 2^{2/3-1}}$$

$$= \pi a\sqrt{\left(\frac{3\rho}{2\Pi}\right)}.\frac{\Gamma\frac{2}{3}}{\left(\Gamma\frac{1}{3}\right)^2 2^{-1/3}}$$

Multiplying Nr and Dr with $\Gamma\frac{1}{3}$, we have

$$= \pi a\sqrt{\left(\frac{3\rho}{2\Pi}\right)}.\frac{\Gamma\left(1-\frac{1}{3}\right)\Gamma\frac{1}{3}}{2^{-1/3}\left(\Gamma\frac{1}{3}\right)_3^3} \quad \left[\text{as}\Gamma n\Gamma(1-n)=\frac{\pi}{\sin n\pi}\right]$$

$$= \pi a\sqrt{\left(\frac{3\rho}{2\Pi}\right).2^{1/3}\Gamma\left(\frac{1}{3}\right)^{-3}}\cdot\frac{\pi}{\sin\pi/3}$$

$$= \pi^2 a \sqrt{\left(\frac{3\rho}{2\Pi}\right)} \cdot \left(\Gamma\frac{1}{3}\right)^{-3} .2^{1/3}.\frac{2}{\sqrt{3}}$$

$$= \pi^2 a(\rho/\Pi)^{1/2} 2^{5/6} \left(\Gamma\frac{1}{3}\right)^{-3}$$ **Proved.**

Example 16:

An infinite mass of fluid is acted on by a force $\mu r^{-3/2}$ per unit mass directed to the origin. If initially the fluid is at rest and there is a cavity to the form of the sphere r = c in it. Show that the cavity will be filled up after an interval of time

$(2/5\mu)^{1/2}\ c^{5/4}$.

Solution:

Let v′ be the velocity at a distance r′ from the origin at any time t and p be the pressure, then the equation of continuity is

$$r'^2 v' = f(t) = r^2 v. \qquad ...(1)$$

Equation of motion is

$$\frac{\partial v'}{\partial t} + v\frac{\partial v'}{\partial r'} = \mu r'^{-3/2} - \frac{1}{\rho}\frac{\partial p}{\partial r'}$$

$$\Rightarrow \ -\frac{f'(t)}{r'2} + v'\frac{\partial v'}{\partial r'} = -\ \mu r t'^{3/2} - \frac{1}{\rho}\frac{\partial p}{\partial r'}$$

By integrating with regard to r¢, we have

$$-\frac{f'(t)}{r'} + \frac{1}{2}v'^2 = \frac{2\mu}{r'^{1/2}} - \frac{p}{\rho} + A. \qquad ...(2)$$

Initially r′ = ∞, v′ = 0,

p = 0; A = 0.

Equation (2) reduces to

$$-\frac{f'(t)}{r'} + \frac{1}{2}v'^2 = \frac{2\mu}{r'^{1/2}} - \frac{p}{\rho}. \qquad ...(3)$$

Let r be the radius of the cavity and v be the velocity at any time t. The motion of the cavity is

$$-\frac{f'(t)}{r} + \frac{1}{2}v^2 = \frac{2\mu}{r^{1/2}}$$

$$\Rightarrow -\left(rv\frac{dv}{dr}+2v^2\right)+\frac{1}{2}v^2=\frac{2\mu}{r^{1/2}}.$$

$$\Rightarrow rv\frac{dv}{dr}+\frac{3}{2}v^2=\frac{2\mu}{r^{1/2}}.$$

Multiplying both the sides with $2r^2$ dr and integrating, we get

$$2r^3vdv + 3r^2v^2\,dr = 4\mu r^{3/2}dr \qquad ...(4)$$

$$r^3v^2 = -\,(8\mu/5)\,r^{5/2} + B.$$

Now $r = c$, $v = 0$; $B = (8\mu/5)c^{5/2}$.

$$\Rightarrow r^3v^2 = \frac{8\mu}{5}(c^{5/2}-r^{5/2})$$

$$\Rightarrow v = \frac{dr}{dt} = \pm\sqrt{\left(\frac{8\mu}{5}\right)}.\sqrt{\left(\frac{c^{5/2}-r^{5/2}}{r^3}\right)},$$

– ve sign is taken since velocity increases as r decreases.

If t be the time of filling up the cavity then

$$t = -\sqrt{\left(\frac{5}{8\mu}\right)}\int_c^0\frac{r^{3/2}dr}{\sqrt{(c^{5/2}-r^{5/2})}}$$

Let $r^{5/2} = c^{5/2}\sin^2\theta$, $\frac{5}{2}r^{3/2}\,dr = c^{5/2}\sin\theta\cos\theta\,d\theta$

$$\text{or } t = \frac{4}{5}\sqrt{\left(\frac{5}{8\mu}\right)}\int_0^{\pi/2}\frac{c^{5/2}\sin\theta\cos\theta}{c^{5/4}\cos\theta}d\theta$$

$$\Rightarrow t = \frac{4}{5}c^{5/4}\sqrt{\left(\frac{5}{8\mu}\right)}\int_0^{\pi/2}\sin\theta d\theta = \sqrt{\left(\frac{2}{5\mu}\right)}c^{564}.$$

Example 17:

A solid sphere of radius a is surrounded by a mass of liquid whose volume is $\frac{4}{5}\pi c^3$ and its centre is a centre of attractive force varying directly as the square of the distance. If the solid sphere be suddenly annihilated, show that the velocity of the inner surface, when its radius is x, is given by

$$x^2x^3\,[x^3 + c^3)^{1/3} - x] = \left[\frac{2\Pi}{3\rho}+\frac{2\mu c^3}{9}\right](a^3-x^3)(c^3-x^3)^{1/3},$$

where ρ is the density, P The external pressure and m the absolute force.

Solution:

Let v′ be the velocity at a distance r′ at any time t and p be the pressure there, then the equation of continuity is given by

$$r'^2 v' = f(t). \qquad ...(1)$$

The equation of motion is given by

$$\frac{\partial v'}{\partial t} + v'\frac{\partial v'}{\partial r'} = -\mu r'^2 - \frac{1}{\rho}\frac{\partial p}{\partial r'},$$

where $\mu r'^2$ is the attractive force.

$$\frac{f'(t)}{r'^2} + v'\frac{\partial v'}{\partial r'} = -\mu r'^2 - \frac{1}{\rho}\frac{\partial p}{\partial r'}$$

Integrating with regard to r′, we have

$$-\frac{f'(t)}{r'^2} + \frac{1}{2}v'^2 = -\frac{1}{3}\mu r'3 - \frac{p}{\rho} + A \qquad ...(1)$$

Let r and R be the internal and external radii of the shell and v and V be the velocities there at any subsequent time t.

Initially I. r′ = r, v′ = v and p = 0

II. r′ = R v′ = v and p = II

Using the initial conditions I and II, the equation (1) reduces to

$$-\frac{f'(t)}{r} + \frac{1}{2}v^2 = \frac{1}{3}\mu r^3 + A,$$

$$-\frac{f'(t)}{R} + \frac{1}{2}V^2 = \frac{1}{3}\mu R^3 - \frac{II}{\rho} + A. \qquad ...(2, 3)$$

Eliminating the constant A, by subtracting (2) and (3), we have

$$f'(t)\left[\frac{1}{r} - \frac{1}{R}\right] - \frac{1}{2}(v^2 - V^2) = \frac{\mu}{3}(r^3 - R^3) - \frac{\Pi}{\rho} \qquad ...(4)$$

Since the volume of the liquid is constant, so

$$\frac{4}{3}\pi R^3 - \frac{4}{3}\pi r^3 = \frac{4}{3}\pi c^3 \Rightarrow R^3 - r^3 = c^3. \qquad ...(5)$$

We know that

$$r^2 v = f(t) = R^2 V$$

$$\Rightarrow r^2\frac{dr}{dt} = f(t) = R^2\frac{dR}{dt}$$

$\Rightarrow r^2\, dr = f(t)\, dt = R^2 dR$

$$f'(t) = \frac{d}{dt}(r^2 v) = \frac{d}{dt}(r^2 v)\frac{dr}{dt} = v\frac{d}{dr}(r^2 v);\ V = \frac{r^2 v}{R^2}. \qquad ...(6)$$

From (4) and (5), we have

$$v\frac{d}{dr}(r^2 v).\left(\frac{1}{r}-\frac{1}{R}\right)-\frac{1}{2}\left(v^2-\frac{r^4 v^2}{R^4}\right) = -\frac{1}{3}\mu c^3 - \frac{\Pi}{\rho}$$

Multiplying both the sides by r^2, we have

$$2r^2 v\frac{d}{dr}(r^2 v).\left(\frac{1}{r}-\frac{1}{R}\right)-v^2 r^4\left(\frac{1}{r^2}-\frac{r^2}{R^4}\right) = \frac{2}{3}\mu c^3 r^2 - \frac{2\Pi}{\rho}r^2$$

$$\Rightarrow \frac{d}{dr}(r^2 v)^2\left(\frac{1}{r}-\frac{1}{R}\right)-(r^2 v)^2\left(\frac{1}{r^2}\frac{r^2}{R^4}\right) = \frac{2}{3\rho}(\mu\rho c^3 + 3\Pi)r^2$$

By integrating, we have

$$r^4 v^2\left(\frac{1}{r}-\frac{1}{R}\right) = -\frac{2}{3\rho}(\mu\rho c^3 + 3\Pi).\frac{r^3}{3} + B$$

when $r = a$, $v = 0$

$$\Rightarrow B = \frac{2}{9\rho}(\mu\rho c^3 + 3\Pi).a^3$$

$$\Rightarrow r^4 v^2\left(\frac{1}{2}-\frac{1}{R}\right) = \frac{2}{9\rho}(\mu\rho c^3 + 3\Pi)(a^3 - r^3)$$

$$\Rightarrow r^4 v^2\left[\frac{1}{r}-\frac{1}{(r^3+c^3)^{1/3}}\right] = \left(\frac{2\mu c^3}{9}+\frac{2\Pi}{3\rho}\right)(a^3 - r^3)$$

For the inner surface $r = x$, $v = x$, we have

$$x^2 x^3\ [(x^3 + c^3) - x] = \left[\frac{2\mu c^3}{9}+\frac{2\Pi}{3\rho}\right](a^3 - x^3)(x^3 + c^3)^{1/3}.$$

Proved.

Example 18:

A sphere is at rest in an infinite mass of homogeneous liquid of density ρ, the pressure at infinity being Π. If the radius R of the sphere varies in such a way that $R = a + b\cos nt$ where $b < a$, show that pressure at the surface of the sphere at any time is

$$P + \frac{1}{4}bn^2\rho(b - 4a\cos nt - 5b\cos 2nt).$$

Solution:

Let v′ be the velocity at a distance r′ at any time t and p′ be the pressure there. Let v be the velocity on the surface of the sphere of radius R where R = a + b cos nt.

The equation of continuity becomes

$$r' v' = f(t) = R^2 v.$$

The equation of motion is given by ...(1)

$$\frac{\partial v'}{\partial t} + v' \frac{\partial v'}{\partial r'} = \frac{1}{\rho}\frac{\partial p'}{\partial r'}$$

$$\frac{f'(t)}{r'^2} + v' \frac{\partial v'}{\partial r'} = -\frac{1}{\rho}\frac{\partial p'}{\partial r'}$$

Integrating with regard to r′, we have

$$-\frac{f'(t)}{r} + \frac{1}{2}v'^2 = \frac{p'}{\rho} + A$$

Initially $r' = \infty$, $v = 0$, $p' = P$ then $A = \dfrac{P}{\rho}$

$$\frac{-f'(t)}{r'} + \frac{1}{2}v'^2 = \frac{P - p'}{\rho}$$

Again $r' = R$, $p' = p$ and $v' = v$

then $$-\frac{f'(t)}{R} + \frac{1}{2}v^2 = \frac{1}{\rho}(P - p)$$

$$p = P + \rho\left[\frac{f'(t)}{R} - \frac{1}{2}v^2\right] \qquad \text{...(2)}$$

R. H. S. $$\frac{f'(t)}{R} + \frac{1}{2}v^2 = 2\left(\frac{dR}{dt}\right)^2 + R\frac{d^2R}{dt^2} - \frac{1}{2}\left(\frac{dR}{dt}\right)^2$$

$$= \frac{3}{2}\left(\frac{dR}{dt}\right)^2 + R\frac{d^2R}{dt^2}$$

$$= \frac{3}{2}(-bn \sin nt)^2 + (a + b\cos nt)(-bn^2 \cos nt)$$

$$= \frac{1}{2} bn^2 (3b \sin^2 nt - 2b \cos^2 nt - 2a \cos nt)$$

$$= \frac{1}{4} bn^2 [3b(1 - \cos 2nt) - 2b(1 + \cos 2nt) - 4a \cos nt]$$

$$= \frac{1}{4} bn^2 \, [b - 4a \cos nt - 5 \, b \cos 2nt]$$

Hence $p = P + \frac{1}{4} \, bn^2\rho \, [b - 4a \cos nt - 5b \cos 2nt]$. **Proved.**

Example 19:

A mass of fluid of density ρ and volume $4/3\pi c^3$ is in the form of a spherical shell. A constant pressure Π is exerted on the external surface of the shell. There is no pressure on the internal surface and no other forces act on liquid. Initially the liquid is at rest and the internal radius of the shell is 2c. Prove the velocity of the internal surface when its radius is c is

$$\sqrt{\left(\frac{14\Pi}{3\rho} \cdot \frac{2^{1/3}}{2^{1/3} - 1}\right)}.$$

Solution:

Let v′ be the velocity at a distance r′ at any time t and p be the pressure there then the equation of continuity is

$$r'^2 v' = f(t) \qquad \text{...(1)}$$

The equation of motion is

$$\frac{\partial v'}{\partial t} + v' \frac{\partial v'}{\partial r'} = -\frac{1}{\rho} \frac{\partial p}{\partial r'} \qquad \text{...(2)}$$

or $$\frac{f'(t)}{r'^2} + v' \frac{\partial v'}{\partial r'} = -\frac{1}{\rho} \frac{\partial p}{\partial r'}$$

or $$-\frac{f'(t)}{r'} + \frac{1}{2} v'^2 = -\frac{p}{\rho} + A \qquad \text{...(3)}$$

Let r and R be the internal and external radii of the shell and v and V be the velocities there at any subsequent time t.

initially (i) $r' = R$, $v' = V$, $p = \Pi$ and (ii) $r' = r$, $v' = v$, $p = 0$. (Since there is no pressure on the internal surface)

The value of the constant A may be determined by using the above conditions with the relation (3). Thus, we get

$$\frac{f'(t)}{R} + \frac{1}{2} V^2 = -\frac{\Pi}{\rho} + A,$$

and $$\frac{f'(t)}{r} + \frac{1}{2} v^2 = A.$$

Eliminating the constant, we have

$$-f'(t)\left\{\frac{1}{r}-\frac{1}{R}\right\}+\frac{1}{2}\{v^2-V^2\}=\frac{\Pi}{\rho} \quad ...(4)$$

We know that

$$r^2v = R^2V = f(t) \quad ...(5)$$

$$\Rightarrow \quad 2r^2dr = 2R^2dR = 2f(t)\,dt \quad ...(6)$$

From (4) and (5), we get

$$-f'(t)\left\{\frac{1}{r}-\frac{1}{R}\right\}+\frac{1}{2}\left\{\frac{f^2(t)}{r^4}-\frac{f^2(t)}{R^4}\right\}=\frac{\Pi}{\rho}.$$

Multiplying both sides by the relation (6) and integrating, we have

$$-2f(t)\,f'(t)\left\{\frac{1}{r}-\frac{1}{R}\right\}+\frac{1}{2}f^2(t)\left\{\frac{2r^2dr}{r^4}-\frac{2R^2dR}{R^4}\right\}=\frac{\Pi}{\rho}.\,2r^2\,dr$$

$$\Rightarrow -2f(t)\,f'(t)\left\{\frac{1}{r}-\frac{1}{R}\right\}+\frac{1}{2}f^2(t)\left\{\frac{2dr}{r^2}-\frac{2dR}{R^2}\right\}=\frac{2\Pi}{\rho}r^2dr$$

$$\Rightarrow f^2(t)\left\{\frac{1}{r}-\frac{1}{R}\right\}=-\frac{2\Pi}{3\rho}r^3+B.$$

Initially $r = 2c$, $v = 0$ *i.e.*, $f(t) = 0 \Rightarrow B = \dfrac{2\Pi}{3\rho}8c^3.$

$$\Rightarrow f^2(t)\left\{\frac{1}{r}-\frac{1}{R}\right\}=\frac{2\Pi}{3\rho}(8c^3-r^3)$$

$$\Rightarrow r^4v^2\left\{\frac{1}{r}-\frac{1}{R}\right\}=\frac{2\Pi}{3\rho}(8c^3-r^3).$$

Since the volume of the liquid is constant,

$$\frac{4}{3}\pi R^3-\frac{4}{3}\pi r^3=\frac{4}{3}\pi c^3,\ R^3-r^3=c^3.$$

$$\Rightarrow r^4v^2\left\{\frac{1}{r}-\frac{1}{(r^3+c^3)^{1/3}}\right\}=\frac{2\Pi}{3\rho}(8c^3-r^3)$$

$$\Rightarrow v^2=\frac{2\Pi}{3\rho}\frac{8c^3-r^3}{r^3\left\{\frac{1}{r}-\frac{1}{(r^3+c^3)^{1/3}}\right\}} \quad ...(7)$$

which gives the velocity in terms of radius r at the inner surface of the cavity.

Hence the velocity of the internal surface (when r = c) is given by

$$v^2 = \frac{2\Pi}{3\rho}\frac{7c^3}{c^4\left\{\frac{1}{c}-\frac{1}{c.2^{1/3}}\right\}} \Rightarrow v = \sqrt{\left(\frac{14\Pi}{3\rho}\cdot\frac{2^{1/3}}{2^{1/3}-1}\right)}$$ **Proved.**

Example 20:

A volume $\frac{4}{3}\pi c^3$ *of gravitating liquid, of density r is initially in the form of a spherical shell of infinitely great radius. If the liquid shell contract under the influence of its own attraction, there being no external or internal pressure, show that when the radius of the inner spherical surface is x, its velocity will be given by*

$$V^2 = \frac{4\pi\gamma\rho z}{15x^3}(2z^4 + 2z^2x^2 - 3zx^3 - 3x^4),$$

where γ is the constant of gravitation, and $z^3 = x^3 + c^3$.

Solution:

Let v′ be the velocity at a distance r′ from the centre of the spherical shell at any time t and p be the pressure there. The equation of continuity is

$$\frac{1}{r'^2}\cdot\frac{d}{dr}(r'^2 v') = 0,\ r'^2 v' = f(t).$$

Let r be the radius of the inner surface. The attraction at the point distance r′ from the centre due to liquid is given by

$$= \frac{\frac{4}{3}\gamma\pi\rho(r'^3 - r^3)}{r'^2} = \frac{4}{3}\gamma\pi\rho\left(r' - \frac{r^3}{r'^2}\right)$$

The equation of motion is

$$\frac{\partial v'}{\partial t} + v'\frac{\partial v'}{\partial r'} = -\frac{4}{3}\gamma\pi\rho\left(r' - \frac{r^3}{r'^2}\right) - \frac{1}{\rho}\frac{\partial p}{\partial r'}.$$

$$\text{or}\quad \frac{f'(t)}{r'^2} + v'\frac{\partial v'}{\partial r'} = -\frac{4}{3}\gamma\pi\rho\left(r' - \frac{r^3}{r'^2}\right) - \frac{1}{\rho}\frac{\partial p}{\partial r'}.$$

Integrating with regard to r′, we get

$$-\frac{f'(t)}{r}+\frac{1}{2}v^2=\frac{4}{3}\gamma\pi\rho\left(\frac{1}{2}r'^2+\frac{r^3}{r'}\right)-\frac{p}{\rho}+A. \quad ...(1)$$

The initial conditions are given as

(i) r′ = r, v′ = v, p = 0,

(ii) r′ = z, v′ = v_1, p = 0.

Using these conditions into (1), we have

$$-\frac{f'(t)}{r}+\frac{1}{2}v^2=\frac{4}{3}\gamma\pi\rho\left(\frac{1}{2}r^2+r^2\right)+A$$

$$-\frac{f'(t)}{z}+\frac{1}{2}v_1^2=\frac{4}{3}\gamma\pi\rho\left(\frac{1}{2}z^2+\frac{r^3}{z}\right)+A.$$

Eliminating the constant, we get

$$-f'(t)\left\{\frac{1}{r}-\frac{1}{z}\right\}+\frac{1}{2}\left\{v^2-v_1^2\right\}=-\frac{4}{3}\gamma\pi\rho\left\{\frac{3r^2}{2}-\frac{z^2}{2}-\frac{r^3}{z}\right\} \quad ...(2)$$

$$\Rightarrow -f'(t)\left\{\frac{1}{r}-\frac{1}{z}\right\}+\frac{1}{2}\left\{\frac{f^2(t)}{r^4}-\frac{f^2(t)}{z^4}\right\}=-\frac{4}{3}\gamma\pi\rho\left\{\frac{3r^2}{2}-\frac{z^2}{2}-\frac{r^3}{z}\right\}$$

Since $r^2v = z^2v_1 = f(t)$

$$\Rightarrow \quad 2r^2\,dr = 2z^2dz = 2f(t)\,dt. \quad ...(3)$$

Multiplying both the the sides of (2) by the relation (3) and integrating, we have

$$f^2(t)\left\{\frac{1}{r}-\frac{1}{z}\right\}=\frac{4}{3}\gamma\pi\rho\int(3r^4dr-dz\frac{r^3}{z}2z^2dz),z^3-r^3=c^3.$$

$$\Rightarrow f^2(t)\left\{\frac{1}{r}-\frac{1}{z}\right\}=\frac{4}{3}\gamma\pi\rho\left\{\frac{3}{5}r^5-\frac{1}{5}z^5-2\int z(z^3-c^3)dz\right\}$$

$$\Rightarrow f^2(t)\left\{\frac{1}{r}-\frac{1}{z}\right\}=\frac{4}{3}\gamma\pi\rho\left\{\frac{3}{5}r^5-\frac{1}{5}z^5-\frac{2}{5}z^5+c^3z^2\right\}$$

$$\Rightarrow f^2(t)\left\{\frac{1}{r}-\frac{1}{z}\right\}=\frac{4}{15}\gamma\pi\rho\left\{3(r^5-z^5)\right\}+5_c^3z^2\}. \quad ...(4)$$

If the radius of the inner spherical surface be x (*i.e.*, r = 1) then the equation of continuity becomes $x^2V = f(t)$ and equation (4) reduces

$$\Rightarrow x^4V^2\left\{\frac{1}{x}-\frac{1}{z}\right\} = \frac{4}{15}\gamma\pi\rho\{3(x^5-z^5)+5c^3z^2\}$$

$$\Rightarrow V^2 = \frac{4}{15}\frac{\pi\gamma\rho z}{x^3}\left\{\frac{3(x^5-z^5)+5c^3z^2}{z-x}\right\}$$

$$\Rightarrow V^2 = \frac{4}{15}\frac{\pi\gamma\rho z}{x^3}\{-3(x^4+x^3z+x^2z^2+xz^3+z^4)+5z^2(z^2+zx+x^2)\}$$

$$\Rightarrow V^2 = \frac{4\pi\gamma\rho z}{15x^3}\{2z^2+2z^3x+2z^2x^2-3zx^3-3x^4\}.$$

Proved.

Example 21:

A mass of gravitating fluid is at rest under its own attraction only, the free surface being a sphere of radius b and the inner surface a rigid concentric shell of radius a. Show that if the shell suddenly disappears, the initial pressure at any point of the fluid at distance from the centre is

$$\frac{2}{3}\pi\gamma\rho^2(b-a)(r-a)\left\{\frac{a+b}{r}+1\right\}.$$

Solution:

Let v′ be the velocity at a distance r′ from the centre of the spherical shell at any time t and p be the pressure there. The equation of continuity is

$$\frac{1}{r'^2}\frac{d}{dr}(r'^2v') = 0$$

$$\Rightarrow r'^2v' = f(t).$$

Let r be the radius of the inner surface. The attraction at the point distance r′ from the centre due to liquid is given by

$$\frac{\frac{4}{3}\pi\gamma\rho(r'^3-r^3)}{r'^2} = \frac{4}{3}\pi\gamma\rho\left(r'-\frac{r^3}{r'^2}\right)$$

The equation of motion is given by

$$\frac{\partial v'}{\partial t}+v'\frac{\partial v'}{\partial r'} = -\frac{4}{3}\pi\gamma\rho\left(r'-\frac{r^3}{r'^2}\right)-\frac{1}{\rho}\frac{\partial p}{\partial r'}$$

$$\Rightarrow \frac{f'(t)}{r'^2}+v'\frac{\partial v'}{\partial r'} = -\frac{4}{3}\pi\gamma\rho\left(r'-\frac{r^3}{r'^2}\right)-\frac{1}{\rho}\frac{\partial p}{\partial r'}$$

Integrating with regard to r′, we have

$$-\frac{f'(t)}{r'}+\frac{1}{2}v'^2=\frac{4}{3}\pi\gamma\rho\left(\frac{1}{2}r'^2+\frac{r^3}{r'}\right)-\frac{p}{\rho}+A \qquad ...(1)$$

Initially t = 0, v′ = 0, r = a

and p = P (let)

$$-\frac{f'(0)}{a}=-\frac{4}{3}\pi\gamma\rho\left(\frac{1}{2}r'^2+\frac{a^3}{r'}\right)-\frac{P}{\rho}+A \qquad ...(2)$$

Again P = 0, r′ = a and r′ = a

and r′ = b, v′ = 0

From (2), we have

$$-\frac{f'(0)}{a}=-\frac{4}{3}\pi\gamma\rho\left(\frac{1}{2}a^2+a^2\right)+A$$

$$-\frac{f'(0)}{b}=-\frac{4}{3}\pi\gamma\rho\left(\frac{1}{2}b^2+\frac{a^3}{b}\right)+A. \qquad ...(3, 4)$$

Subtracting (3) and (4), we have

$$f'(0)\left[\frac{1}{b}-\frac{1}{a}\right]=\frac{4}{3}\pi\gamma\rho\left\{\frac{1}{2}(b^2-a^2)+a^2\left(\frac{a}{b}-1\right)\right\}$$

$$\Rightarrow f'(0)=-\frac{2}{3}\pi\gamma\rho ab(a+b)+\frac{4}{3}\pi\gamma\rho a^3 \qquad ...(5)$$

Multiplying (3) by a and (4) by b and subtracting, we have

$$0=\frac{4}{3}\pi\gamma\rho.\frac{1}{2}(b^3-a^3)+A(a-b)$$

$$\Rightarrow A=\frac{2}{3}\pi\gamma\rho(a^2+b^2+ab) \qquad ...(6)$$

Substituting the values of f′ (0) and the constant A in (2), we have

$$-\frac{1}{r'}\left\{-\frac{2}{3}\pi\gamma\rho ab(a+b)+\frac{4}{3}\pi\gamma\rho a^3\right\}=-\frac{4}{3}\pi\gamma\rho\left(\frac{1}{2}r'^2+\frac{a^3}{r'}\right)-\frac{P}{\rho}$$

$$+\frac{2}{3}\pi\gamma\rho(a^2+b^2+ab)$$

$$\Rightarrow P=\frac{2}{3}\pi\gamma\rho\left\{a^2+b^2+ab-\frac{ab(a+b)}{r'}-r'^2\right\}$$

Replacing r′ by r the initial pressure at any point of the fluid at a distance r from the centre is obtained

$$P = \frac{2}{3}\pi\gamma\rho^2(r-a)(b-r)\left\{\frac{a+b}{r}+1\right\}$$ **Proved.**

Example 22:

Liquid is contained between two parallel planes; the free surfaceis a circular cylinder of radius a whose axis is perpendicular to the planes. All the liquid within a concentric circular cylinder of radius b is suddenly annihilated. Prove that if Π be the pressure at the outer surface, the initial pressure at any point of the liquid distance r from the centre is

$$\Pi\frac{\log r - \log b}{\log a - \log b}.$$

Solution:

In an incompressible fluid, the fluid velocity will be radial outside the cylinder $|z| = b$. It will be a function of r and the time t only. Let v′ be the velocity at a distance r′ and p be the pressure there at any time t, then equation of continuity is

$$r'v' = f(t). \qquad ...(1)$$

Equation of motion is

$$\frac{\partial v'}{\partial t} + v'\frac{\partial v'}{\partial r} = -\frac{1}{\rho}\frac{\partial p}{\partial r'},$$

or $$\frac{f'(t)}{r'} + v'\frac{\partial v'}{\partial r} = -\frac{1}{\rho}\frac{\partial p}{\partial r'},$$

Integrating with regard to r′, we have

$$f'(t)\log r' + \frac{1}{2}v'^2 = -\frac{p}{\rho} + A. \qquad ...(2)$$

Initially t = 0, v′ = 0,

r′ = a, we have

$$f'(0)\log r = \frac{p}{\rho} + A. \qquad ...(3)$$

Again p = Π, when r = a,

and p = 0, when r = b.

By using the above condition, equation (3) becomes

$$f'(0)\log a = \frac{\Pi}{\rho} + A \qquad ...(4)$$

and $$f'(0)\log b = -A. \qquad ...(5)$$

From (3) and (5), we have

$$f'(0)\{\log r - \log b\}. = \frac{p}{\rho}.$$

Also from (4) and (5), we get

$$f'(0)\{\log a - \log b\} = -\frac{\Pi}{\rho}.$$

Eliminating f ′ (0), we have

$$\therefore \frac{\log r - \log b}{\log a - \log b} = \frac{p}{\Pi}.$$

Thus the initial pressure at any point of the liquid at a distance r from the centre is

$$p = \Pi\,\frac{\log r - \log b}{\log a - \log b}.$$ **Proved.**

Example 23:

A mass of liquid of density ρ whose external surface is a ling circular cylinder of radius a, which is subject to a constant pressure cylinder is suddenly destroyed. Show that if V is the velocity at the internal surface when the radius is r, then

$$V^2 = \frac{2\Pi(b^2 - r^2)}{\rho r^2 \log\{(r^2 + a^2 - b^2)/r^2\}}.$$

Solution:

When the internal cylinder is destroyed, suddenly the motion of the liquid will be along the radii of the normal sections of the cylinder. The velocity and p the pressure at any point distance r′ from the axis of the cylinder. The equation of continuity is

$$r'\,v' = f(t). \qquad ...(1)$$

Equation of motion is

$$\frac{\partial v'}{\partial t} + v'\frac{\partial v'}{\partial r'} = \frac{1}{\rho}\frac{\partial p}{\partial r'},$$

or $\frac{f'(t)}{r'} + v'\frac{\partial v'}{\partial r'} = -\frac{1}{\rho}\frac{\partial p}{\partial r'}$.

Integrating with regard to r′, we have

$$f'(t)\log r' + \frac{1}{2}v'^2 = \frac{p}{\rho} + A. \qquad ...(2)$$

Let r and R be the radii of the internal and external surface of the cylinder and v and v_1, the velocities there at any time t.

The conditions are defined as

$$r' = r,\ v' = v,\ p = 0$$

and $\quad r' = R,\ v' = V,\ p = \Pi.$

$$f'(t)\log r + \frac{1}{2}v^2 = A$$

$$f'(t)\log R + \frac{1}{2}V^2 = -\frac{\Pi}{\rho} + A,$$

Eliminating the constant A, we have

$$f'(t)(\log r - \log R) + \frac{1}{2}(v^2 - V^2) = \frac{\Pi}{\rho} \qquad ...(3)$$

From the equation of continuity, we have

$$rv = RV = f(t)$$

$$2r\,dr = 2R\,dR = 2f(t)\,dt \qquad ...(4)$$

and $\quad R^2 - r^2 = a^2 - b^2.$

Multiplying both the side of (3) by the relation (4), we have

$$2f(t)f'(t)(\log r - \log R) + \frac{1}{2}f^2(t)\left\{\frac{2r\,dr}{r^2} - \frac{2R\,dR}{R^2}\right\} = \frac{2\Pi}{\rho}r\,dr$$

$$\Rightarrow 2f(t)f'(t)(\log r - \log R) + f^2(t)\left\{\frac{dr}{r} - \frac{dR}{R}\right\} = \frac{2\Pi}{\rho}r\,dr.$$

By integrating, we have

$$f^2(t)(\log r - \log R) = \frac{\Pi}{\rho}r^2\,B. \qquad ...(5)$$

At $r = b$, $v = 0$; $B = -\frac{\Pi}{\rho}b^2$.

Substituting the value of B into (5), we have

$$f^2(t)(\log r - \log R) = \frac{\Pi}{\rho}(r^2 - b^2)$$

$$\Rightarrow f^2(t) = \frac{\Pi}{\rho}\frac{r^2 - b^2}{\log r - \log R},$$

$$\Rightarrow v^2r^2 = \frac{\Pi}{\rho}\frac{2(r^2 - b^2)}{\log(r^2/R^2)},$$

$$\Rightarrow v^2 = \frac{2\Pi(r^2 - b^2)}{\rho r^2 \log\{r^2/(r^2 + a^2 - b^2)\}},$$

$$\Rightarrow v^2 = \frac{2\Pi(p^2 - r^2 - b^2)}{\rho r^2 \log\{(r^2 + a^2 - b^2)/r^2\}}.$$ **Proved.**

EXERCISES

1. A spherical hollow of radius a initially exists in an infinite fluid subject to constant pressure at infinity. Shew that the pressure at distance r from the centre when the radius of the cavity is x to pressure at infinity as

$$\{3x^2r^4 + (a^3 - 4x^3)\, r^3 - (a^3 - x^3\}; 3x^2r^4.$$

2. A mass of homogeneous liquid is moving so that the velocity at any point is proportional to the time, and that the pressure is given by

$$\frac{p}{\rho} = \mu xyz - \frac{1}{2}t^2(y^2z^2 + z^2x^2 + x^2y^2).$$

Prove that this motion may have been generated from rest by finite natural forces independent of the time, and show that if the direction of motion at every point coincide with the direction of the acting force, each particle of the liquid describe a curve which is the intersection of the two hyperbolic cylinders.

Hint : Equation of motion is

$$\frac{p}{\rho} - \frac{\partial\phi}{\partial t} + \frac{1}{2}q^2 + V = f\,(t). \qquad ...(i)$$

Here $q = \lambda t$

and $$p = \rho\left\{\mu xyz - \frac{1}{2}t^2(y^2z^2 + z^2x^2 + x^2y^2)\right\}. \qquad ...(ii)$$

From (i) and (ii) $\lambda^2 = y^2z^2 + z^2x^2 + x^2y^2$. ...(iii)

$$\text{So } q^2 = t^2 (y^2z^2 + z^2 x^2 + x^2 y^2).$$

Let $\phi = txyz$.

From (i) we have

$$\frac{p}{\rho} = xyz - \frac{1}{2} t^2 (y^2z^2 + z^2x^2 + x^2y^2) - V + f(t). \qquad ...(iv)$$

From (i) and (iv), we have

$$f(t) = 0, \; V = xyz(1 - \mu).$$

Let, u, v, are the component velocities and X, Y, Z be the component forces, then

$$u = -\frac{\partial\phi}{\partial x} = -tyz \text{ etc.}$$

and $$X = -\frac{\partial V}{\partial x} = (\mu - 1)\, yz \text{ etc.}$$

Since the direction of motion coincides with the direction of acting forces, so

$$\frac{u}{X} = \frac{v}{Y} = \frac{w}{Z}.$$

Thus the equation to the path is given by

$$\frac{dx}{X} = \frac{dy}{Y} = \frac{dz}{Z}.$$

3. A sphere whose radius at time t is b + a cos nt is surrounded by liquid extending to infinity under no forces. Prove that the pressure at distance r from the centre is less than the pressure at an infinite distance by

$$\frac{\rho n^2 a}{r}(b+a)\cos nt \left\{a(1-3\sin^2 nt) + b\cos nt + \frac{a}{2r^3}\sin^2 nt (b + a\cos nt)^3\right\},$$

4. A sphere of radius a is alone in an unbounded liquid which is at rest at a great distance from the sphere and is subject to no external forces. The sphere is forced to vibrate radially keeping its spherical shape, the radius r at any time being by r = a + b cos nt. Show that if Π is the pressure in the liquid at a great distance from the sphere, the least pressure (assumed positive)

at the surface of the sphere during the motion is

$$\Pi - n^2\rho b\,(a + b).$$

5. A mass of uniform liquid is in the form of a thick spherical shell bounded by concentric spheres of radii a and b (a < b). The cavity is filled with gas the pressure of which varies according to Boyle's law, and is initially equal to the atmospheric pressure Π, and the mass of which may be neglected. The outer surface of the shell is exposed to atmospheric pressure. Prove that if the system is symmetrically disturbed, so that each particle moves along particle moves along the line joining it to the centre, the time of an oscillation is

$$2pa\left\{\rho\frac{b-a}{3\Pi b}\right\}^{1/2},$$

where ρ is the density of the liquid.

6. A mass of perfect in compressible fluid, of density ρ, is bounded by concentric surface. The outer surface i contained by a flexible envelope which exerts continuously a uniform pressure Π and contracts from radius R_1 to radius R_2. The hollow is filled with a gas obeying Boyle's law, its radius contracts from C_1 to C_2, and the pressure of the gas is initially p_1. Initially the whole mass is at rest. Prove that, neglecting the mass of the gas, the velocity v of the inner surface when the configuration (R_2, C_2) is reached is given by

$$\frac{1}{2}v^2 = \frac{C_1^3}{C_2^1}\left\{\frac{1}{3}\left(1-\frac{C_2^3}{C_1^3}\right)\frac{\Pi}{\rho}-\frac{p_1}{p}\log\frac{C_1}{C_2}\right\}/\left(1-\frac{C_2}{R_2}\right),$$

7. A homogeneous liquid is contained between two concentric spherical the radius of the inner being a and that of the outer indefinitely great. The fluid is attracted to the centre of these surfaces by a force ϕ (r), and constant pressure Π is exerted at the outer surface. Suppose

$$\int\phi(r)dr = \chi\,(r),$$

and that χ (r) vanishes when r is infinite. Show that if the inner surface is suddenly removed, the pressure at the distance r is suddenly diminished by

$$\Pi\left(\frac{a}{r}\right)-\left(\frac{\sigma\rho}{r}\right)\chi(a).$$

Find ϕ (r) so that the pressure immediately after the inner surface is removed may be the same as it would be if no attractive force existed. Also, with this value of f (r), find the velocity of the inner boundary of the fluid at any period of the motion.

IMPULSIVE MOTION OF A FLUID

If at any instant impulsive forces act bodily on the fluid, or if the boundary conditions. Suddenly change, a sudden alternation in the motion may occur. Body forces of large magnitude do not act directly on the fluid, but a sudden change of the motion of the boundaries will give large pressure gradients which produce a change in the velocity at every point of the fluid. The impulse is to be regarded, as an infinitely great force acting for an infinitely short time, the effects of all forces are to be neglected during this interval.

Consider the incompressible fluid particle of volume V bounded by a closed surface S. Let the particle is moving initially with velocity q and is subjected to a sudden external impulsive body force I per unit mass. The impulse changes the velocity of the particle instantaneously form $\mathbf{q}$ to $\mathbf{q}_1$. Let $\overline{\omega}$ denotes the impulsive pressure at the elements δS of S ($\delta S << S$) then the total impulse applied to the fluid particle is

$$= \int_V \rho I dv - \int_s n\overline{\omega} dS,$$

where $\rho I\ \delta v$ is the impulse on a volume element δv of V and that on δS is $(-\ n\overline{\omega}\ \delta S)$, $\mathbf{n}$ is the unit outward drawn normal.

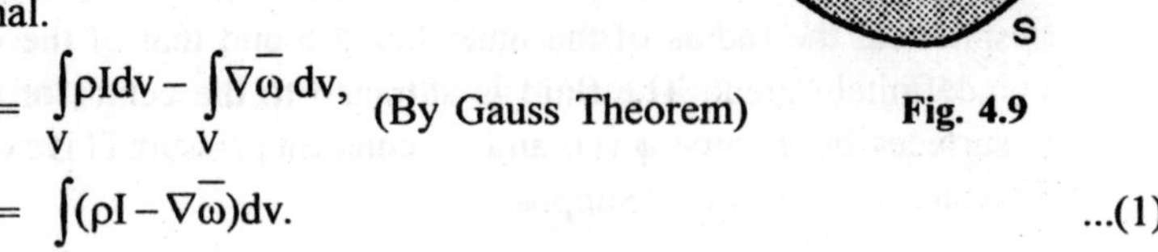

$$= \int_V \rho I dv - \int_V \nabla\overline{\omega}\, dv, \quad \text{(By Gauss Theorem)}$$

Fig. 4.9

$$= \int_V (\rho I - \nabla\overline{\omega}) dv. \qquad ...(1)$$

The change in momentum is given by

$$\int_V \rho(\mathbf{q}_1 - \mathbf{q}) dv. \qquad ...(2)$$

From (1) and (2), we have

$$\int_V \rho(\mathbf{q}_1 - \mathbf{q})\, dv = \int_V (\rho I - \nabla\overline{\omega})dv$$

Since dv is an arbitrary small volume, then

$$\rho\ (\mathbf{q}_1 - \mathbf{q}) = \rho I - \nabla\ \overline{\omega}$$

$$\Rightarrow \quad \mathbf{q}_1 - \mathbf{q} = I - \frac{1}{\rho}\nabla\overline{\omega}, \qquad \text{...(3)}$$

at each point of the fluid.

(i) when there are no externally applied impulses but only impulsive pressure occurs *i.e.*, I = 0 then (3) reduces to

$$\mathbf{q}_1 - \mathbf{q} = -\frac{1}{\rho}\nabla\overline{\omega}$$

(ii) If the velocity distribution in the fluid before the impulse is irrotational with potential ϕ (*i.e.*, $\mathbf{q} = -\nabla\phi$), then equation (4) becomes

$$\nabla\phi_1 - \nabla\phi = (1/\rho)\ \nabla\ \overline{\omega} \Rightarrow \overline{\omega} = \rho\ (\phi_1 - \phi). \qquad \text{...(5)}$$

The constant may be omitted, as an extra pressure remains the same throughout the fluid which will not affect the motion. The relation (5) gives a physical interpretation of the velocity potential. The potential ϕ of given irrotational velocity distribution may be interpreted as $(-1/\rho)$ times the pressure impulse required to set up the given motion from rest or $(1/\rho)$ times the pressure impulse required to set up the given motion from rest or $(1/\rho)$ times the pressure impulse required to reduce the given motion to rest. No rotational motion can be generated from rest or reduced to rest be the action of a pressure impulse.

(iii) If the fluid is incompressible *i.e.*, ρ = const, and there are no extraneous impulses then taking the divergence of both the sides of 1.02 and making use of the equation of continuity, we have

$$\nabla\ (q1 - q) = \nabla.\left[-\frac{1}{\rho}\nabla\overline{\omega}\right]$$

$$\nabla\mathbf{q}_1 - \nabla.\ \mathbf{q} = -\ -\frac{1}{\rho}\nabla\overline{\omega} \Rightarrow \nabla^2\overline{\omega} = 0,$$

Which represents the Laplace equation whose solution can be obtained by solving it with the proper boundary conditions.

IMPULSIVE MOTION OF A FLUID (CARTESIAN COORDINATE)

Consider ρ be the density of the fluid at the point P (x, y,z). Let u_1, v_1, w_1 and u_2, v_2, w_2 be the velocity components at the point P just before and just after the impulsive action. Let I_x, I_y, I_z be the components of the external impulsive forces per unit mass of the fluid. Construct a small parallelopiped with edges of lengths δx, δy, δz parallel to the respective coordinates axes. Let $\bar{\omega}$ be the impulsive pressure at P.

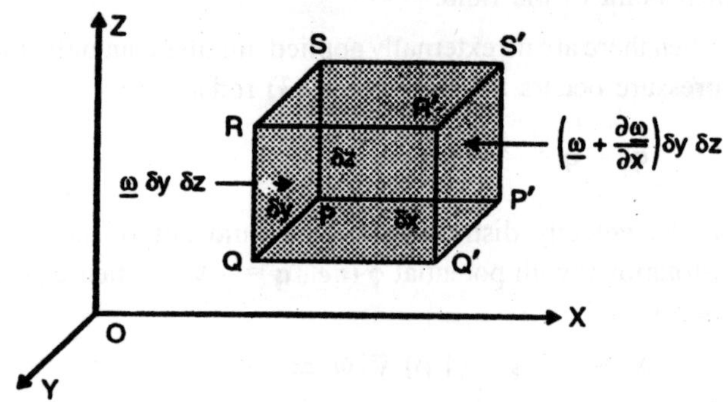

Fig. 4.10

Force on the face PQRS $= \bar{\omega}\,\delta y \delta z$

$$= f\,(x,\ y,\ z)\ \text{(let)}$$

Force on the opposite face P'Q'R'S',

$$= f\,(x,\ \delta,\ y,\ z)\ \text{(let)}$$

$$= f(x, y, z) - f(x, y, z) + \delta x.\frac{\partial}{\partial x} f(x, y, z) + ...$$

$$= f\,(x,\ y,\ z) - f\,(x,\ y,\ z) - \delta x.\ \frac{\partial}{\partial x} f(x, y, z)...$$

$$= -\ \delta x\ \frac{\partial}{\partial x} f(x, y, z)$$

$$= -\ \delta x\ \frac{\partial}{\partial x}(\bar{\omega}\ \delta y\ \delta z)$$

$$= -\ \frac{\partial \bar{\omega}}{\partial x}\delta x \delta y \delta z \text{ along X-axis}$$

The impulse on the parallelopiped along X-axis due to external impulsive body force I_x

$= \rho I_x \, \delta x \, \delta y \, \delta z$

Change in momentum along X-axis

$= \rho \, \delta x \, \delta y \, \delta z \, (u_2 - u_1)$

By Newton's second law of motion, we have

Total impulse applied along X-axis

Change of momentum along X-axis

$$\Rightarrow \rho I_x \delta x \delta y \delta z - \frac{\partial \bar{\omega}}{\partial x} \delta x \delta y \delta z = \rho \, (u_2 - u_1) \, \delta x \delta y \delta z$$

$$\Rightarrow \rho \, (u_2 - u_1) = \rho I_x - \frac{\partial \bar{\omega}}{\partial x}$$

$$\Rightarrow u_2 - u_1 = I_x - \frac{1}{\rho} \frac{\partial \bar{\omega}}{\partial x}, \text{ along X-axis}$$

$$\text{Similarly } v_2 - v_1 = I_y - \frac{1}{\rho} \frac{\partial \bar{\omega}}{\partial y}, \text{ along Y-axis}$$

$$\text{and } w_2 - w_1 = I_z - \frac{1}{\rho} \frac{\partial \bar{\omega}}{\partial z}, \text{ along Z-axis}$$

known as the equations of motion of an inviscid fluid under the action of impulsive forces.

ENERGY EQUATION

The principle of energy enunciates that the change in total energy is equal to work done by the extraneous forces. The potential due to external forces is supposed to be independent of time.

The Euler's equation of motion is

$$\frac{\partial \mathbf{q}}{\partial t} = - \nabla \Omega - \frac{1}{\rho} \nabla p. \quad \text{...(1)}$$

where the extraneous forces are conservative.

$$\rho \frac{\partial \mathbf{q}}{\partial t} = \rho \, (\nabla \Omega) - \nabla p,$$

Multiplying the equation (1) scalarly by **q**, we get

$$\rho \mathbf{q} \frac{\partial \mathbf{q}}{\partial t} = \rho \mathbf{q} \, (\nabla \Omega) - \mathbf{q} . \nabla p$$

$$\text{or } \frac{1}{2}\rho\frac{d}{dt}(\mathbf{q}^2) + \rho\mathbf{q}.(\nabla\Omega) = -\mathbf{q}.\nabla p$$

$$\frac{1}{2}\rho\frac{d}{dt}(\mathbf{q}^2) + \rho\frac{d\Omega}{dt} = -\mathbf{q}.\nabla p$$

$$\text{or } \rho\frac{d}{dt}\left(\frac{1}{2}\mathbf{q}^2 + \Omega\right) = -\mathbf{q}.\nabla p \quad ...(2)$$

Integrating the relation (2) over V, we have

$$\int_V \rho\frac{d}{dt}\left(\frac{1}{2}\mathbf{q}^2 + \Omega\right)dv = \int_V (\mathbf{q}.\nabla p)\, dv$$

$$\text{or } \frac{d}{dt}\left[\int_V \frac{1}{2}\rho\mathbf{q}^2 dv + \int_V \rho\Omega dv\right] = -\int_V (\mathbf{q}.\nabla p)\, dv$$

$$\text{or } \frac{d}{dt}\ (T + W) = -\int_V \nabla.(p\mathbf{q})dv - \int_V \frac{p}{\rho}\frac{d\rho}{dt}dv.$$

By virtue of divergence theorem and the equation of continuity the R. H. S. may be expressed as

$$\frac{d}{dt}\ (T + W) = -\int_S p.\mathbf{q}.\mathbf{n}dS - \frac{dI}{dt}$$

$$\text{or } \frac{d}{dt}\ (T + W + I)^{++} = \int_S p.\mathbf{q}.\mathbf{n}dS^{\dagger}$$

which is known as energy equation.

Thus, the rate of change of total energy (kinetic, potential and intrinsic) of a portion of perfect fluid is equal to the rate at which the work is being done by the pressure on the boundary.

Example 24:

A portion of homogeneous fluid is confined between two concentric spheres of radii A and a, and is attracted towards their centre by a force varying inversely as square of the distance. The inner spherical surface is suddenly annihilated, and when the radii of the inner and outer surface of the fluid are r and R, the fluid impinges on a solid ball concentric with their surfaces. Prove that the impulsive pressure at any point of the ball for different values of R and r varies as

$$\sqrt{\left\{(a^2 - r^2 - A^2 + R^2)\left[\frac{1}{r} - \frac{1}{R}\right]\right\}}.$$

Solution:

The equation of continuity is

$$\frac{1}{r'}\frac{d}{dr'}(r'^2 v') = 0$$

$$\Rightarrow \qquad r'^2 v' = f(t). \qquad ...(1)$$

and the equation of motion is

$$\frac{\partial v'}{\partial t} + v'\frac{\partial v'}{\partial r'} = -\frac{\mu}{r'^2} - \frac{1}{\rho}\frac{\partial p}{\partial r'},$$

$$\text{or} \quad \frac{f'(t)}{r'^2} + v'\frac{\partial v'}{\partial r'} = -\frac{\mu}{r'^2} - \frac{1}{\rho}\frac{\partial p}{\partial r'}. \qquad ...(2)$$

Integrating (2) with regard to r′, we have

$$-\frac{f'(t)}{r'} + \frac{1}{2}v'^2 = \frac{\mu}{r'} - \frac{p}{\rho} + A \qquad ...(3)$$

Let r and R be the internal and outer radii and v and V be the velocities there at any time t. The conditions are defined as

(i) $r' = r, v' = v, p = 0$

(ii) $r' = R, v' = V, p = 0.$

The integration constant can be determined by applying the conditions. Therefore

$$-\frac{f'(t)}{r} + \frac{1}{2}v^2 = A + \frac{\mu}{r}.$$

$$\text{and} \ -\frac{f'(t)}{R} + \frac{1}{2}V^2 = A + \frac{\mu}{r}.$$

By eliminating the constant, we get

$$-f¢\ (t)\left(\frac{1}{r} - \frac{1}{R}\right) + \frac{1}{2}(v^2 - V^2) = m\left(\frac{1}{r} - \frac{1}{R}\right). \qquad ...(4)$$

From the equation of continuity, we have

$$r^2 v = R^2 V = f(t)$$

$$2r^2 dr = 2R^2\, dR = 2f\,(t)\, dt \qquad ...(5)$$

$$\text{or} \ -f'\ (t)\left(\frac{1}{r} - \frac{1}{R}\right) + \frac{1}{2}f^2(t)\left(\frac{1}{r^4} - \frac{1}{R^4}\right)$$

Multiplying by 2f (t) dt both sides and integrating, we have

$$-2f(t)f'(t)\,dt\left(\frac{1}{r}-\frac{1}{R}\right)+\frac{1}{2}f^2(t)\left(\frac{2f(t)dt}{r^4}-\frac{2f(t)dt}{R^4}\right)$$

$$=\mu\left(\frac{2f(t)dt}{r}-\frac{2f(t)dt}{R}\right)$$

or $-2f(t)\,f'(t)\,dt\left(\frac{1}{r}-\frac{1}{R}\right)+f^2(t)\left(\frac{dr}{r^2}-\frac{dR}{R^2}\right)=\mu\,\{2r\,dr-2RdR\}$

or $-f^2(t)\left(\frac{1}{r}-\frac{1}{R}\right)=m\,(r^2-R^2)+B.$ ...(6)

initially $r = a$, $R = A$, $v = 0$; $B = -\mu\,(a^2-A^2)$.

Equation (6) becomes

$$f^2(t)\left(\frac{1}{r}-\frac{1}{R}\right)=m\,(a^2-r^2-A^2+R^2). \qquad ...(7)$$

Let $\bar{\omega}$ be the impulsive pressure at a distance r' then

$$d\bar{\omega}=-\rho v'\,dr'=-\rho\,[f(t)/r'2]\,dr'.$$

By integrating, we have

$\bar{\omega}=\frac{\rho f(t)}{r'}+C$, where C is an arbitrary constant.

But $\bar{\omega}=0$, $r'=R$, then $C=\frac{\rho f(t)}{R}$.

or $\bar{\omega}=r\,f(t)\left\{\frac{1}{r},-\frac{1}{R}\right\}$,

which determines the impulsive pressure at any distance r'.

The impulsive pressure at any point of the ball where $r' = r$ is,

$$\bar{\omega}=\rho f(t)\left\{\frac{1}{r},-\frac{1}{R}\right\}.$$

Substituting the value of f (t) from (7), we have

$$\bar{\omega}=\rho\sqrt{\left\{\frac{\mu(a^2-r^2-A^2+R^2)}{\left(\frac{1}{r}-\frac{1}{R}\right)}\right\}}\left(\frac{1}{r}-\frac{1}{R}\right)$$

$$\text{or } \bar{\omega} = \rho \sqrt{\mu} \sqrt{\left\{(a^2 - r^2 - A^2 - R^2)\left(\frac{1}{r} - \frac{1}{R}\right)\right\}}$$

Hence the impulsive pressure varies as

$$\sqrt{\left\{(a^2 - r^2 - A^2 - R^2)\left(\frac{1}{r} - \frac{1}{R}\right)\right\}}.$$ **Proved.**

Example 25:

A sphere of radius a is surrounded by an infinite liquid of density ρ, the pressure at infinity being Π. The sphere suddenly annihilated. Show that the pressure at distance r from the centre immediately falls to P $\left(I - \frac{a}{r}\right)$.

Show further that if the liquid is brought to rest by impinging on a concęntric sphere of radius, $\frac{1}{2}a$ *, the impulsive pressure sustained by the surface of the sphere is* $\sqrt{\left(\frac{7}{6}\Pi\rho a^2\right)}$.

Solution:

Let v′ be the velocity at a distance r′ from the centre of the sphere at any time t and p be the pressure. The equation of continuity is

$$r'^2 v' = f(t). \quad \text{...(1)}$$

Equation of motion is

$$\frac{\partial v'}{\partial t} + v'\frac{\partial v'}{\partial r'} = -\frac{1}{\rho}\frac{\partial p}{\partial r'}$$

$$\text{or } \frac{f'(t)}{r'^2} + v'\frac{\partial v}{\partial r'} = -\frac{1}{\rho}\frac{\partial p}{\partial r'}.$$

Integrating with regard to r′, we have

$$\frac{f'(t)}{r'} + \frac{1}{2}v'^2 = -\frac{p}{\rho} + A,$$

where A is an arbitrary constant.

$$\text{Since } r' = \infty,\ p = P,\ v' = 0 \text{ so } A = \frac{\Pi}{\rho}. \quad \text{...(2)}$$

When the sphere is suddenly annihilated *i.e.* $r' = a$, $v' = 0$, $p = 0$, then

$$-\frac{f'(t)}{a} = \frac{\Pi}{\rho} \Rightarrow f'(t) = \frac{\Pi a}{\rho}. \qquad ...(3)$$

The velocity v¢ vanishes just after annihilation, so from (2) and (3), we have

$$-\frac{\Pi a}{\rho}\frac{1}{r'} + 0 = \frac{\Pi - p}{\rho}\frac{1}{r'} \Rightarrow \frac{a\Pi}{r'} = \Pi - p.$$

Thus pressure at the time of annihilation ($r' = r$) is

$$\frac{a\Pi}{r} = -\Pi - p \Rightarrow p = \Pi\left(1 - \frac{a}{r}\right). \qquad \textbf{Proved.}$$

Let $\bar{\omega}$ be the impulsive pressure at a distance r′, then

$d\bar{\omega} = -\rho v'\, dr'$.

From the equation of continuity, we have

$r^2 v = r'^2 v' = f(t)$

or $d\bar{\omega} = -\rho v\,(r^2/r'^2)\, dr'$,

where r be the radius of the inner surface and v bethe velocity there

By integrating, we have

$\bar{\omega} = B + \rho v\,(r^2/r')$

When $r' = \infty$, $\bar{\omega} = 0$ then $B = 0$. ...(4)

which determines the impulsive pressure $\bar{\omega}$ at a distance r′.

Since the liquid is brought to rest by impinging on a concentric sphere of radius a/2, then substituting $r = a/2$ in (4), we have

$$\bar{\omega} = \rho v\,(a^2/4r'). \qquad ...(5)$$

The velocity v at the inner surface of the sphere ($p = 0$) is obtained from (2)

$$-\frac{f'(t)}{r} + \frac{1}{2}v^2 = \frac{\Pi}{\rho}.$$

Using the relation (1), we have

$$\left(rv\frac{dv}{dr} + 2v^2\right) - \frac{1}{2}v^2 = \frac{\Pi}{\rho},$$

$$\text{or } rv\frac{dv}{dr}\frac{3}{2}v^2 = -\frac{\Pi}{\rho}.$$

Multiplying by $2r^2$ dr both the sides and integrating, we have

$$2r^3v\,dv + 3v^2r^2\,dr = -\frac{2\Pi}{\rho}r^2 dr$$

$$r^3v^2 = -(2\Pi/3\rho)\,r^3 + C$$

Since r = a, v = 0 then $C = (2\Pi/2\rho)\,a^3$.

$$\text{or } r^3v^2 = \frac{2\Pi}{3\rho}(a^3 - r^3).$$

The velocity v at the surface of the sphere of radius a/2 on which the liquid strikes is

$$v^2 = \frac{2\Pi}{3\rho}\frac{a^3 - r^3}{r^3} = \frac{2\Pi}{3\rho}.\frac{a^3 - a^3/8}{a^3/8} = \frac{14}{3}\frac{\Pi}{\rho}.$$

From the relation (4), we get

$$\bar{\omega} = \frac{1}{4}\rho\sqrt{\left(\frac{14}{3}\frac{\Pi}{\rho}\right)}a^2\frac{1}{r'}, \qquad \text{...(6)}$$

determines the impulsive pressure at a distance r′.

Thus the impulsive pressure at the surface of the sphere of radius a/2 is given by

$$\bar{\omega} = \frac{1}{4}\rho\sqrt{\left(\frac{14}{3}\frac{\Pi}{\rho}\right)}a^2\frac{2}{a} = \sqrt{\left(\frac{7\Pi\rho a^2}{6}\right)}. \textbf{ Proved.}$$

Example 26:

Two equal closed cylinders, of height c, with their base in the same horizontal plane, are filled one with water, and the other with air of such a density as to support a column h of water, h being less than c. If a communication be opened between them at their bases, the height x, to which the water rises, is given by the equation

$$cx - x^2 + ch\log\{(c - x)/x\} = 0.$$

Solution:

Let P, Q be two cylinders containing water andair respectively and k be the cross-section of each cylinder before and after the communication is set up, the air and water are at rest. Thus the initial and final kinetic energies are zero. The intrinsic energies also vanish, because of incompressibility.

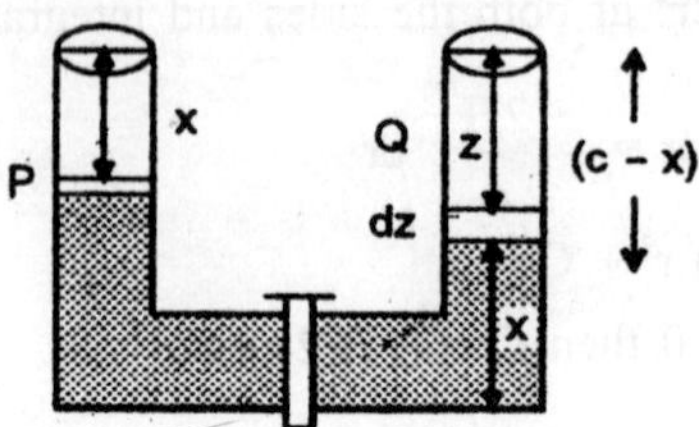

Fig. 4.11

The potential energy due to position of water in the cylinder P is

$$V_a = \int_0^c g\rho kz dz = \frac{1}{2} g\rho kc^2 \qquad ...(1)$$

Let the height x of water rises in cylinder Q then the height (c – x) of water will remain in the cylinder P after communication is opened. The potential energy is

$$V_b = \int_0^{c-x} g\rho\, kz\, dz + \int_0^x g\rho\, kz\, dz$$

$$V_b = \frac{1}{2} g\rho k\, [(c-x)^2) + x^2]. \qquad ...(2)$$

Loss in potential energy of work done by gravity is

$$V_a - V_b = \frac{1}{2} g\rho k\, [c^2 - (c-x)^2) - x^2] \qquad ...(3)$$

Let p be the pressure when the water rises to a height z, then

$g\rho hkc = pk\ (c - z) \Rightarrow p = g\rho hc/\ (c - z)$.

Since the air has been compressed so the work done in compressing this air in the cylinder Q is

$$= -\ g\rho hck \int_0^x \frac{dz}{c-z} = g\rho hck \log\left(\frac{c-x}{c}\right). \qquad ...(4)$$

Total work done = Change in K. E.

Hence $g\rho kx\ (c - x) + g\rho hck \log \{(c - x)/c\} = 0$,

or $g\rho k\ [cx - x^2 + ch \log \{(c - x)/c\}] = 0$,

or $cx - x^2 + ch \log\left(\frac{c-x}{c}\right) = 0$. **Proved.**

Example 27:

A spherical mass of liquid of radius b has a concentric spherical cavity of radius a; which contains gas at pressure p whose mass may be neglected; at every point of the external boundary of the liquid an impulsive pressure $\underline{\omega}$ *per unit area is applied. Assuming that the gas obeys Boyle's law, shew that when the liquid first comes to rest, the radius of the internal surface will be*

$$a \exp.\left\{-\frac{\omega^2 b}{2p\rho a^2(b-a)}\right\},$$

where ρ *is the density of the liquid.*

Solution:

Let v' be the velocity at a distance r' from the centre of the spherical cavity at any time t. The equation of continuity is

$$r'^2 v' = f(t) = b^2 V. \qquad \text{...(1)}$$

Let $\underline{\omega}'$ the impulsive pressure at a distance r' then

$$d\,\underline{\omega}' = -\rho v'\,dr' = -r\,(b^2V/r'^2)\,dr'.$$

Integrating with regard to r', we have

$$\underline{\omega}' = (\rho b^2 V/r') + C, \qquad \text{...(2)}$$

where C is an arbitrary constant.

Again $r' = a$, $\underline{\omega}' = 0$ and $r' = b$, $\underline{\omega}' = \underline{\omega}$. ...(3)

Using (2) and (3), we can determine the arbitrary constant as

$$0 = (\rho b^2 V/a) + C \text{ and } \underline{\omega} = (\rho b^2 V/b) + C,$$

$$\text{or } \underline{\omega} = \rho b^2 V\left(\frac{1}{b}-\frac{1}{a}\right) = \frac{\rho b V}{a}(a-b). \qquad \text{...(4)}$$

The initial kinetic energy is

$$= \frac{1}{2}\int_a^b (4\pi r'^2 \rho\, dr').v'^2 = 2\pi\rho \int_a^b r'^2.\frac{b^4V^2}{r'^4}dr'$$

$$= 2\pi\rho b^4 V^2 \int_a^b \frac{dr'}{r'^2} = \frac{2\pi\rho b^2 V^2}{a}(b-a). \qquad \text{...(5)}$$

Let r be the radius of the internal spherical cavity and p_1 be the pressure of the gas there. Since the gas obeys Boyle's law, we have

$$\frac{4}{3}\pi r^3.p_1 = \frac{4}{3}\pi a^3.p \Rightarrow p_1 = \frac{a^3 p}{r^3}$$

Total work done $= \int_a^r 4\pi r^2.p_1 dr = 4\pi a^3 p \log(r/a)$

Again change in K.E. = total work done

$$\frac{2\pi b^3 V^2}{a}(b-a) = \pi a^3 \rho \log(r/a)$$

$$\text{or } \log(r/a) = \frac{2\pi\rho b^3 V^2(b-a)}{4\pi a^4 p}$$

$$= \frac{2\pi\rho b^3(b-a)}{4\pi p a^4}.\frac{a^2\underline{\omega}^2}{\rho^2 b^2(a-b)^2}$$

$$= -\frac{\underline{\omega}^2 b}{2p\rho a^2(b-a)}$$

$$\text{or } r = a\exp.\left\{\frac{\underline{\omega}^2 b}{2p\rho a^2(b-a)}\right\}.$$ **Proved.**

Example 28:

Show that the rate per unit of time at which work is done by the internal pressures between the parts of a compressible fluid obeying Boyle's law is

$$\iiint p\left(\frac{\partial u}{\partial x}+\frac{\partial v}{\partial y}+\frac{\partial w}{\partial z}\right)dx\,dy\,dz,$$

where p is the pressure and (u,v, w), the velocity at any point and the where p is the pressure through the volume of the fluid.

Solution:

Let W is the work done, p is the pressure and dv an elementary volume. Work done in compressing th fluid is

$$W = \int p(-dv)$$

Rate per unit time of work done is

$$\frac{DW}{Dt} = \iiint \frac{Dp}{Dt}dv. \quad \text{...(1)}$$

Equation of continuity is

$$\frac{D\rho}{Dt} + \rho\left(\frac{\partial u}{\partial x} + \frac{\partial v}{\partial y} + \frac{\partial w}{\partial z}\right) = 0$$

$$\frac{Dp}{Dt} + p\left(\frac{\partial u}{\partial x} + \frac{\partial v}{\partial y} + \frac{\partial w}{\partial z}\right) = 0, \; p = kr \qquad ...(2)$$

From (1) and (2), we have

$$\frac{DW}{Dt} = \iiint p\left(\frac{\partial u}{\partial x} + \frac{\partial v}{\partial y} + \frac{\partial w}{\partial z}\right) dv$$

Hence the rate per unit time of work done is given by

$$\iiint p\left(\frac{\partial u}{\partial x} + \frac{\partial v}{\partial y} + \frac{\partial w}{\partial z}\right) dx\, dy\, dz. \qquad \textbf{Proved.}$$

APPLICATION OF BERNOULLI'S THEOREM

Bernoulli's equation is one of the important tools for solving problems in fluid mechanics. In most of the problems, it is applied in combination with the equation of continuity. The applications of the equation can be classified in two categories : (i) closed flows, and (ii) flows with three surface. Here we shall study some applications of the equation.

Flow from a Tank Through a Small Orifice

When a vessel containing water has a small orifice near its base in one of its walls, the water flows out steadily in the form of a smooth jet. It first converges and then at some distance from the tank its cross-section is minimum. This phenomenon of contraction of the jet is called *vena contracta*.

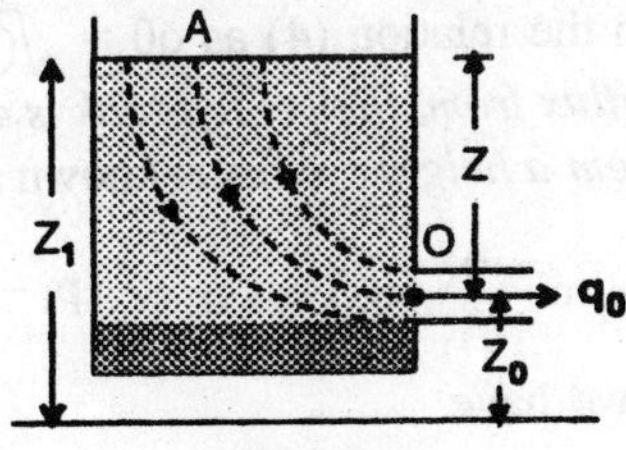

Fig. 4.12

Let A_1 and A_0 be the cross-sectional area of the tank and orifice respectively. All the streamlines passing through the orifice must begin at the free liquid surface, where the velocity is negligibly small. The Bernoulli's equation has the same value for all the emerging streamlines, except those coming from the boundary layer at the wall of the tank which we shall neglect. The equation of motion at a point on the free surface A to the centre O of the vena contracta in the jet gives

$$gz_1 + \frac{p_1}{\rho} + \frac{1}{2}q_1^2 = \frac{p_0}{\rho} + \frac{1}{2}q_0^2, \quad z = z_1 - z_0, \qquad ...(1)$$

where $\mathbf{q}_1$ is velocity with which the level of fluid in the tank is decreasing.

From the continuity equation, we have

$$A_1\mathbf{q}_1 = A_0\mathbf{q}_0 \Rightarrow \mathbf{q}_1 = \frac{A_0 q_0}{A_1} \qquad ...(2)$$

From (1) and (2), we have

$$\frac{1}{2}q_0^2 = g\,(z1 - z_0) + \frac{p_1 - p_0}{\rho} + \frac{1}{2}(A_0^2/A_1^2)\mathbf{q}_0^2$$

or $$\frac{1}{2}\left(1 - \frac{A_0^2}{A_1^2}\right)q_0^2 = g\,(z_1 - z_0) + \frac{1}{\rho}(p_1 - p_0)$$

or $$q_0^2 = \frac{2}{1-(A_0^2/A_1^2)}\left[\frac{1}{\rho}(p_1 - p_0) + gz\right]$$

or $$q_0 = \left[\frac{2}{1-(A_0^2/A_1^2)}\left\{\frac{1}{\rho}(p_1 - p_0) + gz\right\}\right]1/2, \qquad ...(3)$$

which determines the velocity of efflux from the tank through the orifice.

(i) Let $A_0 << A_1$ or $\frac{A_0}{A_1} << 1$. ...(4)

If the tank is vented to the atmosphere so that the pressure is necssarily uniform *i.e.*, $p_1 = p_0$. The velocity of efflux from the vented tank is obeaind from the relation (4) as $q0 = \sqrt{(2gz)}$. *This represents that the velocity of efflux from, the vented tank is equal to that of a right body falling freely from a height z.* This is known as Toricalli's theorem.

(ii) Let $A_0 << A_1$ or $\frac{A_0}{A_1} << 1$ then $-\frac{1}{\rho}(p_1 - p_2) >> GZ$,

Thus from (4), we have

$$q_0 = \left\{\frac{2}{\rho}(p_1 - p_0)\right\}^{1/2} \qquad ...(5)$$

Trajectory of a Free Jet

Consider that a fluid jet is coming out from a small hold of cross-section A0 with velocity q_0 and making an angle q with the horizon. Let A_0 be the exit area of the nozzle. The jet can be considered as a streamline if the atmospheric viscous effects are neglected. Let A is an arbitrary point on it. Since the entire jet is in the atmosphere, so that $p_0 = p_1$. The relations between the velocity components at the point O and A are given by

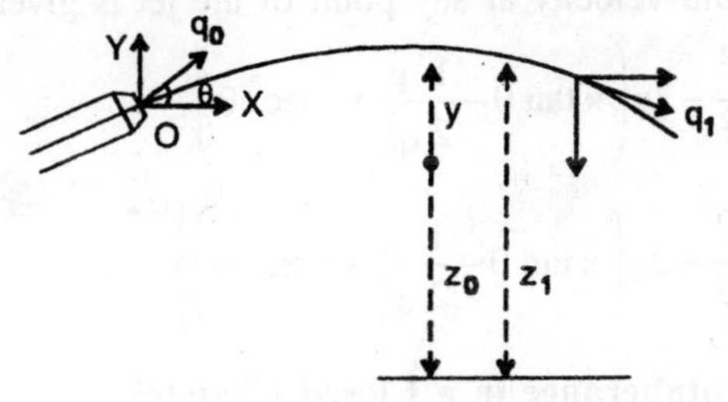

Fig. 4.13

$$q_{1x} = q_{0x}, q_{1y} = q_{0y} - gt. \qquad ...(1)$$

The coordinates of the point A are given by

$$x = q_{0x}t, \; y = q_{0y}t - \frac{1}{2}gt^2,$$

where t is the time for a fluid particle to travel from the vena contracta to the point A.

Eliminating t from (2), we have

$$y = \frac{q_{0y}}{q_{0x}}x - \frac{1}{2}g\frac{x^2}{q_0^2}. \qquad ...(3)$$

The equation of motion at the point of vena contracta O and at an arbitrary point A of the streamline is given by

$$gz_0 + \frac{1}{2}q_0^2 = gz_1 + \frac{1}{2}q_1^2 \qquad ...(4)$$

We shall determine the velocity of the jet at any point on its trajectory as a function of the jet flow rate Q. The volume flow rate is

$Q = A_0 q_0$.

From (4), we have

$$q^2_1 = q^2_0 - 2g\,(z_1 - z_0) = \frac{Q^2}{A_0^2} - 2gy. \qquad ...(5)$$

From (3) and (5), we have

$$q^2{}_1 = \frac{Q^2}{A_0^2} - 2g\left(\frac{q_{0y}}{q_{0x}}x - \frac{1}{2}g\frac{x^2}{q_0^2}\right). \qquad ...(6)$$

Again $\tan\theta = \frac{q_{0y}}{q_{0x}}$ and $\cos\theta = \frac{q_{0x}}{q_0}$.

Thus, the fluid velocity at any point of the jet is given by

$$q_1^2 = \frac{Q^2}{A_0^2} - 2g\left(x\tan\theta - \frac{1}{2}\frac{g}{q_0^2}x^2\sec^2\theta\right),$$

$$\text{or } q_1 = \left\{\frac{Q^2}{A_0^2} - 2g\left(x\tan\theta - \frac{1}{2}\frac{g}{q_0^2}x^2\sec^2\theta\right)\right\}^{1/2}.$$

Flow Over a Protuberance in a Closed Channel

Here we shall determine the pressure distribution on a cylindrical proguberance in a closed channel which is completely filled with the moving fluid.

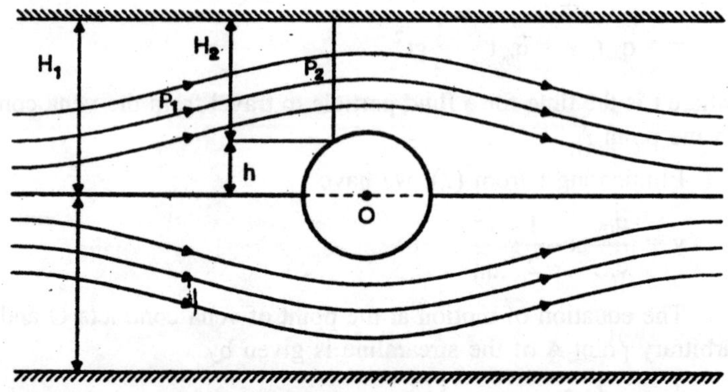

Fig. 4.14

Consider the channel to be of constant width b and depth 2H. Let ρ, q and p be the density, the velocity and the pressure of the fluid at a point far away from the cylindrical protuberance in the upstream direction. The Bernoulli's equation is taken along a stream line as

$$\frac{p_1}{\rho} + \frac{1}{2}q_1^2 = \frac{p_2}{\rho} + \frac{1}{2}q_2^2,$$

$$\Rightarrow \frac{1}{\rho}(p_2 - p_1) = \frac{1}{2}(q_1^2 - q_2^2). \quad ...(1)$$

Equation of continuity gives

$(bH_1)\, q_1 = (bH_2)\, q_2$

$$\Rightarrow q_1 = \left(\frac{bH_2}{bH_1}\right) q_2 \quad ...(2)$$

From (1) and (2), we have

$$\frac{p_2 - p_1}{\rho} = \frac{1}{2}q_1^2\left(1 - \frac{q_2^2}{q_1^2}\right) = \frac{1}{2}q_1^2\left\{1 - \left(\frac{H_1^2}{H_2^2}\right)\right\}$$

$$\Rightarrow \frac{p_2 - p_1}{\rho} = \frac{1}{2}q_1^2\left\{1 - \frac{H_1^2}{(H_1 - h)^2}\right\}$$

$$\Rightarrow p_2 - p_1 = \frac{1}{2}\rho q_1^2\{1 - [1 - (h/H_1)]-^2\}$$

$$p_2 = p_1 + \frac{1}{2}\rho q_1^2\{1 - [1 - (h/H_1)]-^2\},$$

which determines the pressure distribution.

Again the ratio $\dfrac{p_2 - p_1}{\frac{1}{2}\rho q_1^2} = \left\{1 - \left[1 - \dfrac{h}{H_1}\right]^{-2}\right\}$,

is called the ***pressure coefficient***. It is a dimensionless ratio and defines the importance of hydrostatic pressure to the hydrodynamic pressure of the undisturbed flow.

Pitot Tube

Pitot tube is used for the measuring of the fluid velocity of a stream at any point in the fluid region. It consists of a finite tube open at one end, which points upstream, and connected at the other end with a manometer. Fluid elements directly approaching the mouth are brought to rest and a manometer connected to the tube determines the stagnation pressure. The fluid is at rest directly in front of the opening. The stream lines through the point O leads to the point A called the stagnation points, where the fluid is at rest. By using the equation of motion at the point O and A, we have

$$\frac{1}{\rho}p_0 + \frac{1}{2}q_0^2 = \frac{1}{\rho}p_1 + \frac{1}{2}q_1^2. \quad ...(1)$$

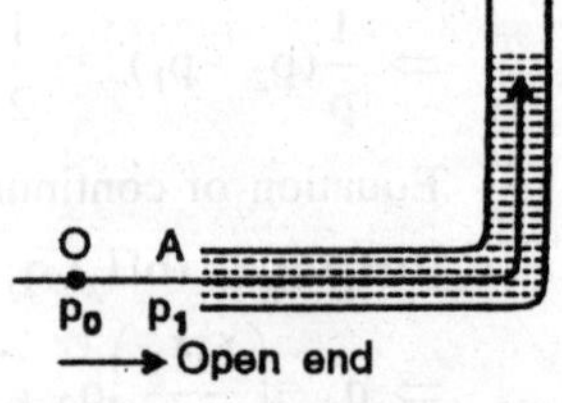

Fig. 4.15

Since the flow is brought to rest at the point A *i.e.*, $q^2 = 0$ and

$p_1 = P$ (say), then we have

$$P = p_0 + \frac{1}{2}\rho q_0^2$$

$$\Rightarrow q_0 = \left\{\frac{2}{\rho}(P - p_0)\right\}^{1/2}. \quad ...(2)$$

Thus, the velocity can be determined by measuring the difference in the stagnation and static pressures.

Let h be the difference in level of the mercury in the U-tube and σ be the density of the mercury then

$$P - p_0 = \sigma g h \Rightarrow q_0 = \sqrt{\left(\frac{2\sigma g h}{\rho}\right)},$$

The measurement of static pressure in a uniform stream can be determined by the use of pitot-static tube. The form of static tube consists of a right-angled tube, one arm of which is sealed at the end and aligned with the wind direction of flow, the holes farther downstream transmit the local external surface pressure to a manometer. The acceleration of the flow over the arm causes the pressure at the holes to fall below its value to an extent which depends on the shape of the nose, on the other hand the presence of the steam causes a deceleration of the flow upstream of itself and a consequent in pressure. The pressure tappings are usually made possibly close to the points where there two opposing effects balance. Pitot and static tubes are often combined into a single instrument so that the velocity can be determined by the measurement of a difference between the Pitot pressure and the pressure at the static holes.

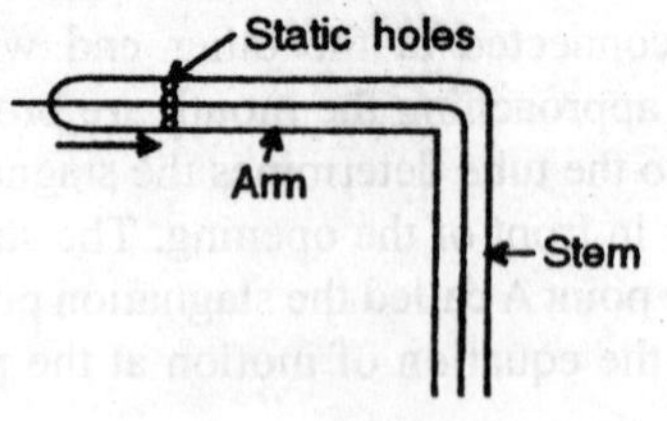

Fig. 4.16

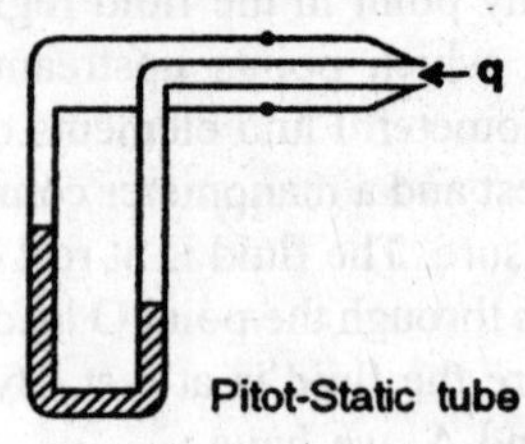

Fig. 4.17

Venturi Tube

The Venturi -tube is used to measure the flow rate of a fluid in a pipe. It consists of upstream section which is of the same cross-section as that of the pipe of which it is attached. It contains a pipe of constant cross-section tapering to a section of smaller cross -section and then gradually expanding to the original cross-section. The cross-section of a venturi tube decreases to a minimum and then increases gradually so that the flow along the tube has a minimum pressure at the throat of the tube. In the flow from the pipe to the throat the velocity is increased and correspondingly the pressure decrease. The flow is accelerated when it moves from the larger to the smaller cross-sectional area and a static pressure variation is produced on it. Consider S_1, S_2 be the cross-sectional areas and p_1, q_1; p_2, q_2 be the pressures and average velocities at section A and B respectively. Then by the equation of continuity, we have

$$q_1S_1 = q_2S_2. \qquad ...(1)$$

Bernoulli's equation for sections A and B can be written as

$$\frac{p_1}{\rho} + \frac{1}{2}q_1^2 + gz_1 = \frac{p_2}{\rho} + \frac{1}{2}q_2^2 + gz_2. \qquad ...(2)$$

From (1) and (2), we have

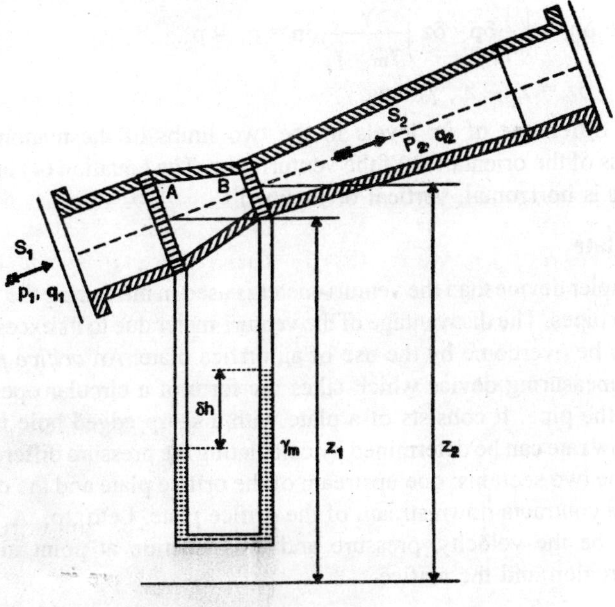

Fig. 4.18

$$\left(1-\frac{S_2^2}{S_1^2}\right)q_2^2 = 2\left[\frac{1}{\rho}(p_1-p_2)+g(z_1-z_2)\right],$$

$$q_2 = \left\{\frac{2\left[\frac{1}{\rho}(p_1-p_2)+g(z_1-z_2)\right]}{1-(S_2^2/S_1^2)}\right\}^{1/2}$$

Thus the flow rate Q at the either section is given by

$$Q = q_2S_2 = \frac{S_2}{\sqrt{\{1-(S_2/S_1)^2\}}}\left[2\left\{\frac{1}{\rho}(p_1-p_2)+g(z_1-z_2)\right\}\right]^{1/2}$$

If $p_1 > p_2$, it follows that the fluid pressure is minimum at a constriction. Also, the flow rate in terms of the manometer deflection is determined as

$$Q_D = \frac{S_2}{\sqrt{\{1-S_2/S_1)^2\}}}\left[2g\delta h\left(\frac{\gamma_m}{\gamma}-1\right)\right]^{1/2} \qquad ...(4)$$

where $\delta h = \left(\frac{1}{\gamma}\delta p - \delta z\right)\frac{\gamma}{\gamma_m - \gamma}$, $\delta p = p_1 - p_2$,

$\delta z = z_2 - z_1$, $\gamma = \rho g$,

δh is the difference of its levels in the two limbs of the manometer reegardless of the orientation of the venturi tube. The equation (4) holds, whether it is horizontal, vertical or inclined.

Orifice Plate

A simpler device than the venturi-meter is used in measuring the fluid velocity in pipes. The disavantage of the venturi-meter due to its excessive length can be overcome by the use of an orifice plate. An *orifice plate* is a fluid measuring device which takes the form of a circular opening placed in the pipe. It consists of a plate with a sharp edged hole in its centre. Flow rate can be determined by calculating the pressure difference between the two sections: one upstream of the orifice plate and the other at the vena contracta down stream of the orifice plate. Letq_1, p_1, A_1 and q_2, p_2, A_2 be the velocity, pressure and cross-section at point in the upstream region and the orifice.

By the equation of continuity, we have

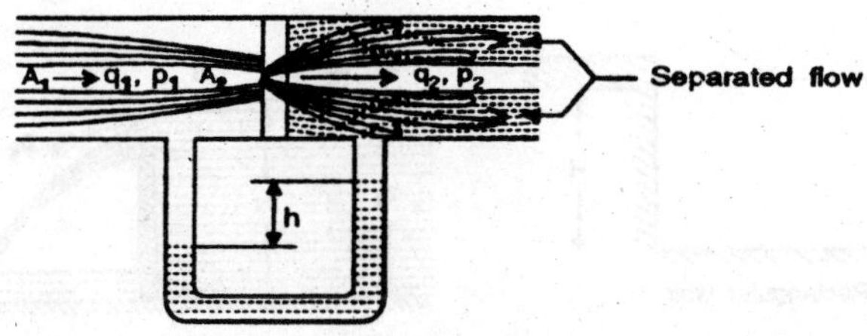

Fig. 4.19

$A_1q_1 = A_2 q_2$ ...(1)

Bernoulli's equation when applied to the stream line, gives

$$\frac{p_1}{\rho}+\frac{1}{2}q_1^2 = \frac{p_2}{\rho}+\frac{1}{2}q_2^2.$$

$$\text{or } p_1 - p_2 = \frac{1}{2}\rho(q_2^2 - q_1^2) \qquad ...(2)$$

Again $p_1 - p_2 = (\rho_1 - \rho)$ gh. ...(3)

From (1), (2) and (3), we have

$$\{A_1^2/A_2^2) - 1\}\ q_1^2 = \frac{2}{\rho}(\rho_0 - \rho)gh$$

$$\Rightarrow q_1 = A_2 \left\{\frac{2gh(\rho_0 - \rho)}{\rho(A_1^2 - A_2^2)}\right\}^{1/2}$$

The actual velocity will be less than the theoretical velocity because we have not considered the los sesdue to viscosity. Experimentally, it is found to be about 10% less than the theoretical velocity. Thus the major disadvantage by using the orifice plates to flow measurement is that a sizeable loss of the pressure is incurred because of the flow separation down stream of the plate.

Weirs

Weirs are used to determine the flow rate of fluid with a free surface such as in open channel or river. The quantity of a fluid flowing steadily along an open channel or through a gate of a reservoir may be determined by a submerged obstacle or weir. The rate of flow is calculated by measuring the height of upstream water surface. Weirs can be classified either according to their shape at right angles to the flow or according to their shape in the direction of flow *e.g.*, rectangular, triangular, parabolic trapezoidal, sharp crested, broad crested, and round crested.

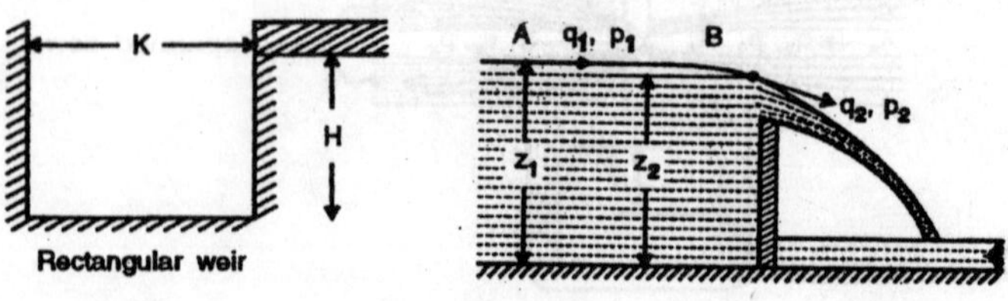

Fig. 4.20

A sharp crested weir is a plate mounted normal to the flow direction and spanning the fluid steam. If the down stream liquid level is below the crest of the weir, the down stream conditions do not affect the flow over the weir, such flow is known as *free flow*. If the down stream liquid level is above the crest, the discharge over the weir depends on height, such flow is known as *submerged flow*.

The sharp-crested rectangular weir has a horizontal crest. The nappe is contracted at top and bottom. By applying the Bernoulli's equation for any streamline in the flow between A and B, we have

$$\frac{p_1}{\rho} + \frac{1}{2}q_1^2 + gz_1 = \frac{p_2}{\rho} + \frac{1}{\rho}q_2^2 + gz_2 . \qquad ...(1)$$

The atmospheric pressure is applied to the entire free surface and

$$z = z_1 - z_2. \qquad ...(2)$$

The velocity at the weir plane is given by

$$\frac{1}{2}q_2^2 = \frac{1}{2}q_1^2 + gz \Rightarrow q_2 = (q_1^2 + 2gz)^{1/2} \qquad ...(3)$$

The flow rate dQ over the weir is given by

$$dQ = q_2 dS_2 \qquad ...(4)$$

For a rectangular weir, we have

$$dS_2 = Kdz$$

or $dQ = Kq^2dz = K\,(q_1^2 + 2gz\}^{1/2}\,dz.$

Thus the theoretical discharge is given by

$$Q = K = \int_0^H (q_1^2 + 2gz)^{1/2} dz \;\; \frac{K}{3g}[(q_1^2 + 2gH)^{3/2} - q_1^3] \qquad ...(5)$$

If the body of water-being measured as large in comparison with the weir opening, then, we have

$$Q = \frac{2K}{3}\sqrt{(2g)H^{3/2}}, q_1 \text{ is negligible,} \qquad ...(6)$$

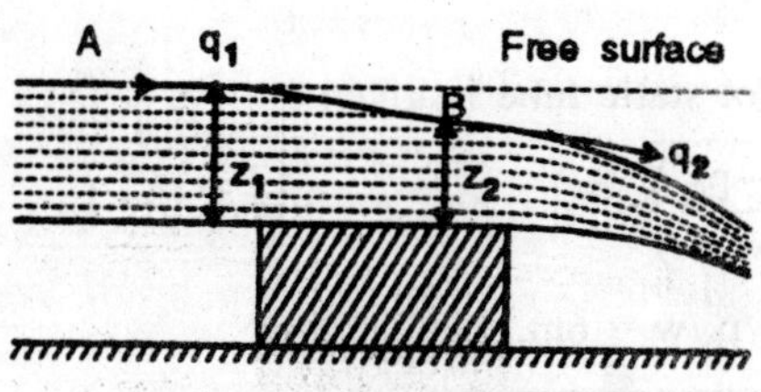

Fig. 4.21

The broad crested weir (*e.g.*, dam spilway) support the flow in a longitudinal direction so that the pressure variation is hydrostatic at the point B and q_2, q_1. The slopes of both the weir and the free surface are small.

By applying the Bernoulli's equation between the points A and B, we have

$$gz_1 = \frac{1}{2}q_2^2 + gz_2 \Rightarrow q_2 = \{2g\,(z_1 - z_2)\}^{1/2}$$

For a weir of width K the theoretical discharge is given by

$$q_2 = Kz_2\,\{2g\,(z_1 - z_2)\}^{1/2}. \qquad ...(7)$$

$$\text{or} \quad \frac{dq_2}{dz_2} = K\sqrt{(2g)}\,[(z^1 - z^2)^{1/2} - \frac{1}{2}z_2(z_1 - z_2)^{-1/2}]$$

For the maximum or minimum rate of discharge $dq_2/dz_2 = 0$

$$K\sqrt{(2g)}\left[(z_1 - z_2)^{1/2} - \frac{1}{2}z_2(z_1 - z_2)^{-1/2}\right] = 0 \text{ Þ } z_2 = \frac{2}{3}z_1. \quad ...(8)$$

For the relation (7) and (8), the maximum rate of discharge is determined as

$$q_2 = K\sqrt{g}\left(\frac{2}{3}z_1\right)^{3/2} \qquad ...(9)$$

The velocity corresponding to the maximum flow rate is given by $\sqrt{(gz_2)}$, which is known as the critical velocity and is equal to the velocity of propagation of a surface wave.

Example 29:

A pitot static tube having a coefficient of 0.98 is used to measure the velocity of water in pipe. The stagnation pressure head is 6 metres and static pressure heads is 5 metres. Determine its velocity

Solution:

Velocity by pitot static tube is given as

$$q = c\sqrt{\left(2g\frac{p_1 - p_0}{w}\right)}$$

Here c = 0.98, p_1/w = 6m., p_0/w = 5m.

$$\therefore\ q = 0.98\ \sqrt{\{2 \times 9.81(6-5)\}}\ = 4.35 \text{ m/sec. } \textbf{Ans.}$$

Example 30:

If the water jet is discharged from a nozzle, inclined at angle π/ 3 with the horizontal, at 20 ft/sec. Calculate the horizontal distance required for the jet striking the ground which is 3 ft below the horizontal line of the nozzle. What is the velocity of the jet just before reaching the ground?

Solution:

The horizontal and vertical components of the velocity at the nozzle exit are, respectively

$$q_{1x} = q_1 \cos\alpha = 20 \cos \pi/3 = 10 \text{ ft/sec.}$$

$$q_{1y} = q_2 \sin\alpha = 20 \cos \pi/3 = 17.32 \text{ ft/sec.}$$

We know that

$$y = \frac{q_{1y}}{q_{1x}}x - \frac{1}{2}g\frac{x^2}{q_1^2}$$

$$\Rightarrow -3 = \frac{17.32}{10}x - \frac{1}{2}g\frac{x^2}{(10)^2}.$$

or $x^2 - 10.74x - 18.61 = 0$.

The horizontal distance between the nozzle and jet striking the ground is

$$x = \frac{1}{2}[10.74 + \sqrt{\{(10.74)^2 - 4 \times 18.62\}}\ = 12.27\text{ft}].$$

The jet velocity at x = 12. 27

and y = – 3 is given by

$$q_2 = \sqrt{\{(Q/A_1)^2 - 2gy\}}\ = \sqrt{(20)^2 + 6g\}} = 24.24 \text{ ft./sec.} \qquad \textbf{Ans.}$$

Example 31:

Fluid is coming out from a small hole of cross-section σ_1 in a tank. If the minimum cross-section of the stream coming out of the hole is σ_2, then show that $\sigma_2/\sigma_1 = \frac{1}{2}$.

Solution:

Let PQ be the hole and P′ Q′ be its image on the opposite wall of the tank. Let p_1 be the pressure at PQ when the hole is closed. Let p_2 be the pressure and q_2 be the velocity at the minimum cross-section. The velocity of the fluid coming out from minimum cross-section is at right angles to the hole and the direction of velocity will be horizontal there. Equation of motion is

$$\sigma_1 (p_1 - p_2) = \sigma_2 \rho q_2^2$$

$$\Rightarrow (p_1 - p_2) = (\sigma_2/\sigma_1)\, \rho q_2^2 \qquad ...(1)$$

Bernoulli's equation for the stream line connecting a point of P′Q′ and a point of minimum cross-section of the jet, becomes

$$\frac{p_1}{\rho} = \frac{p_2}{\rho} + \frac{1}{2} q_2^2 \Rightarrow p_1 - p_2 = \frac{1}{2} \rho q_2^2 \qquad ...(2)$$

From (1) and (2), We have

$$\frac{\sigma_2}{\sigma_1} = \frac{1}{2}.$$

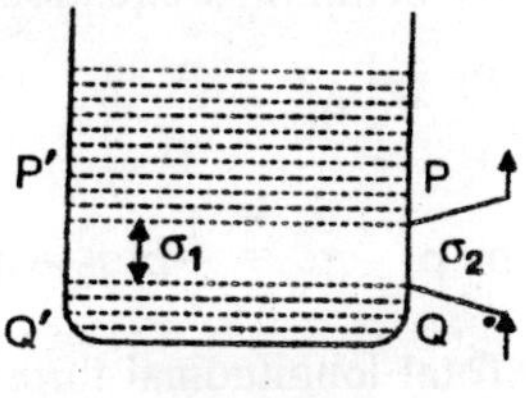

Fig. 4.22

Example 32:

If a jet issuing from an orifice under a constant head D travels a horizontal distance x from the plane of the vena contracta while it drops through a height h, obtain the coefficient of velocity.

Solution:

If a particle falls a vertical distance h under the influence of gravity, then

$$h = ut + \frac{1}{2} gt^2, \qquad ...(1)$$

where u is the initial vertical velocity and t is time taken by the particle to reach a distance h. Here initial velocity is zero, so

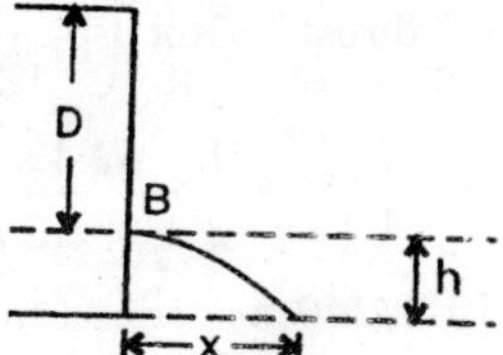

Fig. 4.23

$t = \sqrt{(2h/g)}$.

Horizontal distance $x = V \times t$, where V is the horizontal velocity at vena contracta.

$$V = \frac{x}{t} = \frac{x/\sqrt{(2h/g)}}{\sqrt{2gD}} = \frac{x}{\sqrt{4hD}}. \qquad \textbf{Ans.}$$

Example 33:

A horizontal straight pipe gradually reduces in diameter from 24 in. to 12 in. Determine the total longitudinal thrust exerted on the pipe if the pressure at the larger end is 50 Ibf/in² and the velocity of the water is 8ft/sec.

Solution:

Let A_1 and A_2 be the cross-section of the larger and the smaller end. Let q_1 and q_2 be the velocity and p_1 and p_2 be the pressure at the larger and the smaller end of the pipe. From the equation of continuity, we have

$$A_1q_1 = A_2q_2$$

or $p\ (12)^2q_1 = p\ (6)^2q_2 \Rightarrow 4q_1 = q_2$. ...(1)

By bernoulli's equation, we have

$$\frac{p_1}{\rho} + \frac{1}{2}q_1^2 = \frac{p_2}{\rho} + \frac{1}{2}q_2^2$$

$$\text{or } p_1 - p_2 = \frac{1}{2}\rho(q_2^2 - q_1^2) = \frac{1}{2}\rho \times 15 \times (96)^2. \qquad ...(2)$$

Total longitudinal thrust exerted on the pipe

$= p_1A_1 - p_2A_2$

$= p\ (12)^2\ p_1 - p\ (6)^2p_2$ ...(3)

From (2), we have

$$p_2 = p_1 - \frac{1}{2}\rho \times 15 \times (96)^2 \qquad ...(4)$$

From (3) and (4), we have

$$\text{Total thrust} = 36\pi\left(p_1 + \frac{1}{2}\rho \times 15 \times 96 \times 96\right)$$

$$= 36\pi\left(150 + \frac{1}{2} \times \frac{62.4 \times 15 \times 96 \times 96}{12 \times 12 \times 12}\right)$$

$$= 36 \times 240\ \pi \qquad \textbf{Ans.}$$

Example 34:

Liquid is discharged at the rate of 3.86 ft3/sec from a siphon in the reservoir. The siphon has a diameter of 6 in. Find the elevation z and the fluid pressure at the top of the siphon.

Solution:

Bernoulli's equation for the three points on the same streamline can be written as

$$\frac{q_0^2}{2g}+\frac{p_0}{\rho g}+z_0=\frac{q_1^2}{2g}+\frac{p_1}{\rho g}+z_1=\frac{q_2^2}{2g}+\frac{p_2}{\rho g}+z_2$$

Here $q_0 = 0$, $p_1 = p_2$, $z = z_0 - z_2$ (let)

$$q_2 = \frac{3.86}{\pi\left(\frac{1}{4}\right)^2} = \frac{3.86\times16\times7}{22} = 19.6 \text{ ft./sec.}$$

and $q_2^2 = 2gz$

$$\Rightarrow \quad z = \frac{19.62\times19.62}{2\times32} = 6 \text{ ft. (app).}$$

Since the velocity at the top is the same as that at the bottom. Bernoulli's equation written between these two level gives

$p_1/\rho g = -8$ ft. of liqid

i.e., below the atmospheric pressure. **Ans.**

Example 35:

Water flows through a pipe of length l which tapers from theentrance radius r_1 to the exit radius r_2. If the entrance velocity is v_1, and the relation between r_1 and r_2 is given by $r_2 = r_1 \pm m_l$, where m is the slope, prove that the exit velocity v_2 is

$$v_2 = v_1\left[1-\frac{\pm2m(l/r_1)+m^2(l/r_1)^2}{1\pm2m(l/r_1)+m^2(l/r_1)^2}\right].$$

Solution:

Since r_1 and r_2 be the radius of the pipe of length l_1 at the entrance and the exit. Let A_1 and A_2 be the cross-section of the pipe at the entrance and the exit. From continuity equation, we have

$$A_1v_1 = A_2v_2 \Rightarrow (\pi r_1^2)\ v_2$$

$$\text{or } v_2 = \frac{r_1^2}{r_2^2} v_1 = \frac{r_1^2}{(r_1 \pm ml)^2} v_1$$

$$\text{or } v_2 = v_1 \left[\frac{1}{\{1 \pm (ml/r_1)\}^2} \right].$$

$$\text{or } v_2 = v_1 \left[\frac{1}{1 \pm 2m(l/r_1) + m_2(l^2/r_1^2)} \right]$$

$$\text{or } v_2 = v_1 \left[1 - \frac{\pm 2m(l/r_1) + m^2(l^2/r_1^2)}{1 + 2m(l/r_1) + m^2(l^2/r_1^2)} \right].$$ **Proved.**

Example 36:

A conical pipe has diameters of 10 cm. and 15 cm. at the two ends. If the velocity at the smaller end is 2m/sec, what is the velocity at the other end and the discharge through the pipe?

Solution:

Let q_1 and q_2 be the velocity at the smaller and larger ends. From continuity equation, we have

$q_1A_1 = q_2A_2$.

Here q_2 = 2 m/sec, $A_1 = (\pi/4)\ (0.1)^2$, $A_2 = (p/4)\ (0.15)^2$,

$$q_2 = q_1 \frac{A_1}{A_2} = 2 \frac{(0.1)^2}{(0.15)^2} = 0.89 \text{ m/sec.}$$ **Ans.**

Discharge through the pipe

$Q = q_1A_1 = 2(p/4)\ (0.1)^2 = 0.0157\ m^2/sec.$ **Ans.**

Example 37:

A horizontal conical pipe has diameters 25 cm. and 40 cm. at the two ends. (a) Calculate the pressure at the larger end if the pressure at the smaller end is 5m. of water and rate of flow is 0.3 m²/sec. (b) Calculate the discharge through the pipe if the manometer connected between the two ends reads 10 cm. of mercury.

Solution:

Let q_1, q_2 be the velocities and p_1, p_2 be the pressure at the larger and smaller ends of a conical pipe. Let Q be the discharge through the pipe then

$$Q = A_1 q_1$$

$$q_1 = \frac{Q}{A_1} = \frac{0.3}{(\pi/4)(0.4)^2} \quad ...(1)$$

From the continuity equation, we have

$$A1q_1 = A_2 q_2$$

$$\text{or } q_2 = \frac{A_1}{A_2} q_1 = z\frac{(0.4)^2}{(0.25)^2} \times 2.38 = 6.10\text{m/sec.} \quad ...(2)$$

(a) Using Bernoulli's equation, we have

$$\frac{p_1}{\rho} + \frac{q_1^2}{2g} = \frac{p_2}{\rho} + \frac{q_2^2}{2g}. \quad (\text{Here } z_1 = z_2)$$

$$\text{or } 5 + \frac{(2.38)^2}{2 \times 9.81} = \frac{p_2}{\rho} + \frac{(6.10)^2}{2 \times 9.81}$$

$$\text{or } \frac{p_2}{\rho} = 5 + \frac{(2.38)^2 - (6.10)^2}{2 \times 9.81} = 3.4 \text{ m} = 0.34 \text{ kg/cm}^2$$

Pressure at the larger end = 0.34 kg/cm².

(b) From manometer, we have

$$\frac{p_1}{\rho} - \frac{p_1}{\rho} = 10\ (13.6 - 1) = 126 \text{ cm.} = 1.26 \text{ m.}$$

From manometer, we have

$$A_1 q_1 = A_2 q_2.$$

$$\Rightarrow q2 = \frac{A_1}{A_2} q_1 = \frac{(0.4)^2}{(0.25)^2} q_1 = 2.56 q_1.$$

Using Bernoulli's equation, we have

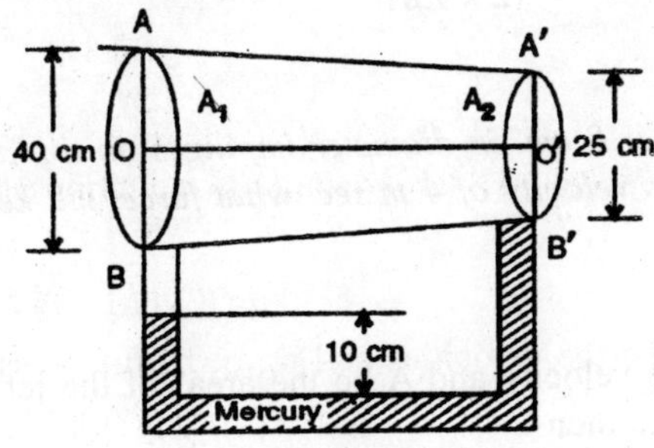

Fig. 4.24

$$\frac{p_1}{\rho.} + \frac{q_1^2}{2g} = \frac{p_2}{\rho} + \frac{q_2^2}{2g}$$

or $\frac{q_1^2}{2g}\left(\frac{q_2^2}{q_1^2} - 1\right) = \frac{p_1 - p_2}{\rho} \Rightarrow \frac{q_1^2}{2g}[(2.56)^2 - 1] = 1.26$

or $q_1 = \sqrt{(1.26 \times 2 \times 9.81/5.55)} = 2.11$ m/sec.

Hence discharge through the pipe is

$Q = A_1 q_1$

$Q = \frac{\pi}{4} \times (0.4)^2 \times 2.11 = 2.56$ m³/sec. **Ans.**

Example 38:

A pipe of 10 cm. diameter is suddenly enlarged to 20 cm. diameter. Find the loss of head when 50 litres/sec. of water is flowing.

Solution:

Let q_1 and q_2 be the velocities at the smaller and larger ends of the pipe then

or $Q = A_1 q_1 = A_2 q_2$

or $q_1 = Q/A_1$, $q_2 = Q/A_2$

or $q_1 = \frac{0.05}{(\pi/4)(0.1)^2}$, $q_2 = \frac{0.05}{(\pi/4)(0.2)^2}$

$q_1 = 6.36$ m/sec, $q_2 = 1.59$ m/sec.

Loss of head due to sudden enlargement

$= \frac{(q_1 - q_2)^2}{2g} = \frac{(6.36 - 1.59)^2}{2 \times 9.81} = 1.6$ m. **Ans.**

Example 39:

A jet of water 8 cm. in diameter impinges on a plate held normal to its axis. For a velocity of 4 m/sec, what force will keep the plate in equilibrium?

Solution:

Let q be the velocity and A be the area of the jet. Let F_x be the force on the plate then

Force on the plate

= change in momentum

$$F_x = \frac{WQV}{g} = \frac{WAV^2}{g}$$

$$= \frac{100 \times \frac{\pi}{4}(0.08)^2 \times 4^2}{9.81} = 32.8\text{kg}.$$

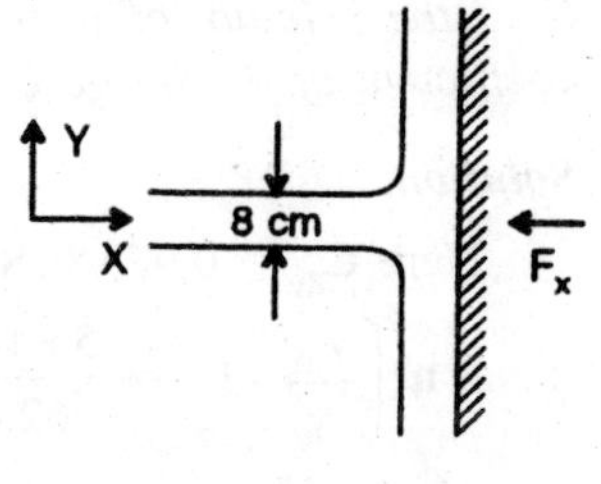

Fig. 4.25

Example 40:

A venturi flume 1 m, wide at entrance and 1 m. wide at the throat-section has its base horizontal. If the depth of water at the entrance and the throat is 1 m. and 0.9 , respectively. Calculate the discharge.

Solution:

The discharge is given by

$$Q = \frac{S_1 S_2}{\sqrt{(S_1^2 - S_2^2)}}\sqrt{(2gh)}$$

Here h = 1.0 = 0.9 = 0.1 m.

$S_1 = 2 \times 1 = 2\text{m}^2$, $S_2 = 1 \times 0.9 = 0.9\text{ m}^2$.

$$\text{or } Q = \frac{2 \times 09}{\sqrt{(2^3 - 0.9^2)}}\sqrt{(2 \times 9.81 \times 0.1)} = 1.41\text{ m}^3/\text{s}.$$ **Ans.**

Example 41:

A venturi meter of throat diameter 4 cm. filled into a pipe line of 10 cm. diameter. The coefficient of discharge is 0.96. Calculate the flow through the meter when the reading on a mercury-water mano-meter connected across the upstream and throat tapping is 25 cm.

Solution : We know that

$$Q = 0.96 \times \frac{\pi}{4} \times (0.10)^2 \sqrt{\left\{\frac{19.62 \times 0.25 \times 12.6}{(10/4)^4 - 1}\right\}}$$

$Q = 9.61 \times 10^{-3}\text{m}^3/\text{s}$. **Ans.**

Example 42:

The venturi meter has an entrance of 6 in. diameter and a throat of 3 in. diameter whose centre is 18 in. above the centre of the entrance.

Find the velocity of water at the throat when $p_1 - p_2 = 5 lbf/in^2$. The coefficient of discharge C_{dv} is 0.97.

Solution:

Here $C_{dv} = 0.97$, $S_2/S_1 = (3/6)^2 = 1/4$,.

$$\delta h\left(\frac{\gamma_m}{\gamma}-1\right) = \frac{5\times 144}{62.4}-\frac{18}{12}=10.04\text{ft}.$$

$$\text{Therefore } q_2 = \frac{0.97}{\sqrt{\left\{1-\left(\frac{1}{4}\right)^2\right\}}}\sqrt{\{2g(10.0.3)\}} = 25.4 \text{ ft/sec.}$$ **Ans.**

Example 43:

A venturi meter measures the flow of oil (sp. gr. 0.85) in a 8cm. diameter pipe. The difference of head between inlet and throat of the meter is measured by a U-tube mercury manometer. What should be the diameter of throat of the meter in order that the difference in level of the mercury in U tube shallbe 25 cm. When the quantity of oil flowing in the pipe is 10 litres/sec. Assume the coefficient of venturi meter as 0.97.

Solution:

Discharge through a venturimeter is given by

$$Q = C_d \frac{S_1}{\sqrt{(r^2-1)}}\sqrt{(2gH)},\ r = s_1/s_2.$$

Here $H = 3.75$ m, $S_1 = p/4\ (0.08)^2\ m^2$

$Q = 10$ litres/sec $= 0.01\ m^3/sec.$, $C_d = 0.97$

$$\text{or } r^2 = \frac{2gHC_d^2S_1^2}{Q^2}+1 = 18.6.$$

Let D_1 be the diameter of the pipe and D_2 be the diameter of the throat, then

$$\frac{S_1^2}{S_2^2} = \frac{D_1^4}{D_2^4} = 18.6$$

or $D_1/D_2 = 2.08$

$\Rightarrow D_2 = 8/2.08 = 3.85$ cm. **Ans.**

Example 44:

Water flows over a re ctangular wire 1m. wide at a depth of 15 cm. and after wards passes thought a triangular right angled weir. Find the depth of water through the triangular weir. The C_d for the rectangular and triangular weirs are 0.62 and 0.59 respectively.

Solution:

The same discharge occurs over the rectangular weir as that over the triangular weir

$$Q = Q_1 = \frac{2}{3} C_{d_1} \sqrt{(2g)LH_1^3 / 2}$$

$$Q = Q_2 = \frac{2}{15} C_{d_2} \sqrt{(2g)H_2^5 /^2 \tan\theta}.$$

Here L = 1m, H_1 = 0.15 m., $\theta = \pi/2$,

C_{d1} = 0.62 and C_{d2} =0.59.

$$H_2^{5/2} = \frac{2}{3} \times \frac{15}{8} \times \frac{0.62}{0.59} \times 1 \times (0.15)^{3/2} = 0.0763$$

or H_2 = 0.3573 m = 35.73 cm. **Ans.**

Example 45:

A reservoir having a surface area of 500 m^2 emptied by a 0.5 m long rectangular weir. How long should it take to empty the reservoir from 0.2 m. height above the sill? Cd = 0.65

Solution:

For`a rectangular weir, we have

$$Q = \frac{2}{3} \sqrt{(2g)H^{3/2} C_d} . \qquad ...(1)$$

Also A dH = – Q dt

(negative sign is taken because h decreases as t increases)

$$dt = - \frac{AH^{-3/2} dH}{\frac{2}{3}\sqrt{(2g)}C_d}$$

$$\text{or} \int_0^t dt = - \frac{3A}{2\sqrt{(2g)}C_d} \int_{H_1}^{H_2} H^{-3/2} dH$$

$$\text{or } t = \frac{3A}{\sqrt{(2g)}C_d}\left(\frac{1}{\sqrt{H_2}}-\frac{1}{\sqrt{H_1}}\right)$$

$$\text{or } t = \frac{3 \times 500}{4.43 \times 0.65}\left(\frac{1}{\sqrt{(0.1)}}-\frac{1}{(0.2)}\right) \Rightarrow t = 965\text{s.}$$ **Ans.**

EXERCISES

1. Determine the horizontal distance required for jet striking the ground which is 2ft above the horizontal line of the nozzle. The jet is inclined at an angle p/6 with the horizontal at a velocity of 25 ft/sec.

   Hint : The horizontal and verticalcomponents of the jet velocity at the nozzle exist are respectively

   $$q_{1x} = q_1 \cos a = 25 \cos \frac{\pi}{6} = 21.65\text{ft/sec.}$$

   $$q_{1y} = q_1 \sin a = 25 \sin \frac{\pi}{6} = 12.5 \text{ ft/sec.}$$

   We know that

   $$y = \frac{q_{1y}}{q_{1x}}x - \frac{1}{2}g\frac{x^2}{q_{1x}^2}$$

   $$\text{or } 2 = \frac{12.5}{21.65}x - \frac{1}{2}g\frac{x^2}{(21.65)^2}.$$

   Find the value of x, which will be the horizontal distance distance required for the jet striking the ground.

2. A jet of water 2 in. in diameter discharge 1.0 ft³/sec. Determine the force required to move flat plate toward jet with a velocity of 25 ft/sec. The jet is perpendicular to the plate.

3. A jet of water having a diameter of 1 in. and a velocity of 20 ft/sec is deflected through an angle of p/3 by a fixed, curved vane. Determine the force components exerted.

4. An orifice plate consists of a circular orifice in a 9 in. diameter pipe and has a coefficient of discharge of 0.65. If the mercury manometer which connects the two sides of the orifice plate has a deflection of 20 in. Calculate the diameter of the orifice hole

when the rate of flow of water in the pipe line is 2.25 ft^3/sec.

5. A horizontal straight pipe gradually reduces in diameter from 30 in to 18 in. Determine the total longitudinal thrust exerted on the pipe if the pressure at the larger end is 75 lbf/in2 and the velocity of the water is 12 ft/sec.

6. A venturi meter has its axis vertical, the inlet and throat diameter ratio being 2.5. The throat is 12 in. above the inlet and the coefficient of discharge is 0.97. Determine the pressure difference between the inlet and throat when the velocity of water at the inlet is 6 ft/sec.

Hint : $\left(\frac{A_1}{A_2}\right)^2 = (2.5)^2$;

$$\frac{6}{0.97} = \sqrt{\left(\frac{(2/\rho)(p_1 - p_2) + 32 \times 1}{1 \times (16.25)}\right)}.$$

Calculate the pressure difference.

7. A venturi meter with an inlet diameter of 10 in. is connected to an alcohol manometer which measures the pressure difference between inlet and throat. If the rate of flow passing through the meter is 100 ft 3/sec of gas (g = 0.062 lbs/ft^3). Calculate the diameter of the throat when the manometer deflection is 3 in. Alcohol has a specific gravity of 0.8.

5

Fluid Pressure and Its Measurement

FLUID PRESSURE AT A POINT

'Pressure' or 'intensity of pressure' may be defined as the force exerted on a unit area. If F represents the total force uniformely distributed over an area A the pressure at any point is p = (F/A). However, if the force is not uniformly distributed, the expression will give the average value only. When the pressured varies from point on an area, the magnitude of pressure at any point can be obtained by the following expression

$$p = \frac{dF}{dA}$$

where dF represents the force acting on an infinitesimal area dA. In SI units pressure is expressed in N/m^2 (or pascal), and in metric gravitational units it is expressed in kg (f) cm^2 or kg (f) m^2.

A fluid is a substance which is capable of flowing. A such when a certain mass of fluid is held in static equilibrium by confining within solid boundaries, it exerts forces against boundary surfaces. The forces so exerted always act in the direction normal to the surface in contact. This is so because a fluid at rest cannot sustain shear stress and hence the forces cannot have tangential components. The normal force exerted by a fluid per area of the surface is called the fluid pressure. It may, however, be noted that even if an imaginary surface is assumed within a fluid body, the fluid pressure and pressure force on the imaginary surface are exactly the same as those acting on any real surface. This is in accordance with Newton's third law of motion, viz., action and reaction exist in pairs.

VARIATION OF PRESSURE IN A FLUID

Consider a small fluid element of size δx × δy × δz at any point in a static mass of fluid as shown in Fig. 5.1. Since the fluid is at rest, the element is in equilibrium under the various forces acting on it. The forces acting on the element are the pressure forces on its faces and the self-weight of the element. Let p be the pressure intensity at the mid point O of the element. Then the pressure intensity on the left hand face of the element is $\left[p-\left(\frac{\partial p}{\partial x}\right)\frac{\delta x}{2}\right]$ and the pressure intensity on the right hand face of the element is $\left[p+\left(\frac{\partial p}{\partial x}\right)\frac{\delta x}{2}\right]$. The corresponding pressure force on the left hand and the right hand faces of the element are

$$\left[p-\left(\frac{\partial p}{\partial x}\right)\frac{\delta x}{2}\right].\ dy\ dz$$

and $$\left[p+\left(\frac{\partial p}{\partial x}\right)\frac{\delta x}{2}\right]\ dy\ dz$$

respectively. Likewise the pressure intensities and the corresponding pressure forces on the other faces of the element may be obtained as shown in Fig. 5.1. Further, if w is the specific weight of the fluid then the weight of the element acting vertically downwards is (wδx δy δz). Since the element is in equilibrium under these froces, the algebraic sum of the forces acting on it in any direction must be zero.

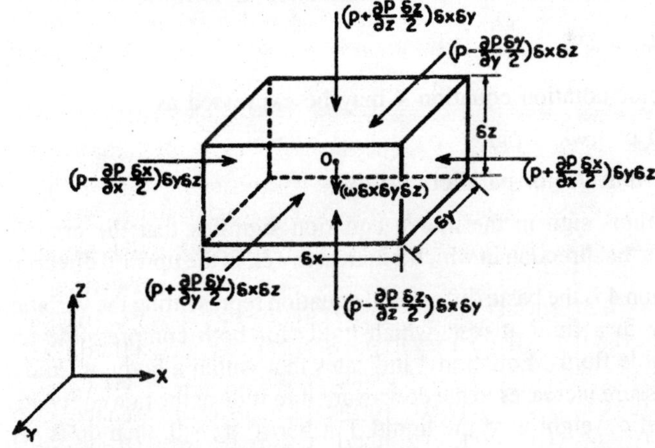

Fig. 5.1 : Fluid element with forces acting on it in a static mass of fluid.

Thus considering the forces acting on the element along x, y and z axes the following equations are obtained,

$$SF_x = 0$$

$$\text{or} \left(p - \frac{\partial p}{\partial x}\frac{\delta x}{2}\right) = dy\ dz - \left(p + \frac{\partial p}{\partial x}\frac{\delta x}{2}\right) dy\ dz = 0$$

$$\text{or} \quad \frac{\partial p}{\partial x} = 0 \qquad \text{...(1)}$$

$$SF_y = 0$$

$$\text{or} \left(p - \frac{\partial p}{\partial y}\frac{\partial y}{2}\right) dx\ dz - \left(p + \frac{\partial p}{\partial y}\frac{\delta x}{2}\right) dx\ dz = 0$$

$$\text{or} \quad \frac{\partial p}{\partial y} = 0 \qquad \text{...(2)}$$

$$\Sigma F_z = 0$$

$$\text{or} \quad \left(p - \frac{\partial p}{\partial x}\frac{\delta z}{2}\right) \delta x\ \delta z - \left(p + \frac{\partial p}{\partial z}\frac{\delta z}{2}\right) \delta x \delta y - (w \delta x\ \delta y\ \delta z) = 0$$

$$\text{or} \quad \frac{\partial p}{\partial z} = -w \qquad \text{...(3)}$$

Equations 1, 2 and 3 indicate that the pressure intensity p at any point in a static mass of fluid does not very in x and y directions and it varies only in z direction. Hence the partial derivative in equation 3 may be reduced to total (or exact) derivative as follows.

$$\frac{dp}{dz} = -w \qquad \text{...(4)}$$

In vector notation equation 4 may be expressed as

$$-\text{grad } p = wk = \rho gk$$

where k is unit vector parallel to z axis.

The minus sign in the above equation signifies that the pressure decreases in the direction in which z increases *i.e.*, in the upward direction.

Equation 4 is the basic differential equation representing the variation of pressure in a fluid at rest, which holds for both compressible and incompresible fluids. Equation 4 indicates that within a body of fluid at rest the pressure increases in the downward direction at the rate equivalent to the specific weight w of the liquid. Further if dz = 0, then dp is also equal to zero; which means that the pressure remains constant over any horizontal plane in a fluid.

Integration of equation 4 yields the pressure at any point in a fluid at reast, which is discussed separately for the incompressible fluids such as liquids having constant density and comprssible fluids such as gases having variable density.

Pressure at a Point in a Liquid

A liquid may be considered as incompressible fluid for which w is constant and hence integration of equation 4 gives

$$p = -wz + C \qquad ...(5)$$

in which p is the pressure at any point at an elevation z in the static mass of liquid and C is the constant of integration. Liquids have a free surface at which the pressure of atmosphere acts. Thus as shown in Fig. 5.2 for a point lying in the free surface of the liquid $z = (H + z_0)$ and if p_a is the atmospheric pressure at the liquid surface then from equation 5 the constant of integration $C = [pa + w(H + z_0)]$ Substituting this value of C in equation 5, it becomes

$$P = wz + [pa + w(H + z_0)] \qquad ...(6)$$

Now if a point is lying in the liquid mass at a vertical depth h below the free surface of the liquid then as shown in Fig. 5.2 for this point $z = (H + z_0 - h)$ and from equation 6

$$p = pa + wh \qquad ...(7)$$

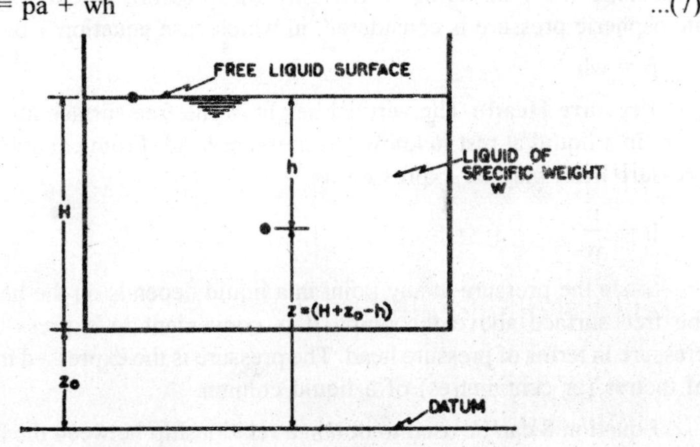

Fig. 5.2 : Pressure at a point in a liquid.

It is evident from equation 7 that the pressure at any point in a static mass of liquid depends only upon the vertical depth of the point below the free surface and the specific weight of the liquid, and it does not depend upon the shape andsize of the bounding container. This fact is illustrated in Fig. 5.3 in which although the containers of different shapes are interconnected, so that the total weight of the liquid in each part differs, yet the pressures at points A, B, C and D lying on the same horizontal level and at the same vertical depth h below the free surface of the liquid will be same.

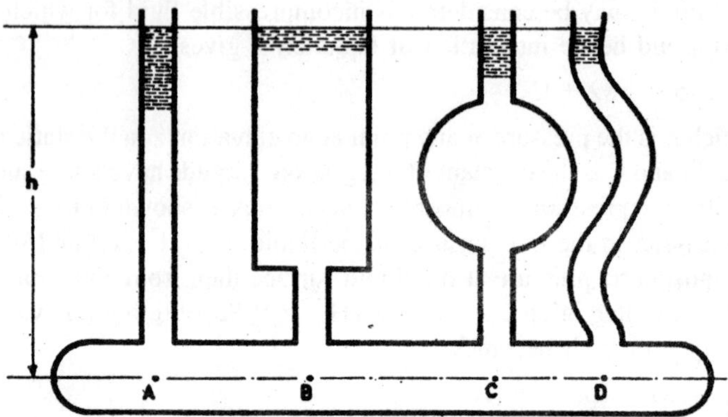

Fig. 5.3 : Interoconnected containers of different shapes.

Since the atmospheric pressure at a place is constant, at any point in a static mass of liquid, often only the pressure in excess of the atmospheric pressure is considered, in which case equation 7 becomes.

$$p = wh \quad \text{...(8)}$$

Pressure Head : The vertical height of the free surface above any point in a liquid at rest in known as pressure head. From equation 8 the pressure head may be expressed as

$$h = \frac{\rho}{w} \quad \text{...(9)}$$

Since the pressure at any point in a liquid depends on the height of the free surface above the point, it is convenient to express a fluid pressure in terms of pressure head. The pressure is the expressed in terms of metres (or centimetres) of a liquid column.

Equation 8 can be used to obtain a relationship between the heights of columns of different liquids which would be develop the same pressure at any point. Thus, if h_1 and h_2 are the heights of the columns of liquids of specific weights w_1 and w_2 required to develop the same pressures p, at any point; then from equation 8.

$$p = w_1 h_1 = w_2 h_2 \quad \text{...(10)}$$

If S_1 and S_2 are the specific gravities of the two liquids and w is the specific weight of water then since $w_1 = S_1$ w and $w_2 = S_2w$, equation 10 may also be written as

$$S_1 h_1 = S_2 h_2 \qquad ...(10a)$$

Further if h_1 and h_2 are the depths of two points below the free surface in static mass of liquid of specific weight w and p_1 and p_2 are the respective pressure intensities at these points, then from equation 2.8 the pressure difference between these points is obtained as

$$(p_1 - p_2) = w\,(h_1 - h_1)$$

Thus it may be stated that the difference in pressure at any two points ina static mass of liquid varies directly as the difference in depth (or elevation) of the two points.

Pressure at any Point in a Compressible Fluid : For a compressible fluid since the density varies with the pressure, equation 4 can be integrated only if the relation between w and p is known. Moreovr, in a compressible fluid there being no free surface, the integration of equation 4 gives the variation of pressure between any two points lying in a static mass of fluid. Thus if p_1 and p_2 are the pressure intensities at two points which are at elevations z_1 and z_2 above an arbitrarily assumed datum, then integration of equation 4 yields

$$\int_{p_1}^{p_2} \frac{dp}{w} = \int_{z_1}^{z_2} dz = (z_1 - z_2)$$

The left hand side of the above expression has been evaluated using different relations between w and p as indicated in the next section.

EQUILIBRIUM OF A COMPRESSIBLE FLUID-ATMOSPHERIC EQUILIBRIUM

Equation 4 expresses the conditions for equilibrium of any fluid, which may be written as

$$dp = -\,wdz = -\,\rho gdz \qquad ...(4)$$

This is a general relationship which can be applied to both incompressible as well as compressible fluids. However, for a compressible fluid the above equation can be integrated, to obtain the value of pressure at any point in the fluid, provided it can be assumed that the mass density ρ is either constant or it is a function of pressure (or elevation) only. The mass ρ is either constant or it is a function of pressure (or elevation) only.

The mass density ρ may be assumed to be constant if the variation in elevation is not great. But if the variation in elevation is great , ρ can no longer be considered constant, in which case the manner in which ρ varies with pressure p (or elevation) must be known. The variation of pressure and mass density of the air in the atmosphere with elevation is one such problem which is encountered in aeronautics and meteorology. Similarly in oceanography too the problem of the variation of pressure and mass density with depth is of concern, since at great depth there is a small increase in the density of the fluid.

In order to obtain the variation of pressure with elevation in the atmosphere, some of the relations between p and ρ which may be used in equation 4 are as indicated below.

(a) **Isothermal State for Atmosphere or Isothermal Atmosphere** : If it is assumed that the variation of p and ρ is according to isothermal condition, *i.e.*, the temperature of atmosphere is assumed to have a constant value (which may however be true only over a relatively small vertical distance), then according to Boyle's law

$$\frac{p}{\rho} = \frac{p_0}{\rho_0} \qquad ...(11)$$

where p_0 and ρ_0 are the values of the pressure and density of the gas at initialcondition at some reference level, for example at the earth's surface *i.e.* at z = 0.

From equation 4

$$\frac{dp}{\rho} = -\,gdz$$

By substituting for ρ from equation 11

$$\frac{dp}{p} = \frac{p_0}{\rho_0} = -\,gdz$$

Since the value of g decreases by only about 0.1% for about 300 m increase in altitude it may be assumed constant. Thus integration of above equation gives

$$-\,gz = \frac{p_0}{\rho_0}\log_e p + C$$

If p_1 is the pressure at height z_1, then

$$-\,gz_1 = \frac{p_0}{\rho_0}\log_e p_1 + C$$

Thus eliminating C from these equations

$$(z - z_1) = \frac{p_0}{\rho_0 g}\log_e\left(\frac{p_1}{p}\right) \quad ...(12)$$

Equation 12 expresses the relation between altitude and pressure when the air is isothermal. The ratio $(p_0/\rho_0 g)$ represents the height of a fluid column of constant specific weight ($\rho o g$). It is called the *equivalent height of a uniform atmosphere.*

Further at z = 0 since $p = p_0$ the integration constant

$$C = \frac{p_0}{\rho_0}\log_e p_0$$

$$\text{Hence } z = \frac{p_0}{\rho_0 g}\log_e\left(\frac{p}{p_0}\right) \quad ...(13)$$

For a perfect gas from equation of state

$$p_0 = r_0 RT_0 \text{ ; or } \frac{p_0}{\rho_0} = RT_0$$

Thus substituting the values of (p_0/ρ_0) in equation (13)

$$z = \frac{RT_0}{g}\log_e\left(\frac{p}{p_0}\right) \quad ...(14)$$

From equation 14 it may be seen that when (gz/RT_0) is small this expression approximates to that for a fluid of constant density.

(b) **Adiabatic State for Atmosphere or Adiabatic (or Isentropic) Atmosphere:** It is well known fact that the temperature in the atmosphere decreases rapidly as one goes to higher altitude. Thus if it is assumed that no heat is added or taken away from a certain column of air, then this column is said to be under adiabatic conditions.

For an adiabatic process

$$\frac{p}{\rho^k} = \frac{p_0}{\rho_0^k} \text{; or } \rho = p_0\left(\frac{p}{p_0}\right)^{1/k} \quad ...(2.15)$$

in which k is the adiabatic exponent (or adiabatic index) defined as the ratio of the specific heat at constant pressure C_p to the specific heat at constant volume C_v

Substituting this value in equation 4 one obtains

$$-\,gdz = \frac{dp}{\rho_0}\left(\frac{p_0}{p}\right)^{1/k}$$

$$\text{or } -dz = \frac{1}{\rho_0 g}\left(\frac{p_0}{p}\right)^{1/k} dp$$

$$\text{or } -dz = \frac{p_0}{\rho_0 g}\left(\frac{p_0}{p}\right)^{1/k} d\left(\frac{p}{p_0}\right)$$

As stated earlier g may be assumed to be constant and hence by integrating the above expression one obtains

$$-z = \frac{p_0}{\rho_0 g}\left(\frac{p}{p_0}\right)^{1-\frac{1}{k}}\left(\frac{1}{1-\frac{1}{k}}\right) + C$$

If $p = p_0$ when $z = z_0$ then integration constant

$$C = z_0 - \left(\frac{k}{k-1}\right)\frac{p_0}{\rho_0 g}$$

$$\text{Hence } (z - z_0) = \left(\frac{k}{k-1}\right)\frac{p_0}{\rho_0 g}\left[1-\left(\frac{p}{p_0}\right)^{\frac{k-1}{k}}\right] \quad \text{...(16)}$$

from which

$$\frac{p}{p_0} = \left[1-\Delta z\frac{\rho_0 g}{p_0}\left(\frac{k-1}{k}\right)\right]^{\frac{k}{k-1}} \quad \text{...(17)}$$

where $(z - z_0) = \Delta z$

Substituting the value of (p/p_0) in equation (15)

$$\frac{\rho}{\rho_0} = \left[1-\Delta z\frac{\rho_0 g}{p_0}\left(\frac{k-1}{k}\right)\right]^{\frac{k}{k-1}} \quad \text{...(18)}$$

Again for a perfect gas from equation of state

$$p = \rho_0 RT_0; \text{ or } \frac{p_0}{\rho_0} = RT_0$$

Substituting the value of (p_0/ρ_0) in equation (17)

$$\frac{p}{p_0} = \left[1-\Delta z\frac{g.}{RT_0}\left(\frac{k-1}{k}\right)\right]\frac{k}{k-1} \quad ...(19)$$

If z is small, all but the first two terms of the binomial expression of the right hand side of equation 17 may be neglected and then

$$\frac{p}{p_0} = 1-\frac{\rho_0 g\Delta z}{p_0}$$

or $p = p_0 - \rho_0\, g\, \Delta z$

This corresponds to the relationship p + wz = constant (equation 2.5) for fluid of constant density. Thus for small difference of height say less than about 300 m in the atmosphere, the fluid may be considered to tbe of constant density without an appreciable error being introuduced.

Further dividing equation 17 by equation 18 one obtains

$$\frac{p}{\rho} = \frac{p_0}{\rho_0}\left[1-\Delta z\frac{\rho_0 g}{p_0}\left(\frac{k-1}{k}\right)\right]$$

$$\text{or } RT = \frac{p_0}{\rho_0}\left[1-\Delta z\frac{\rho_0 g}{p_0}\left(\frac{k-1}{k}\right)\right] \quad ...(20)$$

since $\frac{p}{\rho}$ = RT from the equation of state.

Equation 20 gives the relation between the absolute temperature T and altitude z. Now if $T = T_0$ at $z = z_0$ (or $\Delta z = 0$) then from equation 20

$$RT_0 = \frac{p_0}{\rho_0}$$

Further if $(T_0 - T) = \Delta T$ for a difference in elevation Δz then

$$R\,(\Delta T) = g\,(\Delta z)\left(\frac{k-1}{k}\right) \quad ...(21)$$

For air, R = 29.27 m-kg (f)/kg (m)° C abs and k = 1.405, then for Δz = 100 m, the variation in temperature ΔT may be obtained from equation 2.21 as

$$R\,(\Delta T) = \frac{9.81\times 100\times 0.405}{1.405}$$

R = 29.27 m-kg (f)/kg (m)°C abs

$= 29.27 \times 9.81\ m^2/sec^{2}$°C abs

$$\therefore\ \Delta T = \frac{9.81 \times 100 \times 0.405}{29.27 \times 9.81 \times 1.405} = 0.985°C.$$

This shows that under adiabatic condition, the absolute temperature decreases by about 0.985°C for each 100 m increase in-elevation. However, if somewhat less than that computed above.

(c) Polytropic State for Atmosphere of polytropic Atmosphere : In a more generalized way the variation of atmospheric pressure p with r may be considered according to a polytropic process, in which case, we have

$$\frac{p}{\rho^n} = \frac{p_0}{\rho_0^n} = \text{or } \rho = p_0 \left(\frac{p}{p_0}\right)^{1/n}$$

where n is a positive constant

Again substituting this value in equation 4 and integrating as indicated earlier the following expression may be obtained

$$\frac{P}{P_0} = \left[1 - \frac{g(z - z_0)}{RT_0}\frac{n-1}{n}\right]^{\frac{n}{n-1}} \quad ...(22)$$

Evidently equation 22 is more general being applicable for any value of n, except for the particular case of n = 1.0; and from this, equation 19 can be obtained by considering n = k. Equation 22 represents the variation of pressure with elevation, which has been plotted in Fig. 5.4 for different values of n. The actual variation of pressure with elevation in the atmosphere is also plotted in Fig. 5.4 from which it may be seen that in the atmosphere the actual value of n usually varies between n = 1.2 (wet adiabatic process) to n = 1.4 (dry adiabatic process). As shown below the value of n depends on the *temperature lapse rate* ($\partial T/\partial z$). Combining equation 22 with the equation of state p = ρRT and the equation (p/ρ^n) = constant, we have

$$\frac{T}{T_0} = \left[1 - \frac{g(z - z_0)}{RT_0}\left(\frac{n-1}{n}\right)\right] \quad ...(23)$$

Differentiating with respect to z, we obtain

$$\frac{1}{T_0}\frac{\partial T}{\partial z} = -\frac{g}{RT_0}\left(\frac{n-1}{n}\right)$$

$$\text{or } \frac{g}{R}\left(\frac{n-1}{n}\right) = \frac{\partial T}{\partial z} = \lambda$$

where $\lambda = -\left(\dfrac{\partial T}{\partial z}\right)$

$$\therefore\ n = \frac{1}{1-(R\lambda/g)} \qquad \text{...(24)}$$

For the first 1100 m above the ground the temperature in the atmosphere decreases uniformly, that is – $(\partial T/\partial z) = \lambda$ = constant. The observed value of λ in this region of atmosphere is about 6.56°C/100 m

$$\therefore\ \frac{R\lambda}{g} = \frac{29.27\text{m kg(f)}}{\text{kg(m)}^\circ\text{C}} \times \frac{6.56^\circ\text{C}}{1000\text{ m}} \times \frac{1}{9.81}\ \frac{\sec^2}{\text{m}}$$

$$= \frac{29.27 \times 0.00656}{9.81}\ \frac{\text{kg(f)} - \sec^2}{\text{kg(m)m}}$$

Since 1 kg (f) = 9.81kg (m) $\dfrac{\text{m}}{\sec^2}$

$$\therefore\ \frac{R\lambda}{g} = 29.27 \times 0.00656 = 0.192$$

$$n = \frac{1}{1-0.192} = 1.238$$

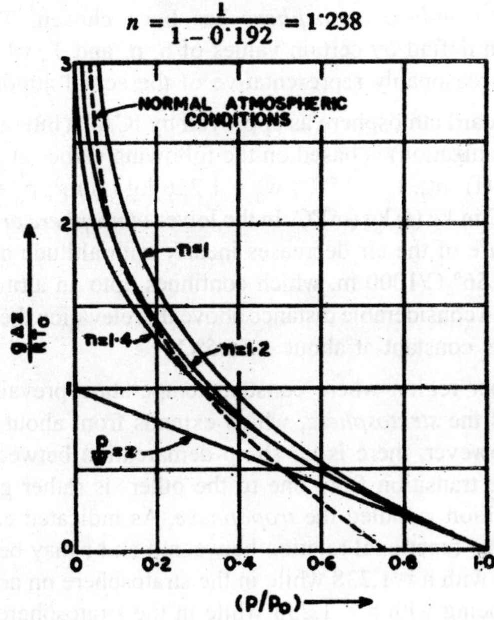

Fig. 5.4 : Variation of pressure with elevation in theatmosphere.

and $n = \dfrac{1}{0.192} = 1.238.$

Form about 11000 m to 32000 m the temperature in theatmosphere remains constant at about – 56.5°C in which case n = 1, that is, isothermal condition may be assumed and then equation 2.14 applies. Beyond 32000 m the temperature rises again. Since n depends on the temperature lapse rate λ, which varies with the altitude, it is obvious that in the atmosphere n varies with altitude. As such the assumption of a constant value of n for all the altitude. As such the assumption of a constant value of n for all the altitudes in the atmosphere may lead to erroneous results.

(d) **Standard Atmosphere :** It is very well known that the densities and the temperatures in the atmosphere vary continuously. They change from day to day and from place to place on the earth. In addition, moisture too plays a part in this variation, though it has a relatively small influence on density ρ and hence it is generally negelected in practical calculations. For aeronautical purpose, particularly for comparison of aircraft performance at different locations and on different days the difficculties that may arise from variations in ρ must be avoided. Therefore, an international *standard atmosphere* has been chosen. The standard atmosphere in defind by certain values of n, p_0 and T_0 which provides a set of data reasonably representative of the actual atmosphere.

The standard atmosphere as approved by ICAO (International Civil Aviation Organization) is based on the following values at sea level : p_0 = 10 332 kg (f) /m_2; t_0 = 15°C; w_0 = 1.226 kg (f) m^3; ρ_0 = 0.125 msl/m^3; R = 29.27 m.kg (f) kg (m)°C. In the lower *atmosphere or troposphere* the temperature of the air decreases inearly with altitude at an average rate of λ = 6.56° C/1000 m, which continues upto an altitude of about 11000 m. For a considerable distance above this elevation the temperature of air remains constant at about – 56.5°C.

This upper region where constant temperature prevails is usually referred to as the *stratosphere*, which extends from about 11000 m to 32000 m. However, there is no sharp demarcation between these two strata, but the transition from one to the other is rather gradual. This zone of transition is called the *tropopause*. As indicated earlier, in the troposhpere the functional relation between p and r may be assumed to be polytropic with n = 1.238 while in the stratosphere on account of the temperature being with n = 1.238 while in the stratosphere on account of the temperature being constant the function relation between p and r is isothernal.

PRESSURE THE SAME IN ALL DIRECTIONS–PASCAL'S LAW

The pressure at any point in a fluid at rest has the same magnitude in all directions. In other words, when a certain pressure is applied at any point ina fluid at rest, the pressure is equally transmitted in all the directions and to every other point in the fluid. This fact was established by B. Pascal, a French Mathematician in 1653, and accordingly it is known as Pascal's Law.

To prove this statement,consider an infinitesimal wedge shaped element of fluid at rest as a free body. The element is arbitrarily chosen and has the dimensions as shown in Fig 5. Since in a fluid at rest there can be no shear forces, the only forces acting on the free body are the normal pressure forces exerted by the surrounding fluid on the plane surfaces, and the weight of the element. As the element is in equilibrium, the sum of the force components on the element in any direction must be equal to zero. So, the equations of equilibrium in the x and z directions are respectively,

$$p_x\ \delta y\ \delta z - ps\ \delta s\ \delta y\ \sin\alpha = 0$$

$$p_x\ \delta x\ \delta y - ps\ \delta s\ \delta y\ \cos\alpha - w\frac{1}{2}\ \delta x\ \delta y \delta z\ = 0$$

in which p_x, p_z, p_s are the average pressures on the three faces and w is the specific weight of the fluid. Since $\delta z = \delta s \sin\alpha$ and $\delta x = \delta s \cos\alpha$ the above equations simplify to

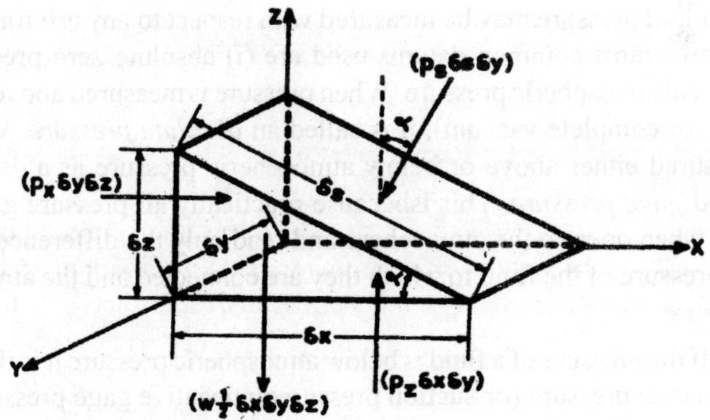

Fig. 5.5 : Free body diagram of a wedge-shaped element of fluid.

$$p_x\ \delta_y \delta_z - p_s\ \delta_y\ \delta_z = 0$$

$$p_x \, \delta x \, \delta y - p_s \, \delta x \, \delta y - \frac{1}{2} w \, \delta x \, \delta y \delta z = 0$$

The third term of the second equation is much smaller than the other two terms (since it involves a product of three infinitesimal quantities), and hence it may be neglected. Now by dividing the equations by $\delta y \, \delta z$ and $\delta x \, \delta y$ respectively and taking the limit, so that the element is reduced to a point, it follows from the equations that $p_s = p_x = p_z$

Since angle α is chosen arbitrarily, this equation proves that the pressure is the same in all directions at a point in a static fluid.

The Pascal's law is made use of in the construction of machines such as hydraulic press, hydraulic jack, hydraulic lift, hydraulic crane, hydraulic rivetter etc. in which by the application of relatively small forces considerably larger forces are developed.

ATMOSPHERIC, ABSOLUTE, GAGE AND VACUUM PRESSURES

The atmospheric air exerts a normal pressure upon all surfaces with which it is contact, and it is known as *atmospheric pressure*. The atmospheric pressure pressure varies with the altitude and it can be measured by means of a barometer. As such it is also called the barometric pressure. At sea level under normal conditions the equivalent values of the atmospheric pressure are 10.1043×10^4 N/m^2; or 1.03 kg (f)/cm^2 ; or 10.3 m of water; or 76 cm of mercury.

Fluid pressure may be measured with respect to any arbitrary datum. The two most common datums used are (i) absolute zero pressure and (ii) local atmospheric pressure. When pressure is measured above absolute zero (or complete vaccum), it is called an *absolute pressure*. When it is measured either above or below atmospheric pressure as a datum, it is called *gage pressure*. This isbecause practically all pressure gages read zero when open to the atmosphere and read only the difference between the pressure of the fluid to which they are connected and the atmospheric pressure.

If the pressure of a fluid is below atmospheric pressure it is designated as vaccum pressure (or suction pressureon negative gage pressure) ; and its gage value is the amount by which it is below that of the atmospheric pressure. A gage which measures vacuum pressure is known as vacuum gage.

All values of absolute pressure are positive, since in the case of fluids the lowest absolute pressure which can possibly exist corresponds to absolute zero or complete vacuum. However, gage pressures are positive if they are above that of the atmosphere and negative if they are vaccum pressures. Fig. 5.6 illustrates the relation between absolute, gage and vaccum pressures.

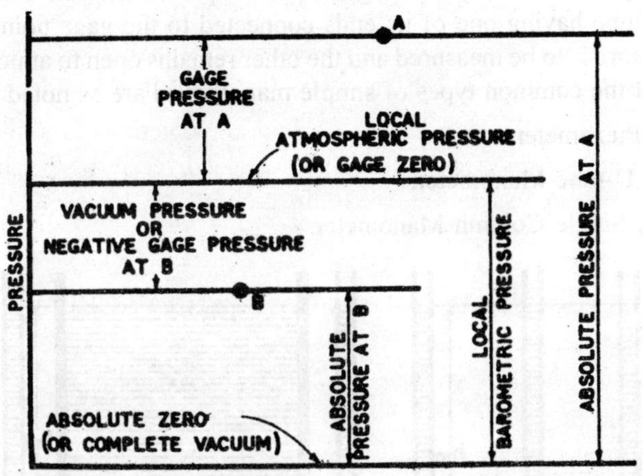

Fig. 5.6 Relationship between absolute, gage and vacuum pressures.

From the foregoing discussion it can be seen that the following relations hold:

Absolute Pressure = Atmospheric Pressure+Gage Pressure ...(25)

Absolute Prssure = Atmospheric Pressure–Vaccum Pressure ...(26)

MEASUREMENT OF PRESSURE

The various devices adopted for measuring fluid pressure may be broadly classified under the following two heads:

(1) Manometers

(2) Mechanical Gages.

Manometers : Manometers are those pressure measuring devices which are based on the principle of balancing the column of liquid (whose pressure is to be found) by the same or another column of liquid. The manometers may be classified as

(a) Simiple Manometers.

(b) Differential Manometers.

Simple Manometers are those which measure pressure at a point in a fluid contained in a pipe or a vessel. On the other hand *Differential Manometers* measure the difference of pressure between any two points in a fluid contained in a pipe or a vessel.

Simple Manometers : In general a simple manometer consists of a glass tube having one of its ends connected to the gage point where the pressure is to be measured and the other remains open to atmosphere. Some of the common types of simple manometers are as noted below :

(i) Piezometer.

(ii) U-tube Manometer.

(iii) Single Column Manometer.

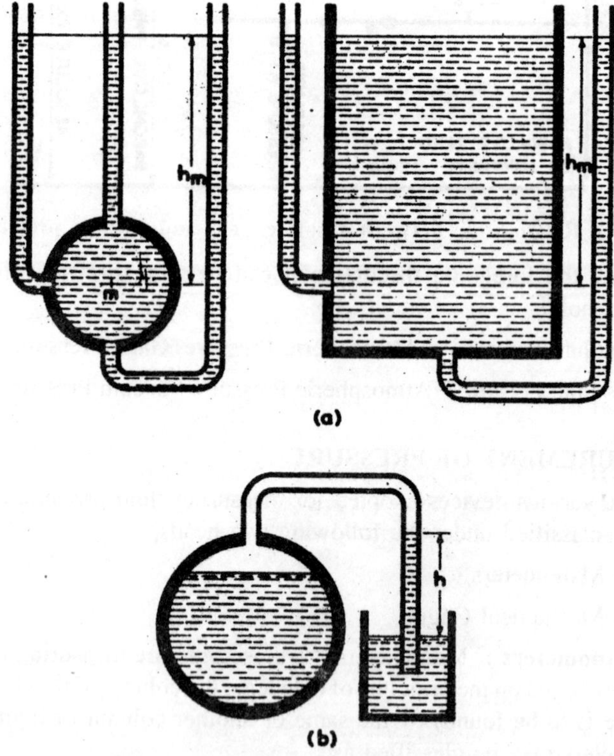

Fig. 5.7 : Piezometers.

(i) **Piezometer** : A piezometer is the simplest form of manometer which can be used measuring moderate pressures of liquids. It consists of a glass tube (Fig. 5.7) inserted in the wall of a pipe or a vessel, containing a liquid whose pressure is to be measured. The tube extends vertically upward to such a height that liquid can freely rise in it without overflowing. The pressure at any point in the liquid is indicated by the height of the liquid in the tube above that point, which can be read on the scale attached to it.

Thus, if w is the specific weigh of the liquid, then the pressure at point m in Fig. 5.7 (a) is pm = whm. In other words, h_m is the pressure head at m. Piezometers measure gage pressure only, since the surface of the liquid in the tube is subjected to atmospheric pressure. From the foregoing principles of pressure in homogeneous liquid at rest, it is obvious that the location of the point of insertion of a piezometer makes no difference. Hence as shown in Fig. 5.7 (a) piezometers may be inserted either in the top, or the side, or the bottom of the container, but the liquid will rise to the same level in the three tubes.

Negative gage pressures (or pressures) less than atmospheric) can be measured by means of the piezometer shown in Fig. 5.7 (b). It is evident that if the pressure in the container is less than the atmospheric no column of liquid will rise in the ordinary piezometer. But if the top of the tube is bent downward and its lower end dipped into a vessel containting water (or same order suitable liquid) [Fig. 5.7 (b)] the atmospherewill cause a column of the liquid to rise to a height h in the tube, from which the magnitube of the pressure of the liquid in the container can be obtained. Neglecting the weight of the air caught in the portion of the tube, the pressure on the free surface in the container is the same as that at free surface in the tube which from equation 2.8 may be expressed as p = – wh, where w is the specific weight of the liquid used in the vessel. Conversly –h is the pressure head at the free surface in the container.

Piezometers are also used to measure pressure heads in pipes where the liquid is in motion. Such tubes should enter the pipe in adirection at right angkes to the direction of flow and the connecting end should be flush with the inner surface of the pipe. All burrs and surface roughness near the hold must be removed, and it is better to round the edge of the hole slightly. Also, the hole should be small, preferably not larger than 3mm.

In order to prevent the capillary action from affecting the height of the column of liquid in a peieometer, the glass tube having an internal diameter less than 12 mm should not be used. Moreover for precise work at low heads the tubes having an internal diameter of 25 mm may be used.

(ii) U-tube Manometer : Piezometers cannot be used when large pressures in the lighter liquids are to be measured, since this would require very long tubes, which can not be handled conveniently. Further more gas pressures can not be measured by means of piezometers because a gas forms no free atmospheric surface. These limitations imposed on the use of piezometers may be overcome by the use of U-tube manometers. A U-tube manometer consists of a glass tube bent in U-shape, one end of which is connected to the gage point and the other end remains open to the atmosphere (Fig. 2.8). The tube contains a liquid of specific gravity greater than that of the fluid of which the pressure is to be measured. Sometimes more than one liquid may also be usedin the manometer. The liquids used in the manometers should be such that they do not get mixed with the fluids of which the pressures are to be measured. Some of the liquids that are frequently used in the manometers are mercury, oil, salt solution, carbon disulphide, carbon tetrachloride, bromoform and alchol. Water may also be usedas a manometric liquid when the pressures of gases or certain coloured liquids (which are immiscible with water) are to be measured.

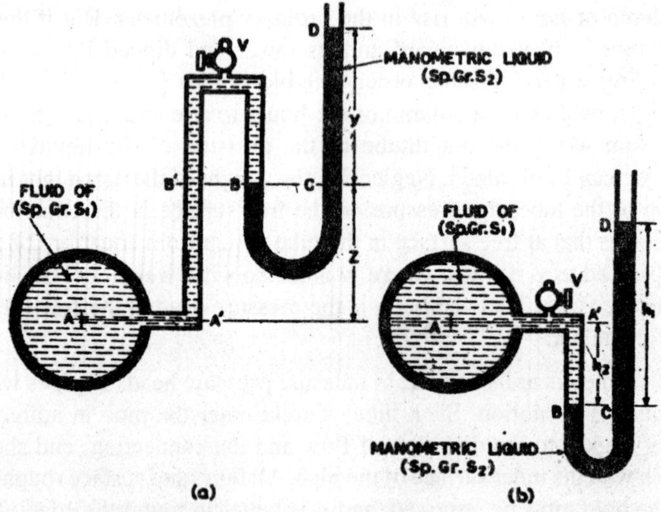

Fig. 5.8 : U-tube simple manometer.

The choice of the manometric liquid, however, depends on the rang of pressure to be measured. For low pressure range, liquids of lower specific gravities are used and for high pressure range, generally mercury is employed.

When one of the limbs of the U-tube manometer is connected to the gage point, the fluid from the container or pipe A will enter the connected limb of the manometer there by causing the manometric liquid to rise in the open limb as shown in Fig. 2.8. An air relief value V is usually provided at the top of the connecting tube which permits the expulsion of all air from the portion A′ B and its place taken by the fluid in A.

This is essential because the presence of even a small air bubble in the portion A′ B would result in an inaccurate pressure measurement.

In order to determine the pressure at A, a gage equation may be written as indicated below. Although any units of pressure or head may be used in the gage equation, it is generally convenient to express all the terms in metres of the fluid whose pressure is to be measured. The following general procedure may be adopted to obtain the gage equation :

(1) Start from either A or form the free surface in the open end of the manometer and write the pressure there in an appropriate unit (say metres of wateror other fluid or N/m^2 or kg (f) cm^2 or kg (f)/m^2). If the pressure is unknown (as at A) it may be expressed in terms of an appropriate symbol. On the other hand the pressure at the free surface in the open end (which is equal to atmospheric pressure) may be taken as zero. So that equation formed in each case will represent the gage pressure.

(2) To the pressure found above, add the change in pressure (in the same units) which will be caused while proceeding from one level to another adjacent level contact of liquids of different specific gravities. Use positive sign if the next level of contact is lower than the first and negative if it is higher. The pressure heads in terms of the heights of columns of same liquid may be obtained by using equation 10 (a).

(3) Continue the process as in (2) until the other end of the gage is reached and equate the expression to the pressure at that point, known or unknown.

The expression will contain only one unknown viz., the pressure at A, which may thus be evaluated.

Thus for the manometer arrangement shown in Fig. 2.8 (a) the gage equation may be written as indicated below.

Starting from A, if pA is the unknown pressure intensity at A, w is the specific weight of water and S_1 is the specific gravity of the liquid in the container, then the pressure head at A = $\frac{p_A}{wS_1}$ (in terms of liquid at A). Since all points lying at the same horizontal level in the same continuous static mass of liquid have same pressure.

Pressure at A = Pressure at A′ :

From A′ to B′ there being increase in elevation, pressure head decreases, so that pressure head at B′ = $\left(\frac{p_A}{wS_1} - z\right)$. Again from the above enunciation,

Pressure at B′ = Pressure at B = Pressrue at C.

From C to D there being gravity of the manometric liquid then from equation 10 (a) the pressure head, in terms of liquid at A, equivalent to the column CD (= y) of the manometric fluid = y $\left(\frac{S_2}{S_1}\right)$. Thus pressure head at D = $\left(\frac{p_A}{wS_1} - z - y\frac{S_2}{S_1}\right)$. But at D there being atmospheric prssure, the pressure head = 0, in terms of the gage pressure. As such equating the pressure heads at D, the gage equation becomes

$$\frac{p_A}{wS_1} - z - \frac{yS_2}{S} = 0$$

$$\text{or } \frac{p_A}{wS_1} = z + \frac{yS_2}{S_1} \qquad \text{...(27)}$$

Equation 27 represents the pressure heads in terems of the liquid at A. However, if the pressure heads are expressed in terms of water of following equation is obtained

$$\frac{p_A}{w} = zS_1 + yS_2 \qquad \text{...(28)}$$

Evidently equation 28 may be obtained directly from equation 27 by multiplying its both the sides by S_1, the specific gravity of liquid at A.

If A contains a gas, its specific weight being quite small, the specific gravity S_1 of the gas (with respect to water) will be so small that zS_1 may be neglected. In which case the pressure head of the gas at A in terms of water is given by

$$\frac{p_A}{w} = yS_2 \qquad ...(29)$$

Fig. 5.8 (b) shows another arrangement for measuring pressure at A by means of a U-tube manometer. By following the same procedure as indicated above the gage equation for this arrangement can also be written, which will be in terms of liquid at A as,

$$\frac{p_A}{wS_1} = h_1\frac{S_2}{S_1} - h_2 \qquad ...(30)$$

and in terms of water as

$$\frac{p_A}{w} = h_1S_2 - h_2S_1 \qquad ...(31)$$

Again if A contains gas, the specific gravity S_1 of the gas (with respect to water) being so small that h_2S_1 may be neglected. In which case equation 31 becomes

$$\frac{p_A}{w} = h_1 S_2 \qquad ...(32)$$

A U-tube manometer can also be used to measure negative or vacuum pressure. For measurement of smalll negative pressure, a U-tube manometer without any manometric liquid may be used, which is as shown in Fig. 5.9 (a). It is evident that for the pressure at A being negative (*i.e.* less than atmospheric prssure) the liquid surface in the open limb of the manometer will be below A. The pressure at A may be determined from the gage equation as obtained below.

At C the pressure head = 0, in terms of gage pressure. Further pressure at C = pressure at B. From to A′, there being increase in elevation, the pressure head decreases, so that pressure head at A′ = 0 – h (in terms of liquid at A). Again,

Pressure at A′ = Pressure at A.

Now if pA is the pressure intensity at A, w is the specific weight of water, and S1 is the specific gravity of the liquid at A, then the pressure head at A = $\frac{p_A}{wS_1}$ (in terms of liquid at A). Thus in terms of liquid at A.

$$\frac{p_A}{wS_1} = -h \qquad ...(33)$$

and in terms of water

$$\frac{p_A}{w} = -S_1h \qquad ...(34)$$

For measuring negative pressures of larger magnitude a manometric liquid having higher specific gravity is employed, for which the arrangement as

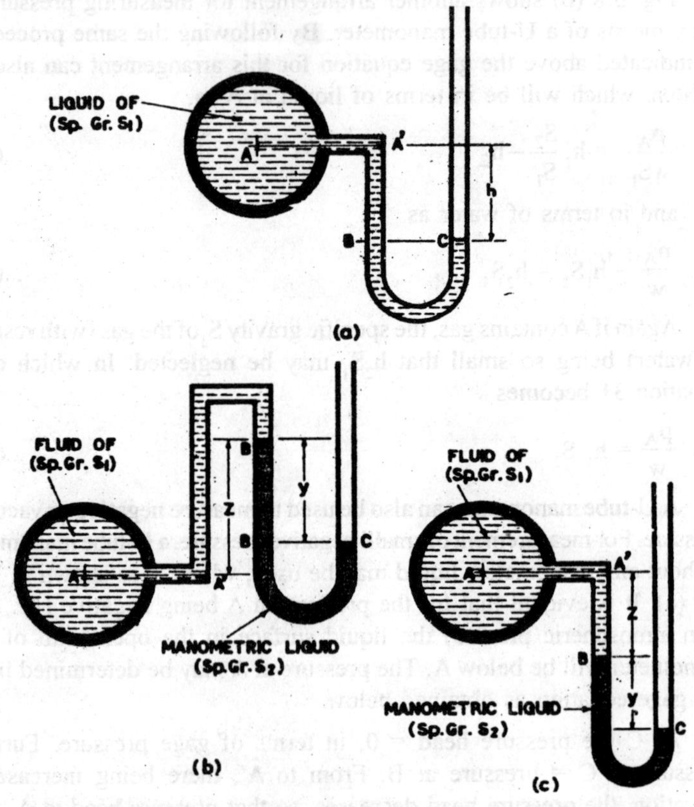

Fig 5.9 : Measurement of negative pressure by U-tube simple manometer.

shown in Fig. 2.9 (b) or Fig, 2.9 (c) may be employed. If the specific gravity of the fluid at A is S_1 (with respect to water) and the specific gravity of the manometric liquid is S_2, the gage equation for the arangement of Fig. 5.9 (b) may be written in terms of liquid at A as,

$$\frac{p_A}{wS_1} = z - y\frac{S_2}{S_1} \qquad ...(35)$$

and in terms of water as

$$\frac{p_A}{w} = zS_1 - yS_2 \qquad ...(36)$$

Similarly, for the arrangement shown in Fig. 5.9 (c) the gage equation may be written in terms of liquid at A as,

$$\frac{p_A}{wS_1} = -z - y\frac{S_2}{S_1} \quad ...(37)$$

and in terms of water as

$$\frac{p_A}{w} = zS_1 - yS_2 \quad ...(38)$$

The arrangement of Fig. 5.9(c) is generally preferred to that of Fig. 5.9(b).

(iii) **Single Column Manometer :** The U-tube manometers described above usually require readings of fluid levels at two or more points, since a change in pressure causes a rise of liquid in one limb of the manometer and a drop in the other. This difficullty may however be overcome by using single column manometers. A single column manometer is a modified form of a U-tube manometer in which a shallow reservoir having a large cross-sectional area (about 100 times) as compared to the area of the tube is introduced into one limb of the manometer, as shown in Fig 5.10. For any variation in pressure, the change in the liquid level in the reservoir will be so small that it may be neglected, and the pressure is indicated approximately by the height of the liquid in the other limb. As sucll only one reading in the narrow limb of the manometer need be taken for all pressure measurements. The narrow limb of the manometer may be vertical as in Fig 5.10 (a) or it may be inclined as in Fig. 5.10 (b). The inclined type is useful for the measurement of small pressures. As indicated later since no reading in required to be taken for the level of liquid in the reservoir, it need not be made of transparent material.

When the manometer is not connected to the container, the surface of the maonometric liquid in the reservoir will stand at level 0 – 0, and since it is subjected to a pressure due to a column of fluid of height y and specific gravity S_1, the surface of the manometric liquid in the tube will stand at B, at a height h_1 above 0 – 0, such that from equation 10 (a)

$$yS_1 = h_1S_2 \quad ...(39)$$

where S_2 is the specific gravity of the manometric liquid. This is known as the normal position of the manometric liquid. On being connected to the container at the gage point, the high pressure fluid will enter the reservoir, due to which there will be a drop in the manometric liquid surface in the reservoir by a distance Δy and a consequent rise in the tube by a distance h_2, above B. If A and a are the cross-sectional areas of the reservoir and the tube respectively, then,

$$A\,(\Delta y) = a h_2 \qquad ...(40)$$

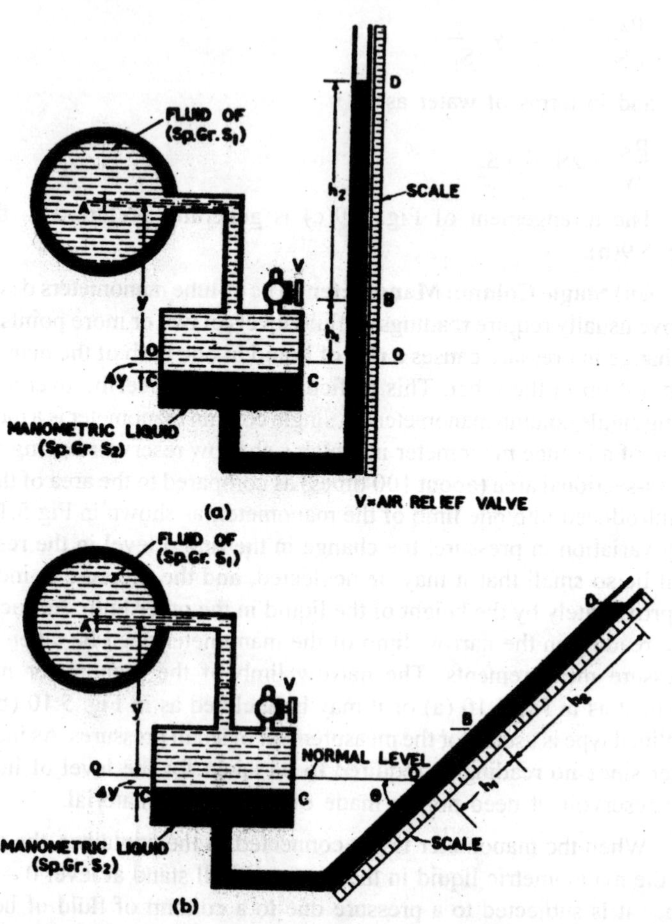

Fig. 5.10 : Single column manometer.

If p_A is the pressure intensity at A and w is the specific weight of water, then starting from D the following gage equation in terms of water is obtained

$$0 + (h_2 + h_1 + \Delta y)\,S_2 - (\Delta y)\,S_1 - yS_1 = \frac{p_A}{w} \qquad ...(41)$$

Introducing equations (39) and (40) in equation (41) it becomes

$$\frac{p_A}{w} = h_2 s_2 \qquad ...(43)$$

so that only one reading on the height of level of liquid in he narrow tube is required to be taken to obtain the pressure head at A. However, if Δy is appreciable, then since the terms within brackets on the right side of equation 42 are constant, the scale on which h_2 is read can be so graduated as to correct for Δy so that again only one reading of the height of liquid level in the narrow tube is required to be taken, which will directly give the pressure head at A.

A single tube manometer can be made more sensitive by making its narrow tube inclined as shown in Fig. 2.10 (b). With this modification the distance moved by the liquid in the narrow tube shall be comparatively more, even for small pressure intensity at A. As before when the manometer is not connected to the container, the manometric liquid surface in the reservoir will stand at level 0 – 0 and that in the tube will stand at B, such that

$$yS_1 = (h_1 \sin \theta)\, S_2 \qquad ...(44)$$

Due to high pressure fluid entering the reservoir, the maonmetric liquid surface will drop to level C – C by a distance Δy, and it will travel a distance BD equal to h_2 in the narrow tube. Thus.

$$A(\Delta y) = ah_2 \qquad ...(45)$$

Again starting from D the gage equation in this case becomes

$$0 + (h_1 + h_2)(\sin \theta)\, S_2 + \Delta y S_2 \;\; \Delta y S_1 - y S_1 = \frac{p_A}{w} \qquad ...(46)$$

Introducingequations 44 and 45 in equation 46 it becomes

$$\frac{p_A}{w} = h_2 \left[S_2 \sin\theta + (S_2 - S_1)\frac{a}{A} \right] \qquad ...(47)$$

Again if the ratio $\frac{a}{A}$ is negligible then equation 47 reduces to

$$\frac{p_A}{w} = (h2 \sin \theta)\, S_2 \qquad ...(48)$$

Single column manometers can also be employed to measure the negative gage pressures. If the pressure at A in the container is negative, the manometric liuquid surface in the reservoir will be raised by a certain distance and consequently there will be drop in the surface in the tube. Again by adopting the same procedure the gage equations for the negative pressure measurement can also be obtained.

Differential Manometers : For measuring the difference of pressure between any two points in a pipeline or in two pipes or containers, a differential manometer is employed. In general a differential manometer consists of a bent glass tube, the two ends of which are connected to each of the two gage points between which the pressure difference is required to be measured. Some of the common types of differential manometers are as noted below :

(i) Two- Piezometer Manometer.

(ii) Inverted U-Tube Manometer.

(iii) U-Tube Differential Manometer.

(iv) Micromanometer.

(i) *Two-Piezometer Manometer :* As the name suggests this manometer consists of two separate piezometers which are inserted at the two gage points between which the difference of pressure is required to be measured. The difference between the two points. Evidently this method is useful only in the pressure at each of the two points is small. Moreover it cannot be used to measure the pressure difference in gases, for which the other types of differential manometers described below may be employed.

(ii) *Inverted U- tube Manometer :* It consists of a glass tube bent in U-shape and held inverted as shown in Fig. 5.11. Thus it is as two piezometers described above are connected with each other at top. When the two ends of the manometer areconnected to the points between which the pressure difference is required to be measured, the liquid under pressure will enter the two limbs of the manometer, thereby causing the air within the manometer to get compressed. The presence of the compressed air results in restricting the heights of the columns of liquids raised in the two limbs of the manometer. An air cock as shown in Fig. 5.11, is usually provided at the top of the inverted U-tube which facilitates the raising of the liquid column to suitable level in both the limbs by driving out a portion of the compressed air. It also permits the expulsion of air bubbles of air bubbles which might have been entrapped some where in the pipeline.

If p_A and p_B are the pressure intensities at points A and B between which the inverted U-tube manometer is connected, then corresponding to these pressure intensities the liquid will rise above points A and B up to C and D in the two limbs of the manometer as shown in Fig. 5.11. Now if w represents the specific weight of water and S_1 represents the specific gravity of the liquid at A or B, then commencing from A, the pressure head at C in terms of water is equal to $\left(\frac{p_A}{w} - yS_1\right)$. Since points C and C′ are at the same horizontal level and in the same continuous static mass of fluid

Pressure at C = Pressure at C′

Between points C′ and D there is a column of compressed air. Since the specific weight of air is negligible as compared with that of liquid, the weight of air column between C′ and D may be neglected, Hence,

Pressure at C′ = Pressure at D

From D to B there being decrease in elevation, the pressure head increase so that the pressure head at B is equal to pressure head at D plus (y – h) S_1. Thus the gage equation may be expressed as

$$\frac{p_A}{w} - yS_1 + (y - h)\,S_1 = \frac{p_B}{w}$$

or $$\frac{p_A}{w} - \frac{p_B}{w} = hS_1 \qquad \text{...(49)}$$

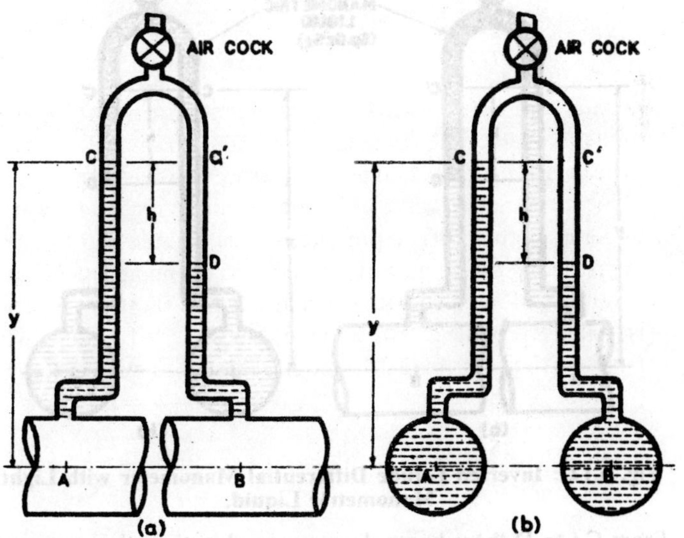

Fig. 5.11 : Inverted U-Tube Differential Manometer.

Inverted U-tube manometers are suitable for the measurement of small pressure difference in liquids. Sometimes instead of air, the upper part of this manometer is filled with a manometric liquid which is lighter than the liquid for which the pressure difference is to be measured and is immiscible with it. Such an arrangement is shown in Fig. 5.12. As indicated later the use of manometric liquid in this manometer results in increasing the sensitivity of the manometer.

Again if P_A and p_B are the pressure intensities at point A and B between which the inverted U-tube manometer is connected, then corresponding to these pressure intensities the liquid will rise above points A and B upto C and D in the two limbs of the manometer as shown in Fig. 5.12. Now if w represent the specific weight of water and S_1 and S_2 are the specific gravities of the liquid at A and B and the manometric liquid (in the upper part of the manometer) respectively, then commencing from A, the pressure head at C in terms of water is equal to $\left(\frac{p_A}{w} - yS_1\right)$. Since points C and C¢ are at the same horizontal level and in the same continuous static mass of liquid

Pressure at C = Pressure at C¢

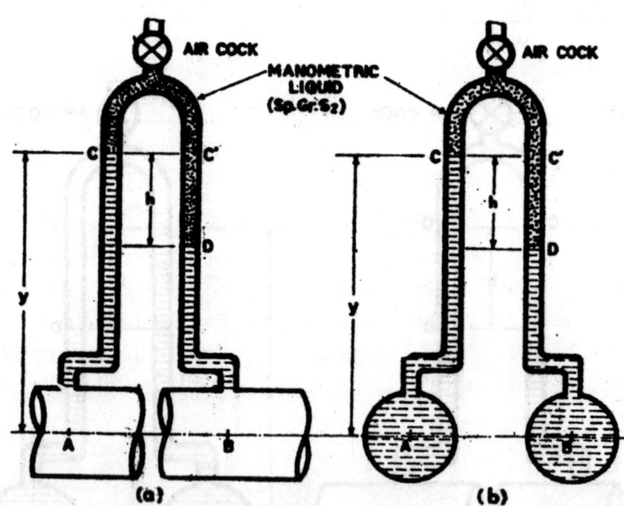

Fig. 5.12 : Inverted U-tube Differential Manometer with Light Manometric Liquid.

From C¢ to D there being decrease in elevation, the pressure head increase,so that the pressure head at D in terms of water is equal to

$$\left[\left(\frac{p_A}{w} - yS_1\right) + hS_2\right].$$

Further from D to B there is decrease in elevation and hence the gage equaion becomes

$$\frac{p_A}{w} - yS_1 + hS_2 + (y - h)\, S_1 = \frac{p_B}{w}$$

$$\text{or } \frac{p_A}{w} - \frac{p_B}{w} = h\,(S_1 - S_2) \qquad ...(50)$$

It is evident from equation 50 that as the specific gravity of the manometric liquid approaches that of the liquid at A or B $(S_1 - /S_2)$ approaches zero and large values of h will be obtained even for small pressure differences, thus increasing the sensitivity of these manometers is to incline the gage tubes so that a vertical gage difference h is transposed into a reading which is magnified by $\frac{1}{\sin\theta}$ where θ is angle of inclination with the horizontal.

(iii) *U-Tube Differential Manometer :* It consists of glass tube bent in U-shape, the two ends of which are connected to the two gage points between which the pressure difference is required to be measured. Fig. 5.13 shows such an arrangement for measuring the pressure difference between any two points A and B. The lower part of the manometer contains a manometric liquid which is heavier than the liquid for which the pressure difference is to be measured and is immiscible with it.

When the two limbs of the manometer are connected to the gage points A and B, then corresponding to the difference in the pressure intensities p_A and p_B the levels of manometric liquid in the two limbs of the manometer will be displaced through a distance x as shown in Fig. 5.13. By measuring this difference in the levels of the manometric liquid, the pressure difference $(p_A - p_B)$ may be computed as indicated below.

If S_1 and S_2 are the specific gravities of the liquid at A or B and the manometric liquid respectively, then by commencing at A where the pressure is p_A, the pressure head at C in terms of water is equal to $\left[\frac{p_A}{w} + (y + x)S_1\right]$, in which w is the specific weight of water. Further, since points C and C′ are at the same level and are lying in the same continuous static mass of liquid,

Pressure at C = Pressure at C′.

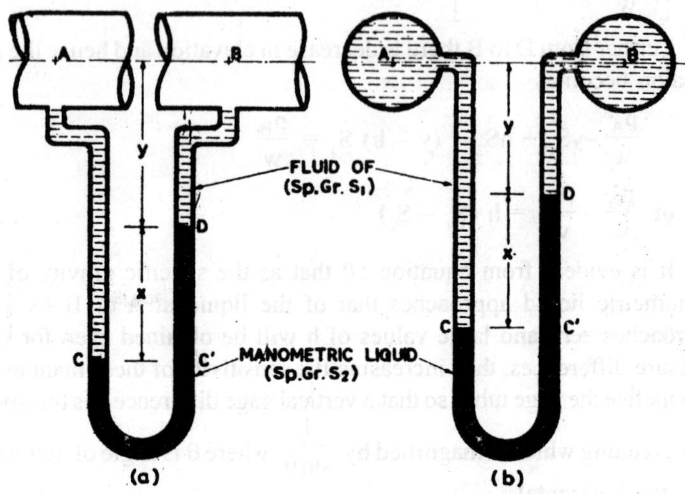

Fig. 5.13 : U-tube Differential Manometer.

Further from C′ to D there being an increases in the elevation, the pressure head decreases, so that the prêssure head at D in terms of water is equal to $\left[\frac{p_A}{w}+(y+x)S_1 - xS_2\right]$. Similarly from D to B there is an increase in elevation and hence the gage equation becomes

$$\left[\frac{p_A}{w}+(y+x)S_1 - xS_2 - yS_1\right] = \frac{p_B}{w}$$

$$\text{or } \frac{p_A}{w}-\frac{p_B}{w} = x\,(S_2 - S_1) \qquad ...(51)$$

If, for example, the manometric liquid is mecury (S2 = 13.6) and the liquid at A or B is water (S1 = 1) then the difference in pressure heads at the points A and B is 12.6 times the deflection x of the manometric liquid in the two limbs of the manometer. As such the use of mercury as manometric liquid in U-tube manometer is suitable for measuring large pressure differences. However, for small pressure differences, mercury makes precise measurement difficult, and hence for such cases it is common to use a liquid which is only slightly heavier than the liquid for which the pressure difference is to be measured.

Often the points A and B between which the pressure difference is to be measured are not at the same level, as shown in Fig. 5.14. For such cases also, by adopting the same procedure, the following gage equation may be obtained in order to compute the pressure difference between the points A and B.

$$\left[\frac{p_A}{w} + (z + y + x)S_1 - xS_2 - yS_3\right] = \frac{p_B}{w}$$

or $$\frac{p_A}{w} - \frac{p_B}{w} = [x\,(S_2 - S_1) + y\,(S_3 - S_1) - zS_1] \quad ...(52)$$

Equation 52 is, however, a general equation, which may be modified to derive the equations for different conditins. Thus, for example, if there is same liquid at A and B, then since $S_1 = S_3$, equation 52 becomes

$$\frac{p_A}{w} - \frac{p_B}{w} = [x\,(S_2 - S_1) - zS_1] \quad ...(53)$$

Further if A and B are at the same level, then since z = 0, equation 53 becomes same as equation 51 which is quite obvious.

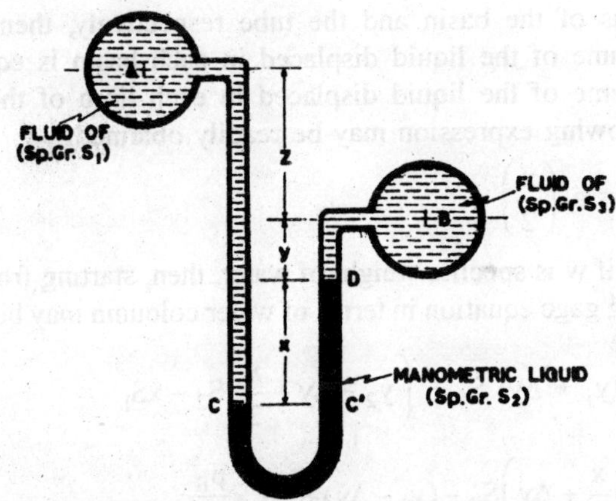

Fig. 5.14 : U-tube Differential Manometer between Two Points at Different Levels.

(iv) **Micromanometers :** For the measurement of very small pressure differences, or for the measurement of pressure differences with very high precision, special forms of manometers called micromanometers are used. A wide variety of micromanometers have been developed, which eigher magnify the readings or permit the readings to be observed with greater accuracy. One simple type of micromanometer consists of a glass U-tube, provided with two transparent basins of wider sections at the top of the two limbs, as shown in Fig. 5.15. The manometer contains two manometric liquids of different specific gravities andimmiscible with each other and with the fluid for which the pressure difference is to be measured.

Before the manometer is connected to the pressure points A and B, both the limbs are subjected to the same pressure. As such the heavier manometeric liquid of sp. gr. S_1 will occupy the level DD′ and the lighter manometric liquid of sp. gr. S_2 will occupy the level CC′. When the manometer is connected to the pressure points A and B where the pressure intensities are p_A and p_B respectively, such that $p_A > p_B$ then the level of the lighter manometric liquid will fall in the left basin and rise in the right basin by the same amount 'Δy'. Similarly the level of the heavier manometric liquid will fall in the left limb to point E and rise in the right limb to point F. If A and a are the cross-sectional areas of the basin and the tube respectively, then since the volume of the liquid displaced in each basin is equal to the volume of the liquid displaced in each limb of the tube the following expression may be readily obtained.

$$A\,(\Delta y) = a\left(\frac{x}{2}\right) \qquad ...(54)$$

Further if w is specific weight of water, then, starting from point A the following gage equation in terms of water coloumn may be obtained.

$$\frac{p_A}{w} + (y_1 + \Delta y)\,S_3 + \left(y_2 - \Delta y + \frac{x}{2}\right)S_2 - xS_1$$

$$-\left(y_2 - \frac{x}{2} + \Delta y\right)S_2 - (y_1 - \Delta y)\,S_3 = \frac{p_B}{w}$$

Substituting the value of Δy from equation 54 and simplifying the above equation it becomes

$$\frac{p_A}{w} - \frac{p_B}{w} = x\left[S_1 - S_2\left(1 - \frac{a}{A}\right) - S_3\frac{a}{A}\right] \qquad ...(55)$$

The quantities within brackets on right side of equation 55 are constant for a particular manometer. Thud by measuring x and substituting in equation 55 the pressure difference between any two points can be known.

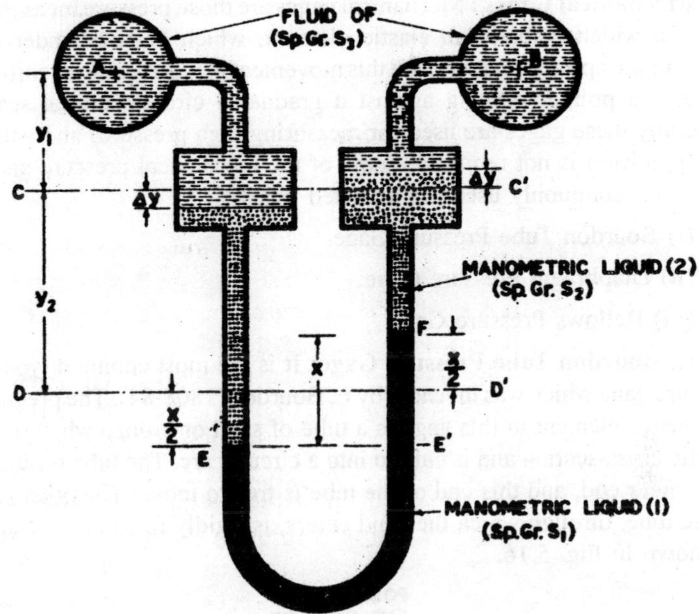

Fig. 5.15 : Micromanometer.

If the cross-sectional area of the basin is larege as compared with the cross-sectional area of the tube, then the ratio a/A is very small and the equation 55 reduces to

$$\frac{p_A}{w} - \frac{p_B}{w} = x\,(S_1 - S_2) \qquad \text{...(56)}$$

By selecting the two manometric liquid such that their specific gravities are very nearly equal then a measurable value of x may be achieved even for a verysmall pressure difference between the two points.

In a number of other types of micromanometers the pressure difference to be measured is balanced by the slight raising or lowering (on a micrometer screw) of one arem of the manometer where by a meniscus is brought back to its original poisition. The micromanometers of this type are those invented by Chattock, Small and Krell, which are sensitive to prssure difference down to less than 0.0025 mm of water. However the disadvantage with such manometers is than an appreciable time is required to take a reading and they are therefore suitable only for completely steady pressures.

Mechanical Gages : Mechanical gages are those pressure measuring devices, which embody an elastic element, which deflects under the action of the applied pressure,and this movement mechanically magnified, operates a pointer moving against a graduated circumferential scale. Generally these gages are used for measuring high pressures and where high precision is not required. Some of the mechanical pressure gages which are commonly used are as noted below:

(i) Bourdon Tube Pressure Gage.

(ii) Diaphragm Pressure Gage.

(iii) Bellows Pressure Gage.

(i) **Bourdon Tube Pressure Gage:** It is the most common type of pressure gage which was invented by E. Bourdon (1808-84). The pressure responsive element in this gage is a tube of steel or bronze which is of elliptic cross-section and is curved into a circular arc. The tube is closed at its outer end, and this end of the tube is free to move. The other end of the tube, through which the fluid enters, is rigidly fixed to the frame as shown in Fig. 5.16.

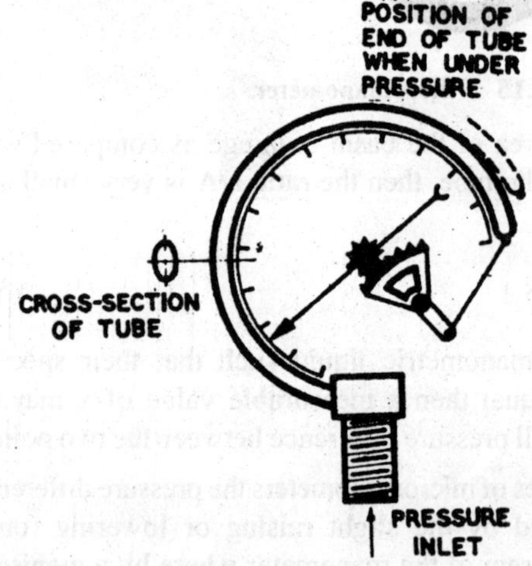

Fig. 5.16 : Bourdon tube pressure gage.

When the gage is connected to the gage point, fluid under pressure eners the tube. Due to increase in internal pressure, the eliptical cross-section of the tube tends to become circular, thus causing the tube to straighten out sightly. The small outward movement of the free end of the tube is transmitted, through a link, quadrant and pinion, to a pointer which by moving clockwise on the graduated circular dial indicated the prssure intensity of the fluid. The dial of the gage is so calibrated that it reads zero wheh the pressure inside the tube equals the local atmospheric pressure, and the elastic deformation of the tube causes the pointer to be displaced on the dial in proportion to the pressure intensity of the fluid. By using tubes of appropriate stiffness, gages for wide range of pressures may be made. Further by suitably modifying the graduations of the dial and adjusting the pointer Bourdon tube vacuum gages can also be made. When a vacuum gage is connected to a partial vacuum, the tube tends to close, thereby moving the pointer in anti clockwise direction, indicating the negative direction, indicating the negative or vacuum pressure. The gage dials are usually calibrated to read newton per square metre (N/m^2), or pascal (pa), or kilogram (f) per square centimetre ($kg(f)/cm^2$]. However other units of pressure, such as metres of water or centimetres of mercury, are also frequently used

(ii) **Diaphragm Pressure Gage:** The pressure responsive element in this gage is an elastic steel corrugated daiphragm. The elastic deformation of the diaphragm under pressure is transmitted to a pointer by a similar arrangement as in the case of Bourdon tube pressure gage (see Fig. 5.17). However, this gage is used to measure relaviely low pressure intensities. The Aneroid barometer operates on a similar principle.

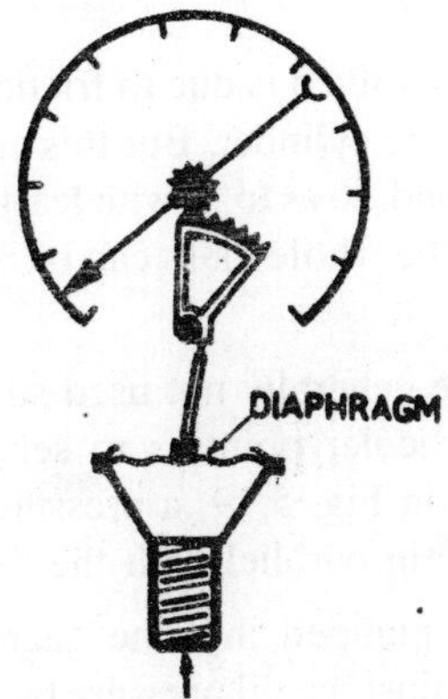

Fig. 5.17 : Diaphragm pressure gage.

(iii) **Bellows Pressure Gage :** In this gage the pressure responsive element is made up of a thin metallic tube having deep circumferential corrugations. In responseto the pressure changes this elastic element expands or contracts, thereby moving the pointer on a graduated circular dial as shown in Fig. 5.18.

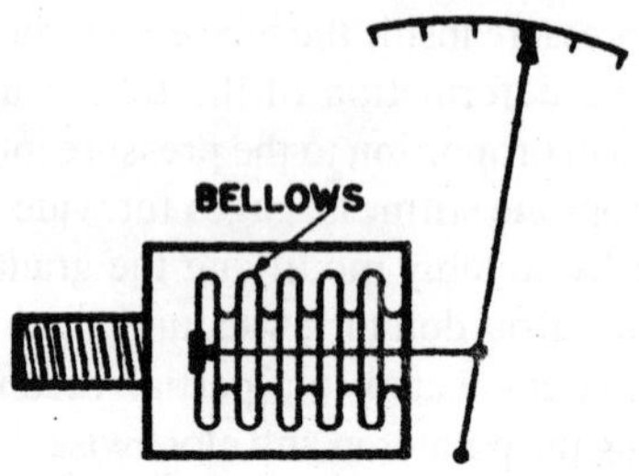

Fig. 5.18 : Bellows pressure gage.

(iv) **Dead-Weight Pressure Gage :** A simple form of a dead-weight pressure gage consistsof a plunger of diameter d, which can side within a verticalcylinder, as shown in Fig. 5.19. The fluid under pressure, entering the cylinder, exerts a force on the plunger, which is balanced by the weights loaded on the top of the plunger. If the weight required to balance the fluid under pressure is W,then the pressure intensity p of the fluid may be determined as.

$$p = W/\left(\frac{\pi}{4}d^2\right)$$

The only that may be involved is due to frictional resistance offered to motion of the pluger in the cylinder. But this error can be avoided if the plunger is carefully ground, so as to fit with least permissible clearance in the cylinder. Moreover, the whole mass can be rotated by hand before final reading are taken.

Dead-weight gages are generally not used so much to measure the pressure intensity at a particular point as to serve as standards of co nparison. Hence as shown in Fig. 5.19, a pressure gage which is to be checked or calibrated is setin parallel with the dead-weight gage.

Oil under pressure is pumped into the gages, thereby lifting the plunger and balancing it against the oil pressure by loading it with known weights. The pressure intensity of the being thus known, the attached pressure gage can either be tested for its accuracy or it can be calibrated.

A dead-weight gage which can be used for measuring pressure at a point with more convenience is also shown in Fig. 5.19. In this gage a lever, same as in some of theweighing machines, is provided to magnify the pull of the weights.

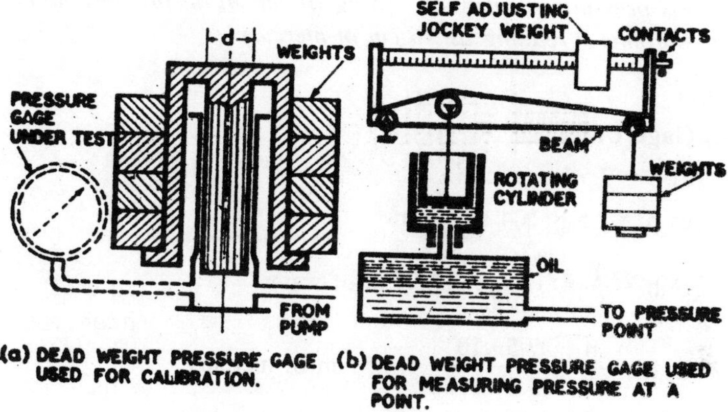

Fig. 5.19 : Dead-weight pressure gage.

The load required to balance the force due to fluid pressure is first roughly adusted by hanging weights from the end of main beam. Then a smaller jockey weight is slided along to give precise balance.

In more precise type of gage the sliding motion may be contrived automatically by an electric motor.

GENERAL COMMENTS ON CONNECTIONS FOR MANOMETERS AND GAGES

The following points should be kept in view while making connections for the various pressure measuring devices :

(i) At the gage point the hole should drilled normal to the surface and it should flush with the inner surface.

(ii) The diameters of the holes at the gage points should be about 3 mm to 6 mm.

(iii) The holes should not disturb the internal surface andno burrs or irregularities must be left.

(iv) There should be no air pockets left over in the connecting tubes, which should be completely filled with the liquid. The presence of air bubbles can easily by detected if the connecting tubes are made of polythene or similar transparent material.

SOLVED EXAMPLES

Example 1:

Express pressure intensity of 7.5 kg (f)/cm² in all pressure units. Take the barometer reading as 76 cm of mercury.

Solution:

(A) **Gage Units :**

(a) $p = 7.5 \text{ kg (f)/cm}^2$

(b) $p = 7.5 \times 104 \text{ kg (f) /m}^2$

(c) $h = \frac{p}{w} = \frac{7.5 \times 10^4}{1000} = 75 \text{ m of water}$

(d) $h = \frac{p}{w} = \frac{7.5 \times 10^4}{13.6 \times 1000} = 5.51 \text{ m of mercury}$

(e) $p = 9810 \times 75 = 73.575 \times 104 \text{ N/m}^2$

(B) **Absolute pr. :**

Absolute pr.= Gage pr + Atmospheric pr.

Atmospheric pr. = 76 cm of mercury

$= \frac{76 \times 13.6}{} = 10.34 \text{ m of water}$

$= \frac{76 \times 13.6 \times 1000}{100} = 1.034 \times 10^4 \text{ kg (f)/ m}^2$

$= \frac{76 \times 13.6 \times 1000}{100 \times 10^4} = 1.034 \text{ kg (f)/cm}^2$

$= \frac{76 \times 13.6 \times 9810}{100} = 10.14 \times 10^4 \text{ N/m}^2$

(a) Absolute pr. = (k 7.5 + 1.034)

= 8.534 kg (f) cm²

(b) Absolute pr. = $(7.5 \times 10^4 + 1.34 \times 10^4)$

$= 8.534 \times 10^4 \text{ kg (f) /m}^2$

(c) Absolute pr.head = (75 + 10.34)

= 85.34 m of water

(d) Absolutepr. head = (5.51 + 0.76)

= 6.27 m of mercury

(e) Absolute pr. = $\frac{6.27}{0.76}$

= 8.52 atmospheres

(f) Absolute pr. = $(73.58 \times 10^4 + 10.14 \times 10^4)$

Example 2:

Find the depth of a point below water surface in sea where pressure intensity is 1.006 MN/m². Specific gravity of sea water = 1.025.

Solution:

Depth of sea water above the point is

$$h = \frac{p}{w}$$

$p = 1.006\ MN/m^2$

$= 1.006 \times 106\ N/m^2$

$w = (1.025 \times 9810)\ N/m^2$

$= 1.006 \times 10^4\ N/m^3$

$$h = \frac{1.006 \times 10^6}{1.006 \times 10^4}$$

= 100 m.

Example 3:

Covert a pressure head of 100 m of water to (a) kerosene of specific gravity 0.81, (b) carbon tetrachloride of specific gravity 1.6.

Solution:

From equation 10 (a)

$$h_1S_1 = h_2\ S_2$$

Thus by substitution

(a) $100 \times 1 = h_2 \times 0.81$

$$h_2 = \frac{100}{0.81}$$

= 123.46 m of kerosene

(b) $100 \times 1 = h_2 \times 1.6$

$$h_2 = \frac{100}{1.6}$$

= 62.5 m of carbon tetrachloride

Example 4:

The left leg of a U-tube mercury manometer is connected to a pipe-line conveying water, the level of mercury in the leg being 0.6 m below the centre of pipe-line, and the right leg is open to atmosphere. The level of mercury in the right leg is 0.45 m above that in the left leg and the space above mercury in the right leg contains Benzene (specific gravity 0.88) to a height of 0.3 m. Find the pressure in the pipe.

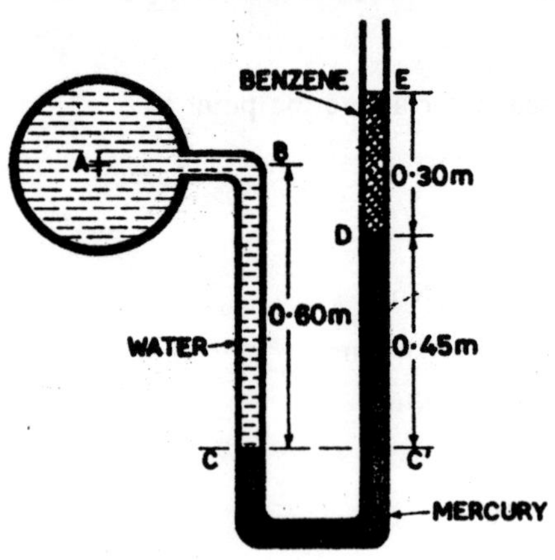

Fig. 5.20

Solution:

In the accompanying figure the pressures at C and C′ are equal. Thus computing the pressure heads at C and C′ from either side and equating the same, we get

$$\frac{p_A}{w} + 0.6 = 0.45 \times 1.36 + 0.3 \times 0.88$$

$$\text{or } \frac{p_A}{w} = 5.784 \text{ m of water}$$

$\therefore\ p_A = (5.784 \times 9810)$

$= 5.674 \times 10^4$ N/m^2

or $p_A = (5.784 \times 1000)$

$= 5.784 \times 10^3$ kg (f)/ m^2

$= 0.5784$ kg (f)/cm^2

Example 5:

As shown in the accompanying figure, pipe M contains carbon tetrachloride of specific gravity 1.594 under a pressure of 1.05 kg (f)/ cm² and pipe N contains oil of specific gravity 0.8. If the pressure in the pipe N is 1.75 kg (f) cm² and the manometric fluid is mercury, find the difference x between the levels of mercury.

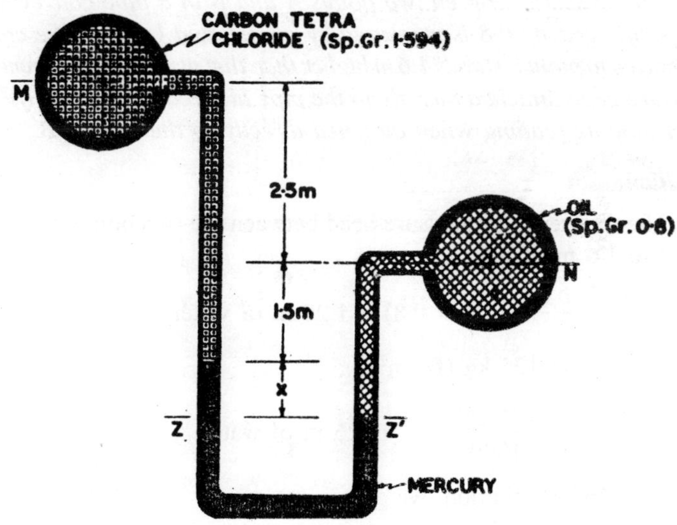

Fig. 5.21

Solution:

Equat the pressure heads at Z and Z′ as shown in the figure.

Pressure head at Z in terms of water column

$$= \left[\frac{1.05 \times 10^4}{1000} + (2.5 + 1.5) \times 1.594 + x(13.6)\right]$$

$$= 16.876 \quad + 13.6x$$

Similarly pressure head at Z′ in terms of water column

$$= \left[\frac{1.75 \times 10^4}{1000} + (1.5 \times 0.8) + x(0.8)\right]$$

Thus equating the two, we get

$$16.876 + 13.6x = 18.7 + 0.8x$$

or $12.8x = 1.824$

$$\therefore x = \frac{1.824}{12.8} = 0.1425 \text{ m} = 14.25 \text{ cm}$$

Example 6:

The pressure between two points A and B in a pipe conveying oil of specific gravity 0.8 is measured by an inverted U-tube. The column connected to point B stands 1.6 m higher than that at point A. A commercial pressure gage attached directly to the pipe at A reads 1.125 kg (f)/ cm². determine its reading when attached directly to the pipe at B.

Solution:

The difference of pressure head between the two points A and B is equal to 1.6 m of oil.

$$\therefore \frac{p_B}{w} - \frac{p_A}{w} = (1.6 \times 0.8) = 1.28 \text{ m of water}$$

But $p_A = 1.125 \text{ kg (f)/cm}^2$

$$\therefore \frac{p_A}{w} = \frac{1.125 \times 10^4}{1000} = 11.25 \text{ m of water}$$

Thus by substitution, we get

$$\frac{p_B}{w} - 11.25 = 1.28$$

or $\frac{p_B}{w} = 12.53$ m of water

$$\therefore p_B = (12.53 \times 1000) \text{ kg (f)/ m}^2$$

$$= \frac{12.53 \times 1000}{10^4} = 1.253 \text{ kg (f) /cm}^2.$$

Example 7:

Two pipes as shown in figure convey toluene of specific gravity 0.875 and water respectively. Both the liquids in the pipes are under pressure. The pipes are connectd to a U-tube manometer and the hoses connecting the pipes to the tubes are filled with the corresponding liquids. Find the difference of pressure in two pipes if the level of manometric liquid having a specific gravity 1.25 is 7.25 m higher in the right limb than the lower level of toluene in the left limb of the manometer.

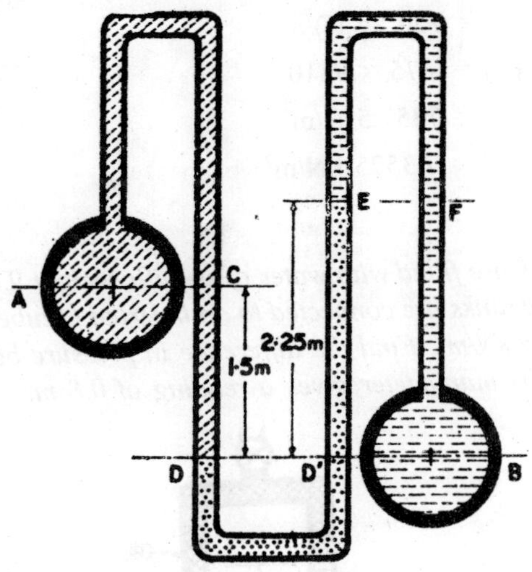

Fig. 5.22

Solution:

Let p_A and p_B be the pressure intensities at the centre of the pipes A and B respectively.

Since points D and D′ lie at the same horizontal plane and in the same continuous static mass of liquid, the pressure heads at these two points will be equal.

$$\text{Pressure head at D} = \frac{p_A}{w} + (1.5 \times 0.875)$$

$$\text{Pressure head at D}' = \frac{p_B}{w} - 2.25 + (2.25 \times 1.25)$$

The pressure heads have been calculated in terms of water column. Equating the pressure heads, we get

$$\frac{p_A}{w} + (1.5 \times 0.875) = \frac{p_B}{w} - 2.2m5 + (2.25 \times 1.25)$$

$$\text{or} \quad \frac{p_A}{w} + 1.3125 = \frac{p_B}{w} + 0.5625$$

$$\text{or} \quad \frac{p_B}{w} - \frac{p_A}{w} = 0.75$$

or $(p_B - P_A) = (0.75 \times 1000)$

$= 750 \text{ kg (f)/m}^2$

or $(p_B - p_A) = 0.75 \times 9810$

$= 7357.5 \text{ N/m}^2$

$= 7.3575 \text{ kN/m}^2.$

Example 8:

Two tanks are filled with water of specific weight 9.81 kN/ m³. The bottoms of the tanks are connected to an inverted U-tube containing oil weighing 7.85 kN/m³. Find the difference in pressure between the two tanks when the manometer gives a reading of 0.8 m.

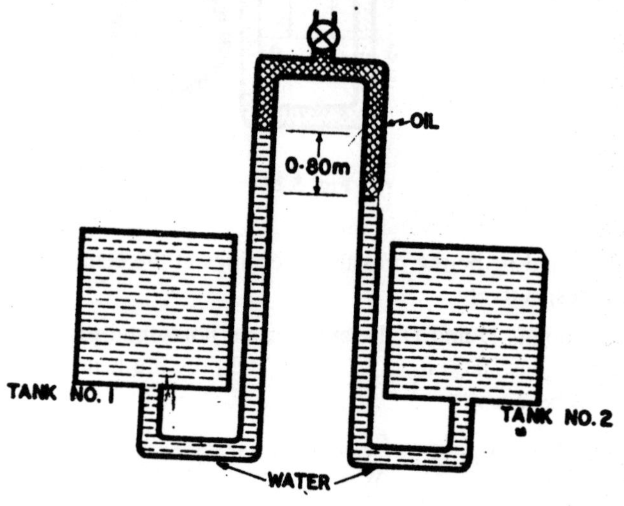

Fig. 5.23

Solution:

The difference of pressure head in the two tanks is given by

$$\frac{p_1}{w} - \frac{p_2}{w} = x\,(S_1 - S_2)$$

$$= 0.8\left(1 - \frac{7.85}{9.81}\right)$$

$= 0.16$ m of water $= 16$ cm of water.

Eample 9:

In the accompanying figure, fluid A is water, fluid B is oil of specific gravity 0.85, Z = 0.7 m and y = 1.5 m. Compute pressure difference between m and n.

Solution:

Let the height of the common surface above the point m be x. Since pressure head at T = pressure head at T; we have

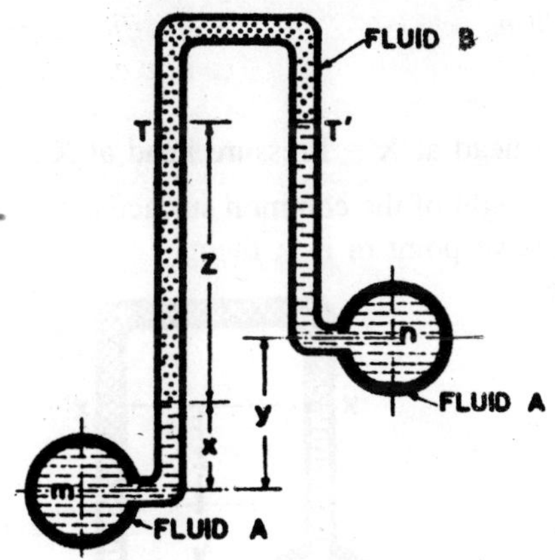

Fig. 5.24

$$\frac{p_m}{w} - x - (Z \times 0.85)$$

$$= \frac{p_n}{w} - (Z + x - y)$$

$$\text{or } \frac{p_m}{w} - \frac{p_n}{w} = y - Z\,(1 - 0.85)$$

$$= 1.5 - 0.7\,(0.15)$$

$$= (1.5 - 0.105)$$

$$= 1.395 \text{ m of water}$$

$$\text{or } (p_m - p_n) = \frac{1.395 \times 1000}{10^4}$$

$$= 0.1395 \text{ kg (f)/cm}_2$$

$$\text{or } (p_m - p_n) = 1.395 \times 9.810$$

$$= 13.685 \text{ kN/m}^2$$

Example 10:

(a) *Water fills the vessels shown in the accompanying figure and a portion of the connecting tube. If the manometric liquid is oil of specific gravity 0.9 find the difference in pressure intensity at m and n when h = 1.25 m and Z = 0.3 m.*

(b) *If in the same figure instead of water there is mercury and the manometric liquid used has a specific gravity of 1.6, find the difference in pressure intensity at m and n when h = 0.6 m and Z = 1.0 m.*

Solution:

(a) Pressure head at X = Pressure head at X′.

Thus if the height of the common surface between oil and water in the right limb above point m is y, then

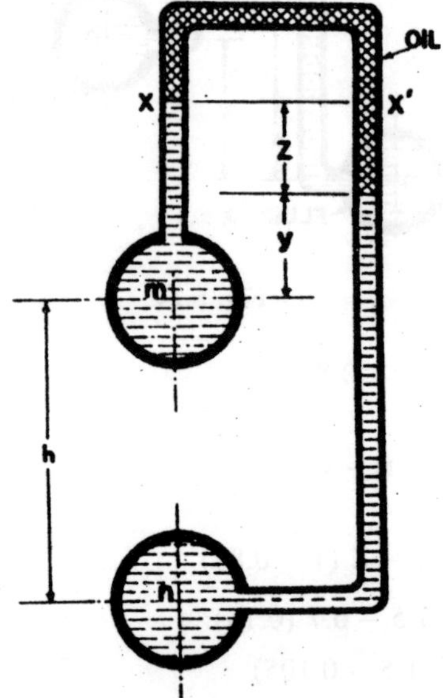

Fig. 5.25

$$\frac{p_m}{w} - Z - y = \frac{p_n}{w} - (h + y) - Z(0.9) \qquad \text{...(i)}$$

or $$\frac{p_n}{w} - \frac{p_m}{w} = h - Z\,(1 - 0.9)$$

$= 1.25 - 0.3\ (1 - 0.9)$

$= 1.22$ m of water

$$\text{or } p_n - p_m = \frac{1.22 \times 1000}{10^4} = 0.122 \text{ kg (f)/cm}^2$$

$$\text{or } p_n - p_m = 1.22 \times 9810$$

$= 11968.2$ N/m^2

$= 11.968$ kN/m^2

(b) substituting the corresponding values in equation (i) noted above, we get

$$\frac{p_m}{w} - (Z + y) \times 13.6 = \frac{p_n}{w} - (h + y) \times 13.6 - Z\ (1.6)$$

$$\text{or } \frac{p_m}{w} - \frac{p_n}{w} = Z\ (13.6 - 1.6) - h\ (13.6)$$

$= 1.0\ (13.6 - 1.6) - 0.6\ (13.6)$

$= (12.0 - 8.16) = 3.84$ m of water

$$\text{or } (p_m - p_n) = \frac{3.84 \times 1000}{10^4} = 0.384 \text{ kg (f) cm}^2$$

$$\text{or } (p_m - p_n) = 3.84 \times 9810$$

$= 37670.4$ N/m^2 $= 37.6704$ kN/m^2

Example 11:

Figure shows a differential gage. X and Y are connected to two different sources of pressure. With the equal pressure at X and Y, tops of kerosene columns stand at the common level J-J and water at 0.0. Points X and Y are at the same level. Find the difference of pressure head between X and Y in mm of water if h is 0.3 m. Take the reservoir cross-section 100 times that of glass tube.

Solution:

Let A and a be the cross- sectional areas of the reservoir and the glass tube respectively.

$\therefore$ (A/a) = 100

Let l be the vertical height of kerosene column above 0-0 in the right limb. Also let Z and Z′ be the two points in the left and the right limbs respectively at the same level and at height (h/2) above J-J as shown in the figure.

$\therefore$ Pressure head at Z = Pressure head at Z′.

The pressure head at Z will be equal to

$$\frac{p_x}{w} - t - y - \left(1 - \frac{y}{2}\right)0.82 - h\,(0.82)$$

where t is the vertical depth of the points X and Y below the common surface between kerosene and water in the right limb.

Similarly the pressure head at point Z′ will be equal to

$$\frac{p_y}{w} - t - \left(1 + \frac{y}{2}\right)0.82 - h\,(0.80)$$

Thus, we have

$$\frac{p_x}{w} - t - y - \left(1 - \frac{y}{2}\right)0.82 - h\,(0.82)$$

$$= \frac{p_y}{w} - t - \left(1 + \frac{y}{2}\right)0.82 - h\,(0.80)$$

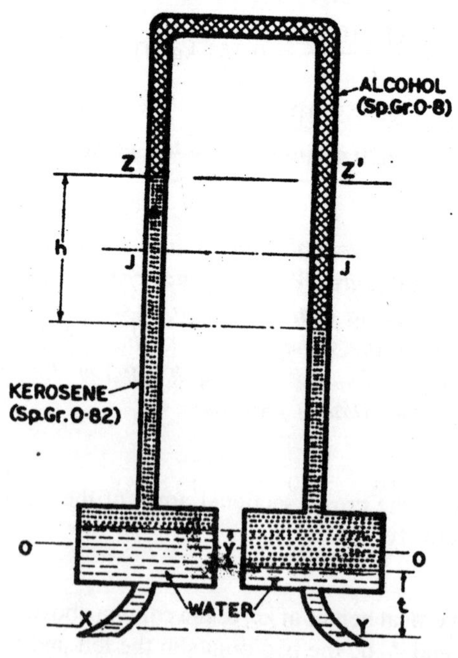

Fig. 5.26

or $\frac{p_x}{w} - \frac{p_y}{w}$

$= h\,(0.82 - 0.80) + y\,(1 - 0.82)$

Further

$A \times y = a \times h$

and $h = 0.3$ m

$\therefore\ y = \frac{a}{A} \times h = \frac{0.3}{100} = 0.03$ m

By Substituting these values, we get

$$\frac{p_x}{w} - \frac{p_y}{w} = 0.3\,(0.82 - 0.80) + 0.003\,(1 - 0.82)$$

$= 6.54 \times 10^{-3}$ m $= 6.54$ mm of water.

Example 12:

A two-liquid double column enlarged ends manometer is used to measure a high precision pressure difference between two points of a pipeline containing gas under pressure. The basins are partly filled with methyl alcohol of specific gravity 0.78 and the lower portion of the U-tube is filled with mercury of specific gravity 13.6. The specific weight of the gas which is methane is 0.47 kg (f) m³. Find the pressure difference if the U-tube reading is 30 mm and the diameter of the basin is 15 times that of the U-tube.

Solution:

Let p_1 and p_2 be pressure intensities at the two points 1 and 2 in the pipeline. The equating the pressure heads at the two points Z and Z′, as shown in the accompanying figure, we get

$$\frac{p_1}{w} + (1 + y)\frac{0.476}{1000} + m\,(0.78) + 30\,(0.78)$$

$$= \frac{p_1}{w} + 1\left(\frac{0.476}{1000}\right) + (m + y)\,0.78 + 30\,(13.6)$$

or $\frac{p_1}{w} - \frac{p_2}{w}$

$= 30\,(13.6 - 0.78) + y\,(0.78 - 0.476 \times 10^{-3})$

$= 384.6 + y\,(0.7795)$

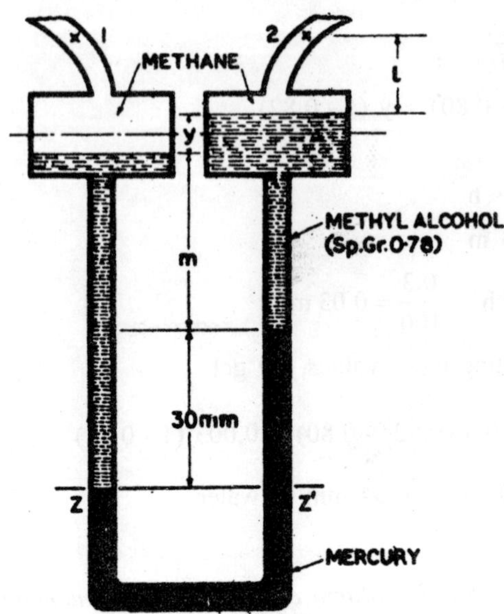

Fig. 5.27

Further

$A \times y = 30 \times a$

or $y = 30 \dfrac{a}{A}$ mm

But $A = (15)^2 \; a = 225 \; a$

$\therefore \; y = \dfrac{30}{225}$

By substituting these values, we get

$$\frac{p_1}{w} - \frac{p_2}{w} = 384.6 + \frac{30}{225}(0.7795)$$

= 384.704 mm of water

= 38.4704 cm of water

Example 13:

A manometer consists of an inclined glass tube which is connected to a metal cylinder standing upright, and manometric liquid fills the apparatus to a fixed zero mark on the tube when both cylinder and the tube are open to the atmosphere. The upper end of the cylinder and the tube are open to the atmosphere. The upper end of the cylinder is then connected to a gas supply at a pressure p and the liquid rises in the tube.

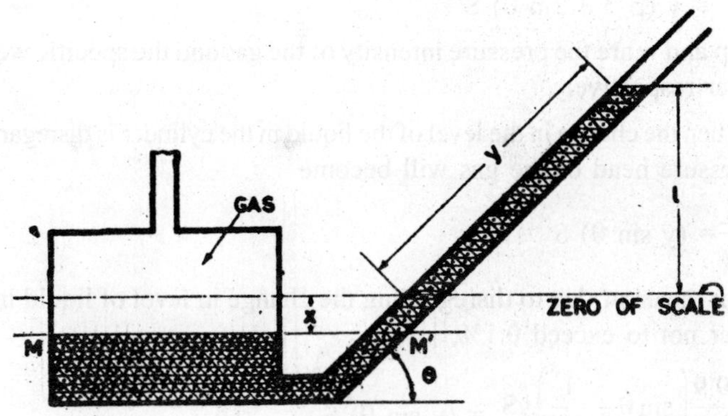

Fig. 5.28

Find an expression for the pressure p in cm of water when the liquid reads y cm in the tube, in terms of inclination θ of the tube, the specific gravity of the liquid S, and the ratio ρ of the diameter of the cylinder to the diameter of the tube. Hence determine the value of ρ so that the error due to disregarding the change in level in the cylinder will not exceed 0.1% when θ = 30°.

Solution:

When the level of liquid rises in the tube by y cm, the level of the liquid will fall in the reservoir by certain amount ; say x°, which will be given by

$$x\left(\frac{\pi D^2}{4}\right) = y\left(\frac{\pi d^2}{4}\right)$$

$$\text{or} \quad x = y\left(\frac{d}{D}\right)^2 = \frac{y}{\rho^2}$$

Further the vertical rise corresponding to y cm of rise = l (y sin θ).

Thus equating the pressure heads at the two points M and M′, and neglecting the height of the gas column, we get

$$\frac{p}{w} = xS + (y \sin \theta)\, S.$$

$$= \left(\frac{y}{\rho^2} + y \sin\theta\right) S$$

$$= y\,(\rho^{-2} + \sin\theta)\,S$$

where p and w are the pressure intensity of the gas and the specific weight of water respectively.

When the change in the level of the liquid in the cylinder is disregarded, the pressure head of the gas will become

$$\frac{p}{w} = (y \sin\theta)\,S$$

For the error due to disregarding the change in level of liquid in the cylinder not to exceed 0.1%,

$$\frac{99.9}{100}\left(\sin\theta + \frac{1}{\rho^2}\right) yS = (y \sin\theta)\,S$$

But $\theta = 30°$

Thus $$\frac{99.9}{100}\left(\frac{1}{2} + \frac{1}{\rho^2}\right) = \frac{1}{2}$$

or $$\frac{0.999}{\rho^2} = \frac{0.1}{2 \times 100}$$

or $\rho^2 = 1998$

$\therefore\ \rho = 44.7.$

Example 14:

For a gage pressure at A of – 0.15 kg (f)/cm², determine the specific gravity of the gage liquid B in the figure given below.

Solution:

Pressure at C = Pressure at D

Thus

$$-(0.15) \times 10^4 + (1000 \times 1.6 \times 0.5) = p_D$$

or $p_D = -0.7 \times 10^3$ kg (f)/m²

Between points D and E, since there is an air column which can be neglected. Thus

$$p_D = p_E$$

Also, pressure at F = pressure at G

But point G being at atmospheric pressure,

$p_G = 0 = p_F$

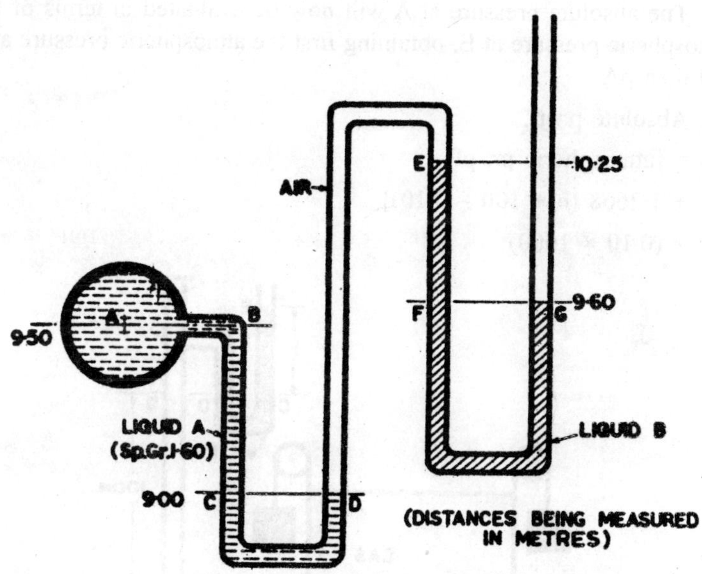

Fig. 5.29

Thus $p_F = p_E + S \times 1000\ (10.25 - 960) = 0$

or $-\ 0.7 \times 103 + S \times 1000\ (0.65) = 0$

$\therefore\ S = \dfrac{0.70}{0.65} = 1.077.$

Example 15:

The pressure head at level A – A is 0.1m of water and the unit weight of gas and air 0.5607 kg (f) m3 and 1.2608 kg (f)/ m³ respectively. Determine the reading of the water in the U-tube gage which measures the gas pressure at level B in the figure given below.

Solution:

Assume that the values of specific weight for air gas remain constant for the 100 m difference of elevation. Because the specific weights of gas and air are of the same order of magnitude, the change in atmosphere pressure with altitude must be taken into account.

Thus absolute pressure at

C = Absolute pressure at D or Atmospheric pr. pE + 1000 × h

= Absolute pr. at A – (0.5607× 100)

The absolute pressure at A will now be evaluated in terms of the atmospheric pressure at E, obtaining first the atmospheric pressure at F and then pA.

Absolute pr. p_A.

= [atmospheric pr. pE

+ 1.2608 (h + 100 – 0.10)]

+ (0.10 × 1000)

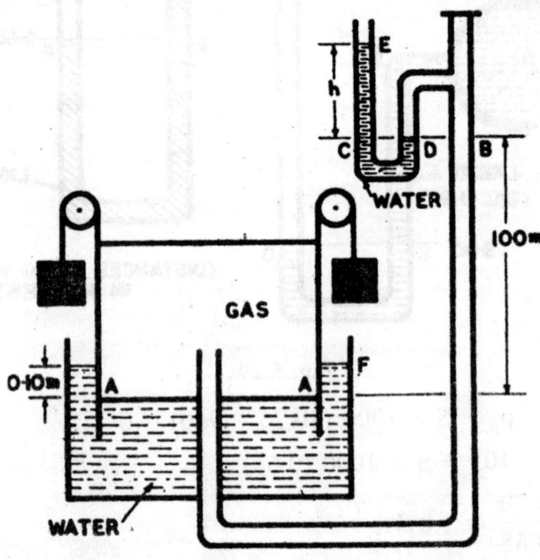

Fig. 5.30

By substituting the value of absolute pr. pA in the above equation and cancelling atmospheric pr. pE, we get

1000 × h = 1.2608 (h + 99.9) + (1000 × 0.10) – (0.5607 × 100)

or (1000 – 1.2608) h = (1.2608 × 99.9) + 100 – 56.07

= 169.884

or h = $\frac{169.884}{998.7392}$ m of water

$$= \frac{169.884 \times 100}{998.7392} \text{ cm of water}$$

Example 16:

Point A is 0.25 m below the surface of the liquid of specific gravity 1.25, in the vessel as shown in the figure. What is the pressure at A if a liquid of specific gravity 1.36 rises 2.1. m the tube?

Solution:

The pressure of the air in the vessel

$= -(2.1 \times 1.36)$ m of water

$= -2.856$ m of water.

Thus pressure at point A in the vessel

$= [-2.856 + 0.25 \times 1.25]$ m of water

$= [-2.856 + 0.3125]$ m of water

$= -2.5435$ m of water

$= -(2.5435 \times 9810)\text{N/m}^2$

$= 24.952 \text{ kN/m}^2 = -2543.5$ kg (f) m^2

$= 0.25435$ kg (f) cm^2

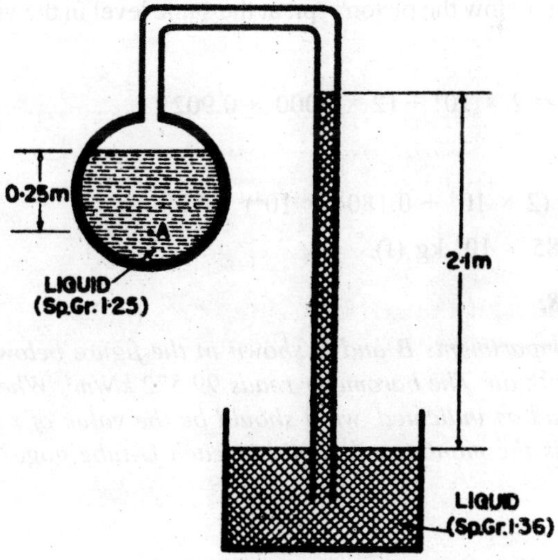

Fig. 5.31

Example 17:

The cylinder and tubing shown in the figure contain oil of specific gravity 0.902. For a gage reading of 2kg (f)/cm². what is the total weight of the piston and the slab placed on it?

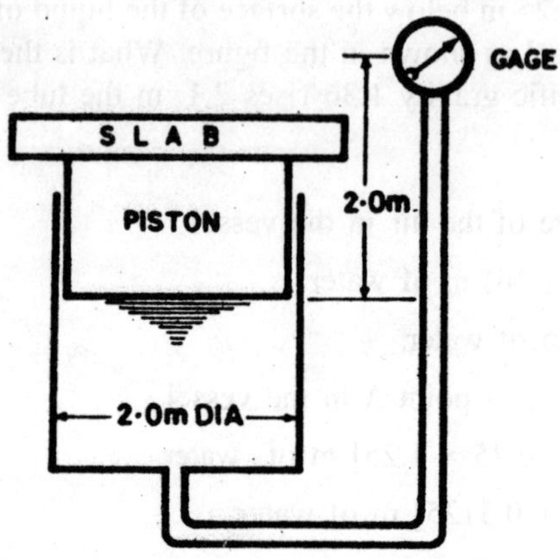

Fig. 5.32

Solution:

Let W be the total weight of the piston and the slab.

Since pr. below the piston = pr. at the same level in the vertical, tube, we have

$$\frac{W}{\frac{\pi}{4}(2)^2} = 2 \times 10^4 + (2 \times 1000 \times 0.902)$$

$$W = \pi\,(2 \times 10^4 + 0.1804 \times 10^4)$$

$$= 6.85 \times 10^4 \text{ kg (f)}.$$

Example 18:

The compartments B and C shown in the figure below are closed and filled with air. The barometer reads 99.572 kN/m². When the gages A and D read as indicated, what should be the value of x for gage E, if mercury is the manometric liquid in each U-tube gage?

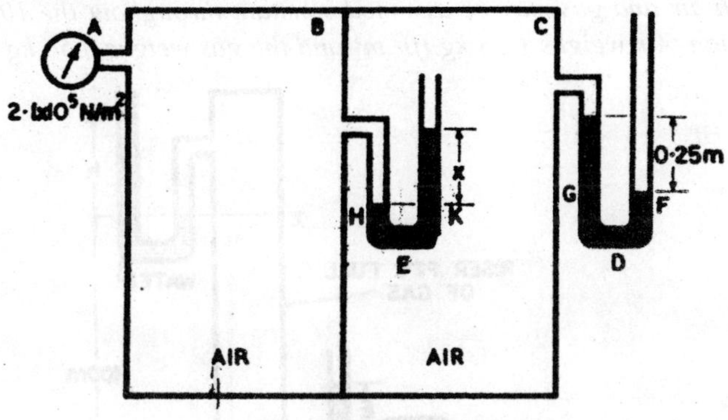

Fig. 5.33

Solution:

Pressure at F = Pressure at G Thus if p is the pressure intensity of air in the compartment C in terms of absolute units, then

$$\frac{p_c}{w} + 0.25 \times 13.6 = \frac{99.572 \times 10^3}{9810}$$

$$\text{or } \frac{p_c}{w} = (10.15 - 3.4)$$

= 6.75 m of water (absolute)

Further Pressure at K = Pressure at H

Thus in terms of absolute units

$$6.75 + x\,(13.6) = \left[\frac{2.1 \times 10^5}{9810} + \frac{99.572 \times 10^3}{9810}\right]$$

$$\text{or } x = \frac{1}{13.6}(31.56 - 6.75)$$

= 1.824 m

Example 19:

The pressure of illuminating gas is measured as shown in the figure below. The U-tube at the gas main shows a pressure of 75 mm of water. What difference should the upper tube show in mm of water? The right hand column of both gages are open to the atmosphere. The density of both air and gas may be assumed constant throughout the 100 m of elevation. Air weighs 1.33 kg (f)/ m³ and the gas weighs 0.56 kg (f)/m³.

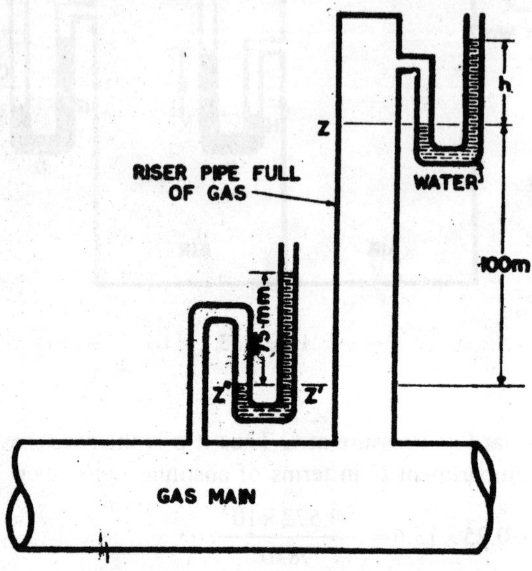

Fig. 5.34

Solution:

Let p_a represent the intensity of the atmospheric pressure in kg (f)/ m² at an elevation of 100 m.

Thus the absolute pressure of the gas at point Z will be equal to

$(p_a + h \times 1000)$ kg (f)/ m²

The pressure at point Z in the riser pipe at 100 m below Z will be given by

Pr. at Z′ = Pr. at Z + (100×0.56)

$= pa + h \times 1000 + (100 \times 0.56)$

Further

Pr. at Z′ = Pr. at Z″

Thus

$(p_a + h \times 1000) + (100 \times 0.56)$

$= [p_a + (1.33 \times 100)] + \frac{75}{1000} \times 1000$

or $1000\ h = (133 + 75 - 56)$

$\therefore\ h = 0.152$ m

Example 20:

For the multiple differential manometer shown in the sketch, if points A, B and C are at the same elevation, what is the difference in pressure heads in terms of water column between A and B, between A and C and between B and C?

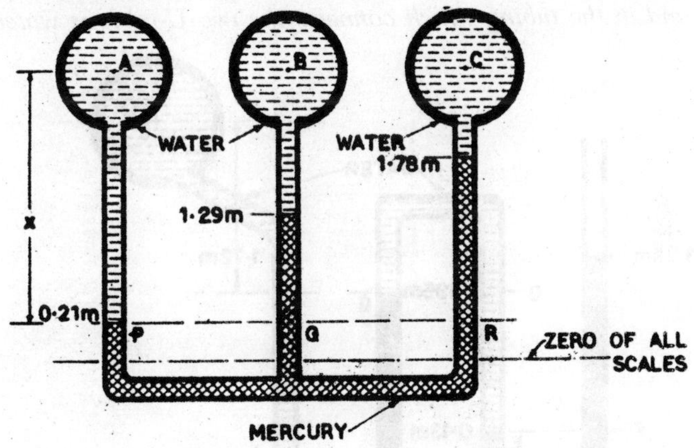

Fig. 5.35

Solution:

Pr. at P = Pr. at Q = Pr. at R

Thus if the depth of the point P below A = x, then

$$\frac{p_A}{w} + x = \frac{p_B}{w} + (x - 1.08) + (1.08 \times 13.6)$$

$$= \frac{p_C}{w} + (x - 1.57) + (1.57 \times 13.6)$$

Thus

$$\frac{p_A}{w} - \frac{p_B}{w} = 1.08\ (13.6 - 1)$$

= 13.608 m of water

Similarly

$$\frac{p_A}{w} - \frac{p_C}{w} = 1.57\ (13.6 - 1)$$

$= 19.782$ m of water

and $\frac{p_B}{w} - \frac{p_C}{w} = (19.782 - 13.608)$

$= 6.174$ m of water

Example 21:

For a compound manometer shown in the figure below, what is gage pressure at C if the manometric fluid is mercury and if the fluid in the pipe and in the tubing which connects the two U-tubes is water?

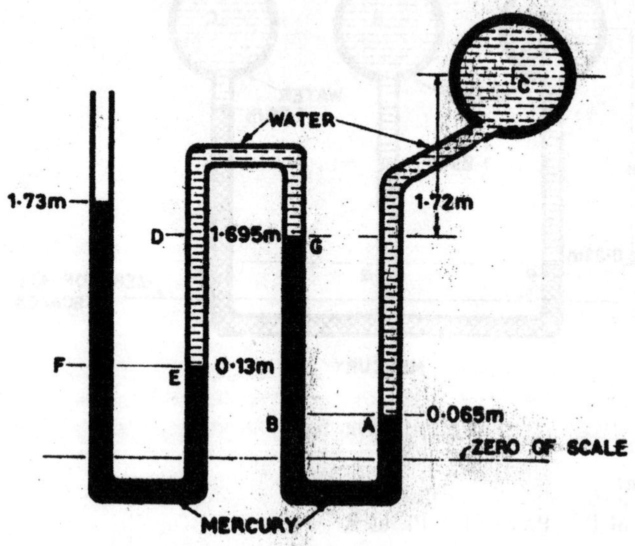

Fig. 5.36

Solution:

Pressure at A, p_A = Pressure at B, p_B

Thus we have

$$\frac{p_A}{w} = \frac{p_C}{w} + 1.72 + 1.695 - 0.065 = \frac{p_B}{w}$$

Further

pressure at G, $\frac{p_G}{w} = \frac{p_B}{w} - (1.695 - 0.065)\ 13.6$

$$= \frac{p_C}{w} + (1.72 + 1.695 - 0.065\ )$$

Also – (1.695 – 0.065) × 13.6

Pressure at G, P_G = Pressure at D, p_D

and Pressure at E, p_E = Pressure at F, p_F

But $\frac{p_E}{w} - \frac{p_D}{w} + (1.695 - 0.13)$

and $\frac{p_F}{w} = (1.73 - 0.13) \times 13.6$

Thus by substitution, we get

$\frac{p_C}{w} + (1.72 + 1.695 - 0.065) - (1.695 - 0.065) \times 13.6$

$= (1.73 - 0.13) \times 13.6 - (1.695 - 0.13)$

or $\frac{p_C}{w} + 3.35 - 22.168 = 21.76 - 1.565$

$\therefore \quad \frac{p_C}{w} = (43.928 - 4.915) = 39.013$ m of water

i.e., pc = 39.013 × 10^3 kg (f) /m^2

= 3.9013 kg (f)/ cm^2

Example 22:

An empty cylindrical bucket, 0.3 m in diameter and 0.5 m long whose wall thickness and weight can be considered as negligible is forced with its open end first into water until its lower edge is 4 m below the surface.

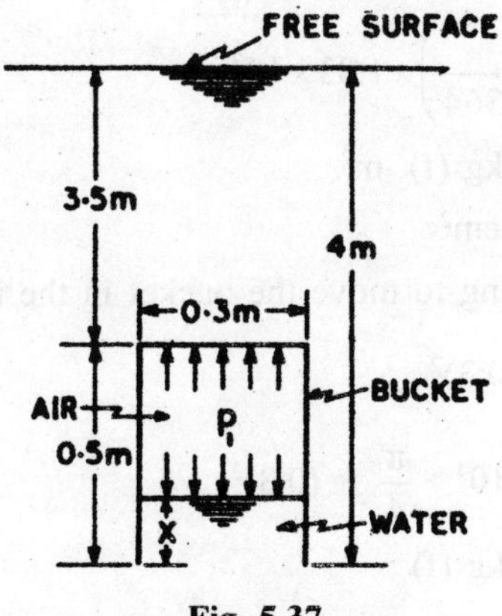

Fig. 5.37

What force will be required to maintain position, assuming the trapped air to remain at constant temperature during the entire operation? Atmospheric pressure = 1.03 kg (f)/cm².

Solution:

Let P_a be the atmospheric pressure and P_1 be the absolute pressure of the compressed air trapped in the bucket. If the depth of water raised in the bucket is x, then since the temperature of the air remains, constant, according to isothermal condition.

$$p_a \times \frac{\pi}{4}(0.3)^2 \times 0.5 = p_1 \times \frac{\pi}{4}(0.3)^2 (0.5 - x)$$

$$\text{or } p_1 = \left(\frac{0.5}{0.5-x}\right)p_a \qquad \text{...(i)}$$

$$\text{Also } p_1 = p_a + (4 - x) \times 1000 \qquad \text{...(ii)}$$

From equations (i) and (ii), we get

$$\left(\frac{0.5}{0.5-x}\right)p_a = pa + (4 - x) \times 1000$$

Since $p_a = 1.03 \times 10^4$ kg (f)/m², we get

$$\text{or } \left(\frac{0.5}{0.5-x}\right) 1.03 \times 10^4 + (4 - x) \times 1000$$

$$\text{or } \left(\frac{x}{0.5-x}\right) 1.03 \times 10^4 = (4 - x) \times 1000$$

$\therefore$ x = 0.1364 m

Substituting the value of x in (i), we get

$$p_1 = \left(\frac{0.5}{0.5-0.1364}\right) \times 1.03 \times 10^4$$

$= 1.416 \times 10^4$ kg (f)/ m²

= 1.416 kg (f) cm²

The force tending to move the bucket in the upward direction

$$P_1 = p_1 \times \frac{\pi}{4}(0.3)^2$$

$$= 1.4164 \times 10^4 \times \frac{\pi}{4} \times (0.3)^2$$

= 1001.194 kg (f)

The force acting on the bucket in the downward direction

$$p^2 = [1.03 \times 10^4 + 1000 \times 3.5] \frac{\pi}{4} \times (0.3)^2$$

$$= 13.8 \times 10^3 \times \frac{\pi}{4} \times (0.3)^2$$

$$= 975.464 \text{ kg (f)}$$

Thus the force required to maintain the bucket in this position

$$F = (p_1 - p_2)$$

$$= (1000.194 - 975.464)$$

$$= 24.73 \text{ kg (f)}$$

Example 23:

Petrol of specific gravity 0.8 flows upwards though a vertical pipe. A and B are two points in the pipe, B being 0.3 m higher than A. Connections are led from A and B to a U-tube containing mercury. If the difference of pressure between A and B is 0.18 kg (f)/ cm², find the reading shown by the differential mercury manometer gage.

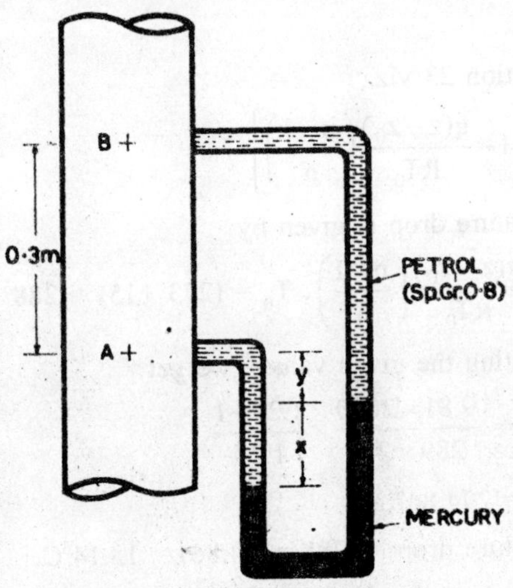

Fig. 5.38

Solution:

Let p_A and p_B be the pressure intensities at the points A and B, if the reading of the gage is x, then as shown in the figure

$$\frac{p_A}{w} + (x + y) \times 0.8 = \frac{p_B}{w} +$$

$$(0.3 + y)\ 0.8 + (x \times 13.6)$$

$$\left(\frac{p_A}{w} - \frac{p_B}{w}\right) = x\ (13.6 - 0.8) + \ (0.3 \times 0.8)$$

$$\text{but } \left(\frac{p_A}{w} - \frac{p_B}{w}\right) = \frac{0.18 \times 10^4}{1000} = 1.8 \text{ m}$$

Hence $1.8 = 12.8\ x + 0.24$

$$\therefore\ x = \frac{1.56}{12.8} = 0.122 \text{ m or } 12.2 \text{ cm.}$$

Example 24:

Assuming adiabatic conditions, find the temperature drop for an altitude of 2000 meters above the earth's surface. Assume the earth's surface temperature as 15°C, the gas constant R for air 289 the atmospheric pressure at the altitude, if it is given that it is 1.033 kg (f)/ cm² at earth's surface.

Solution:

From equation 23 viz.,

$$\frac{T}{T_0} = \left[1 - \frac{g(z - z_0)}{RT_0}\left(\frac{n-1}{n}\right)\right]$$

the temperature drop is given by

$$\frac{T_0 - T}{T_0} = \frac{g(z - z_0)}{RT_0}\left(\frac{n-1}{n}\right); \ T_0 = (273 + 15) = 288° \text{ abs}$$

By substituting the given values, we get

$$\frac{288 - T}{288} = \frac{9.81 \times 2000}{289 \times 288} \times \frac{1.24 - 1}{1.24}$$

$$\therefore \qquad T = 274.86° \text{ abs.}$$

$\therefore$ Temperature drop = (288 –274.86) = 13.14°C.

Further from equation 22 the variation of atmospheric pressure with altitude is given as

$$\frac{p}{p_0} = \left[1 - \frac{g(z - z_0)}{RT_0}\left(\frac{n-1}{n}\right)\right]^{\frac{n}{n-1}}$$

By substituting the given values, we get

$$\frac{p}{1.033} = \left[1 - \frac{9.81 \times 200}{289 \times 288} \times \left(\frac{1.24-1}{1.24}\right)\right]^{\left(\frac{1.24}{1.24-1}\right)}$$

$p = 0.812$ kg (f)/cm^2.

Example 25:

At the top of a mountain the temperature is- 5°C and mercury barometer reads 56.6 cm, whereas the reading at the foot joule/[kg (m) deg C abs], calculate the height of the mountain.

Solution:

From equation 23, we have

$$\frac{T}{T_0} = \left[1 - \frac{g(z-z_0)}{RT_0}\left(\frac{n-1}{n}\right)\right]$$

$R = 287$ joule/[kg (m) deg C abs]

$= 287$ m^2/(sec^2 deg C abs)

$T = (273 - 5) = 268°$ C abs.

For dry abiabatic conditions, $n = 1.4$.

Thus by substitution, we get

$$\frac{268}{T_0} = \left[1 - \frac{9.81(z-z_0)}{287 \times T_0} \times \left(\frac{1.4-1}{1.4}\right)\right]$$

$$\therefore T_0 = \left[268 + \frac{9.81 \times (z-z_0) \times 0.4}{287 \times 1.4}\right]$$

Further from equation 22

$$\frac{p}{p_0} = \left[1 - \frac{g(z-z_0)}{RT_0}\left(\frac{n-1}{n}\right)\right]^{\frac{n}{n-1}}$$

By substitution, we get

$$\frac{56.6}{74.9} = \left[1 - \frac{9.81 \times (z-z_0) \times 0.4 \times 287 \times 1.4}{(287 \times 1.4)(268 \times 287 \times 1.4) + 9.81 \times (z-z_0) \times 0.4}\right]^{\frac{1.4}{0.4}}$$

$$\text{or} \quad \left(\frac{56.6}{74.9}\right)^{\frac{0.4}{1.4}} = \left[1 - \frac{9.81 \times (z-z_0) \times 0.4}{(268 \times 287 \times 1.4) + 9.81 \times (z-z_0) \times 0.4}\right]$$

$$\text{or} \quad 0.923 = \frac{268 \times 287 \times 1.4}{(268 \times 287 \times 1.4) + 9.81 \times (z - z_0) \times 0.4}$$

$\therefore (z - z_0) = 2289$ m.

EXERCISES

1. Explain the terms-intensity of pressure head.
2. Prove that the pressure is the same in all directions at a point in a static fluid.
3. State Pascal's Law and give some examples where this principle is applied.
4. Define the terms gage pressure, vacuum pressure and absolute pressure. Indicate their relative positions on a chart.
5. Briefly explain the principle employed in the manometers used for the measurement of pressure.
6. Differentiate between simple and differential type of manometers.
7. Describe with the help of neat sketches different types of manometers.
8. State the advantages of mechanical pressure gages over the manometers.
9. Explain how vacuum pressure can be measured with the help of a U-tube manometer.
10. Describe with a neat sketch a micro-manometer used for very precise measurement of small pressure difference between two points.
11. Express a pressure intensity of 5 kg (f)/cm^2 in all possible units. Take the barometer reading as 76 cm of mercury.

    [Gage units : 5 kg (f)/cm^2 ; 5 × 104 kg (f)/m^2 ; 50 m of water; 49.05 x × 10^4 N/m^2; 3.68 m of mercury.]

    [Absolute units: 6.034 kg (f)/cm^2; 6.034 × 10^4 kg (f)/m^2 ; 59.19 × 10^4 N/m^2; 60.34 m of water ; 4.44 m of mercury; 5.842 atmospheres]
12. Find the depth of a point below free surface in a tank containing oil where the pressure intensity is 9 kg (f)/cm^2. Specific gravity of oil is 0.9. [100 m]

13. Convert a pressure head of 15 m of water to (a) metres of oil of specific gravity 0.750; (b) metres of mercury of specific gravity 13.6. [(a) 20. m; (b) 1.103 m]

14. A U-tube containing mercury has its right limb open to atmosphere. The left limb is full of water and is connected to a pipe containing water under pressure, the centre of which is in level with the free surface of mercury. Find the pressure of the water in the pipe above atmosphere, if the difference of level of mercury in the limbs is 5.08 cm. [0.064 kg (f)/ cm^2]

15. The pressure between two points M and N in a pipe conveying oil of specific gravity 0.9 is measured by an inverted U-tube and the column connected to point N stands 1.5 m higher than that at point M. A commercial pressure gage attached directly to the pipe at M reads 2 kg (f)/cm^2; determine its reading when attached directly to the pipe at N. [2.135 kg (f)/cm^2]

16. A pipe containing water at 172 kN/m^2 pressure is connected by a differential gage to another pipe 1.5 m lower than the first pipe and containing water at high pressure. If the difference in heights of the two mercury columns of the gage is equal to 75 mm, what is the pressure in the lower pipe? Specific gravity of mercury is 13.6. [196 kN/m^2]

17. Two pressure points in a water pipe are connected to a manometer which has the form of an inverted U-tube. The space above the water in the two limbs of the manometers is filled with toluene of specific gravity 0.875. If the difference of level of water columns in the two limbs is equal to 0.12 m what is the corresponding difference of pressure expressed in (a) kg (f)/cm^2, (b) N/m^2. [(a) 0.0015 kg (f) cm^2; (b) 147.15 N/m^2]

18. A U-tube differential gage is attached to two sections A and B in a horizontal pipe in which oil of specific gravity 0.8 is flowing. The deflection of the mercury in the gage is 60 cm, the level nearer to A being the lower one. Calculate the difference of pressure in kg (f) cm^2 between the sections A and B. [0.768 kg (f)/cm^2]

19. In the accompanying figure the areas of the plunger A and cylinder B are 40 and 4000 cm^2 respectively. The weight of cylinder B is 4 100 kg (f). The vessel and the connecting passage are filled

with oil of specific gravity 0.750. What force F is required for equilibrium, if the weight of A is neglected?

[40.85 kg (f)]

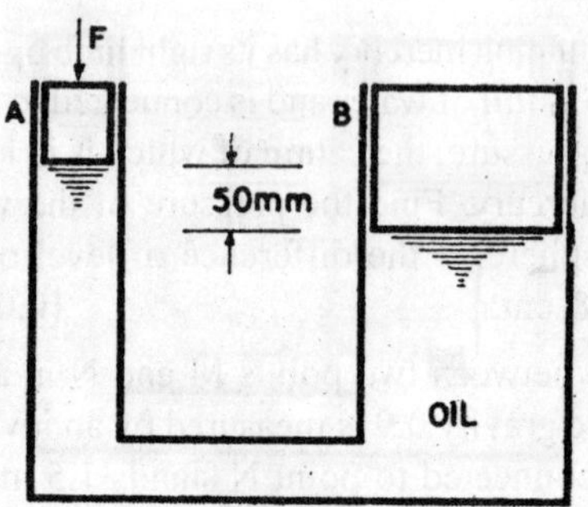

Fig. 5.39

20. Vessels A and B contain water under pressures of 274.68kN/m² [2.8 kg (f)/cm²] and 137.34 kN/m² (1.4 kg (f)/cm²] respectively. What is the deflection of the mercury in the differential gage shown in the accompanying figure?

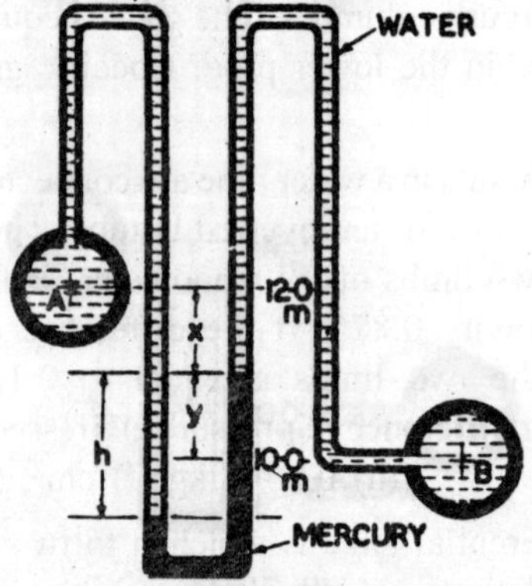

Fig. 5.40

21. The tank in the accompanying figure contains oil of specific gravity 0.750. Determine the reading of pressure gage A in (a) kg (f)/cm²; kN/m².

[(a) – 0.096 kg (f)/cm²; (b) – 9.442 kN/m²]

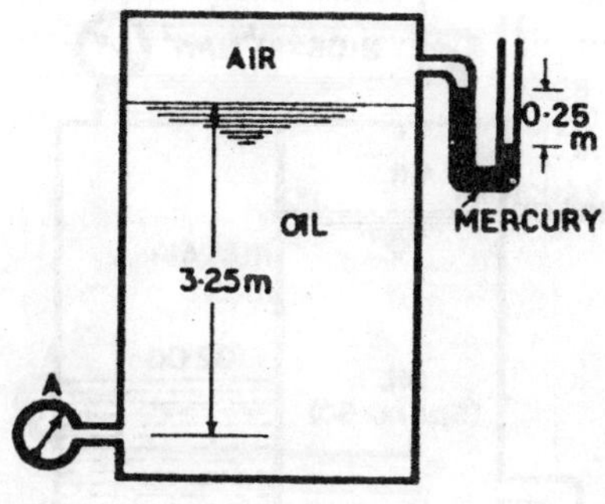

Fig. 5.41

22. In the manometer shows in the accompanying figure the liquid on the left side is carbon-tetrachloride of specific gravity 1.60 and liquid on the right side is mercury. If $(p_A - p_B)$ is 525 kg (f) m² (5 150.25 N/m²), find the specific gravity of the liquid X.

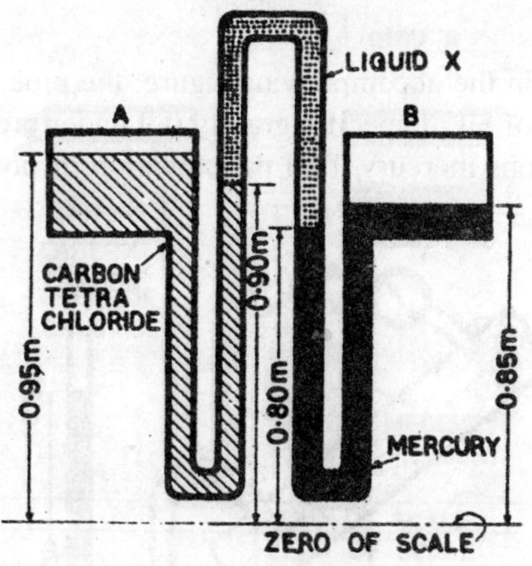

Fig. 5.42

23. In the left hand shown in the accompanying figure, the air pressure is –0.23 m of mercury. Determine the elevation of the gage liquid in the right hand column at A, if the liquid in the right hand tank is water. [94.62 m]

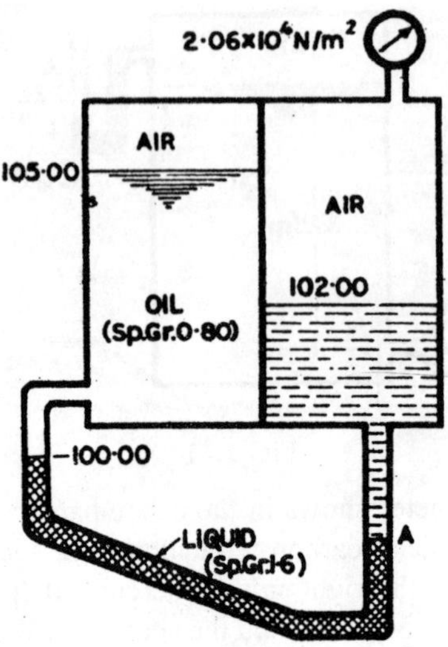

Fig. 5.43

24. As shown in the accompanying figure, the pipe and connection B are full of oil of specific gravity 0.9 under pressure. If the U-tube contains mercury, find the elevation of point A in metres.

[112.76 m]

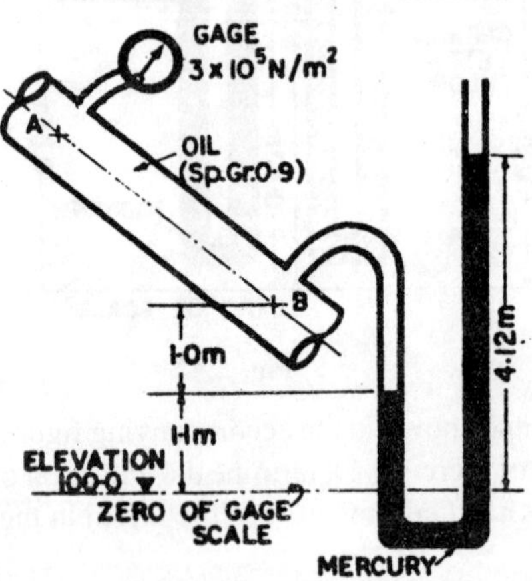

Fig. 5.44

25. Neglecting the friction between the piston A and the gas tank, find the gage reading at B in metres of water. Assuming gas and

air to be of constant specific weight and equal to 5.501 N/m^3 and 12.267 N/m^3 respectively. [0.101 m]

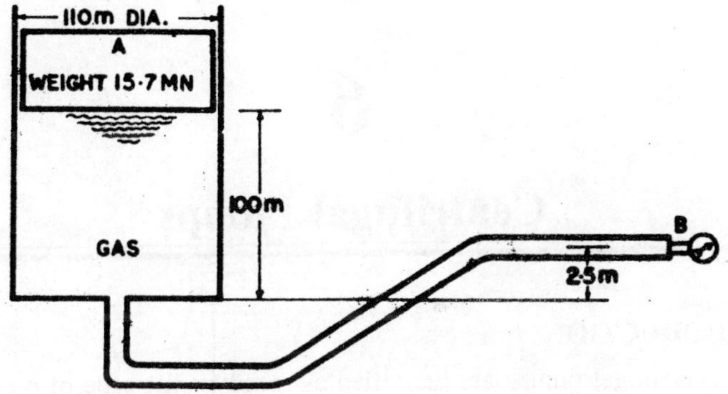

Fig. 5.45

26. A micro-manometer consists of two cylindrical bulbs A and B each 1000 sq. mm cross-sectional area which are connected by a U-tube with vertical limbs each of 25 sq. mm cross-sectional area. A liquid of specific gravity 1.2 in filled in A and another liquid of specific gravity 0.9 is filled in B, the surface of separation being in the limb attached to B. Find the displacement of the surface of separation when the pressure on the surface in B is greater than that in A by an amount equal to 15 mm head of water. [42.6 mm]

6

Centrifugal Pumps

INTRODUCTION

Centrifugal pumps are classified as rotodynamic type of pumps in which a dynamic pressure is developed which enables the lifting of liquids from a lower to a higher level. The basic principle on which a centrifugal pump works is that when a certain mass of liquid is made to rotate by an external force, it is thrown away from the central axis of rotation and a centrifugal head is impressed which enables it to rise to a higher level. Now if more liquid is constantly made available at the centre of rotation, a continuous supply of liquid at a higher level may be ensured. Since in these pumps the lifting of the liquid is due to centrifugal action, these pumps are called '*centrifugal pumps*'. In addition to the centrifugal action, as the liquid passes through the revolving wheel or impeller, its angular momentum changes, which also results in increasing the pressure of the liquid. As such centrifugal pumps behave quite differently from positive displacement pumps. A centrifugal pump does not push the liquid as in the case of a positive displacement pump, but it modifies the hydraulic gradient such that the liquid is lifted to a higher level.

According to the general direction of flow of liquid within the passage of the rotating wheel or impeller the rotodynamic pumps are classified as,

(i) Centrifugal pumps,

(ii) Half axial or screw or mixed flow pumps,

(iii) Axial flow or propeller pumps.

In the impeller of a centrifugal pump the liquid flows in the outward radial direction, while the flow of liquid in a propeller pump impeller is in the axial direction, parallel to the rotating shaft. The mixed flow pump impeller has an intermediate form so that the flow of liquid is in between the radial and axial directions. However, there are no rigid boundaries separating these three types of pumps, and often all the three types of pumps are called centrifugal pumps.

In general all the rotodynamic pumps closely resemble reaction type of hydraulic turbines and they may be regarded as reversed reaction turbines. Thus the action of a centrifugal pump is just the reverse of a radially inward flow reaction turbine. Similarly the axial flow pumps are reverse of propeller or Kaplan turbines and the mixed flow pumps are the reverse of mixed flow type turbines such as Francis turbine. In the present chapter only centrifugal pumps have been described.

ADVANTAGES OF CENTRIFUGAL PUMPS OVER RECIPROCATING PUMPS

The main advantage of a centrifugal pump is that its discharging capacity is very much greater than that of a reciprocating pump which can handle relatively small quantity of liquid only. A centrifugal pump can be used for lifting highly viscous liquids such as oils, muddy and sewage water, paper pulp, sugar molasses, chemicals etc. But a reciprocating pump can handle only pure water or less viscous liquids free from impurities as otherwise its valves may cause frequent trouble. A centrifugal pump can be operated at very high speeds without any danger of separation and cavitation. As such it can be coupled directly through flanged coupling to electric motor. On the other hand as indicated the maximum speed of a reciprocating pump is limited from the considerations of separation and cavitation. As such reciprocating pumps can be operated at low speeds only and for that these pumps are mostly belt driven. The maintenance cost of a centrifugal pump is low and only periodical check up sufficient. But for a reciprocating pump the maintenance cost is high because the parts such as valves etc. may need frequent replacement. However, a reciprocating pump can build up very high pressures as high as 69×10^6 N/m^2 {700 kg(f)/cm^2} or even more and hence these pumps are used for lifting oil from very deep oil wells.

COMPONENT PARTS OF A CENTRIFUGAL PUMP

Fig. 6.1 shows the main component parts of a centrifugal pump which are described below:

(i) *Impeller* : It is a wheel or rotor which is provided with a series of backward curved blades or vanes. It is mounted on a shaft which is coupled to an external source of energy (usually an electric motor) which imparts the required energy to the impeller thereby making it to rotate.

The impellers may be classified as:

(a) Shrouded or closed impeller,

(b) Semi-open impeller; and

(c) Open impeller, which are all shown in Fig. 1.

A 'closed or shrouded impeller' is that whose vanes are provided with metal cover plates or shrouds on both sides. These plates or shrouds are known as crown plate and lower or base plate as shown in Fig. 1. The closed impeller provides better guidance for the liquid and is more efficient. However, this type of impeller is most suited when the liquid to be pumped is pure and comparatively free from debris.

If the vanes have only the base plate and no crown plate, then the impeller is known as 'semi-open type impeller'. Such an impeller is suitable even if the liquids are charged with some debris.

An 'open impeller' is that whose vanes have neither the crown plate nor the base plate. Such impellers are useful in the pumping of liquids containing suspended solid matter, such as paper pulp, sewage and water containing sand or grit. These impellers are less liable to clog when handing liquids charged with a large quantity of debris.

(ii) *Casing :* It is an airtight chamber which surrounds the impeller. It is similar to the casing of a reaction turbine. The different types of casings that are commonly adopted are described later.

(iii) *Suction Pipe :* It is a pipe which is connected at its upper end to the inlet of the pump or to the Centre of the impeller which is commonly known as eye. The lower end of the suction pipe dips into liquid in a suction tank or a sump from which the liquid is to be pumped or lifted up.

The lower end of the suction pipe is fitted with afoot valve and strainer. The liquid first enters the strainer which is provided in order to keep the debris (such as leaves, wooden pieces and other rubbish) away from the pump. It then passes through the foot valve to enter the suction pipe. A 'foot valve' is a non-return or one-way type of valve which opens only in the upward direction. As such the liquid will pass through the foot valve only upwards and it will not allow the liquid to move downwards back to the pump.

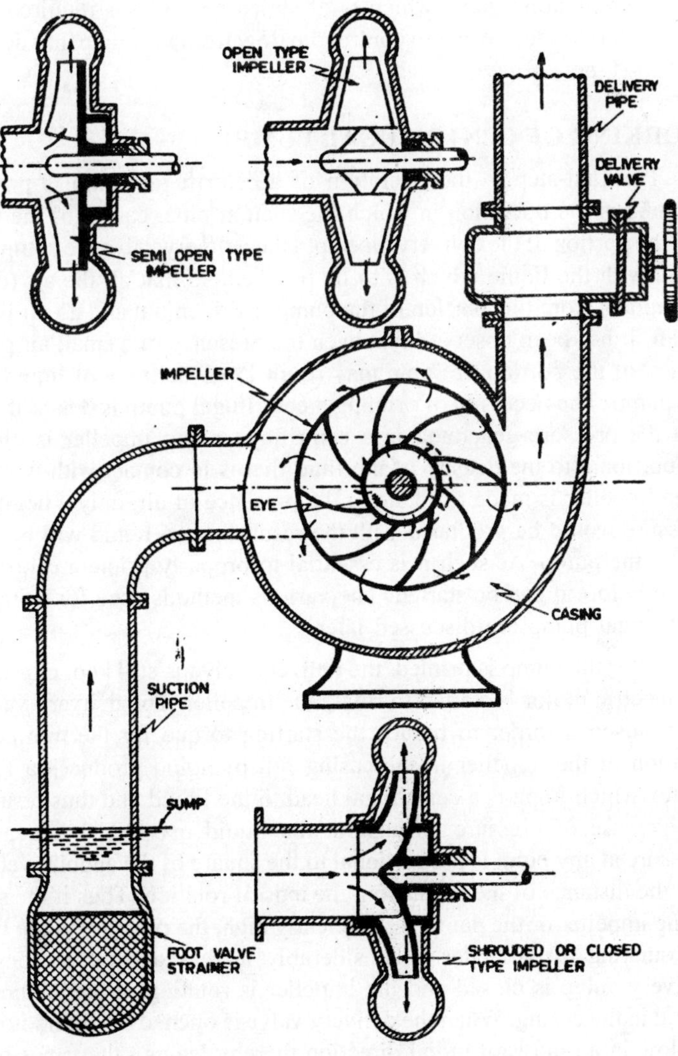

Fig. 6.1 : Component parts of a centrifugal pump.

(iv) *Delivery Pipe :* It is a pipe which is connected at its lower end to the outlet of the pump and it delivers the liquid to the required height. Just near the outlet of the pump on the delivery pipe a delivery valve is invariably provided. A delivery valve is a regulating valve which is of sluice type and is required to be provided in order to control the flow from me pump into delivery Pipe.

WORKING OF CENTRIFUGAL PUMP

The first-step in the operation of a centrifugal pump is priming. Priming is the operation in which the suction pipe, casing of the pump and the portion of the delivery pipe upto the delivery valve are completely filled with the liquid which is to be pumped, so that all the air (or gas or vapour) from this portion of the pump is driven out and no air pocket is left. It has been observed that even the presence of a small air pocket in any of the portion of pump may result in no delivery of liquid from the pump. The necessity of priming a centrifugal pump is due to the fact that the pressure generated in a centrifugal pump impeller is directly proportional to the density of the fluid that is in contact with it. Hence If an impeller is made to rotate in the presence of air, only a negligible pressure would be produced with the result that no liquid will be lifted up by the pump. As such it is essential to properly prime a centrifugal pump before it can be started. The various methods used for priming a centrifugal pump are discussed later.

After the pump is primed, the delivery valve is still kept closed and the electric motor is started to rotate the impeller. The delivery valve is kept closed in order to reduce the starting torque for the motor, The rotation of the impeller in the casing full of liquid produces a forced vortex which imparts a centrifugal head to the liquid and thus results in an increase of pressure throughout the liquid mass. The increase of pressure at any point is proportional to the square of the angular velocity and the distance of the point from the axis of rotation. Thus if the speed of the impeller or the pump is sufficiently high, the pressure in the liquid surrounding the impeller is considerably increased. Now as long the delivery valve is closed and the impeller is rotating, it just chums the liquid in the casing. When the delivery valve is opened the liquid is made to flow in an outward radial direction thereby leaving the vanes of the impeller at the outer circumference with high velocity and pressure. At the eye of the impeller due to the centrifugal action a partial vacuum is created. This causes the liquid from the sump, which is at atmospheric pressure, to rush through the suction pipe to the eye of the impeller thereby replacing the liquid which is being discharged from the entire circumference of the impeller. The high pressure of the liquid leaving the impeller is utilized in lifting the liquid to the required height through the delivery pipe.

As the liquid flows through the rotating impeller it receives energy from the vanes which results in an increase in both pressure and velocity energy. As such the liquid leaves the impeller with a high absolute velocity. In order that the kinetic energy corresponding to the high velocity of the leaving liquid is not wasted in eddies and efficiency of the pump thereby lowered, it is essential that this high velocity of the leaving liquid is gradually reduced to a lower velocity of the delivery pipe, so that the larger portion of the kinetic energy is converted into useful pressure energy. Usually this is achieved by shaping the casing such that the leaving liquid flows through a passage of gradually expanded area. The gradually increased cross-sectional area of the casing also helps in maintaining uniform velocity of flow throughout, because as the flow proceeds from the tongue T (Fig. 6.2) to the delivery pipe, more and more liquid is added from the impeller. There are different types of casings that are adopted for this purpose and on the basis of the type of casing used, the centrifugal pumps are classified into different types as described in the next section.

TYPES OF CENTRIFUGAL PUMPS

According to the type of casing provided, centrifugal pumps are classified into the following two classes :

(1) Volute pump.

(2) Diffuser or turbine pump.

Volute Pump

In a volute pump the impeller is surrounded by a spiral shaped casing which is known as *volute chamber*. As shown in Fig. 6.2 the shape of the casing is such that the sectional area of flow around the periphery of the impeller gradually increases from the tongue T towards the delivery pipe. This increase in the cross-sectional area results in developing a uniform velocity throughout the casing, because as the flow progresses from the tongue T towards the delivery pipe, more and more liquid is added to the stream from the periphery of the impeller. The volute casing may be designed to have the velocity of flow approximately equal to that of the liquid leaving the impeller. If the casing is designed according to this consideration then the loss of energy is considerably reduced, but the conversion of kinetic energy into useful pressure energy will not be possible. On the other hand if the casing is so designed that the casing velocity may be kept down to the value of the velocity in the delivery pipe, then there will be considerable loss of energy due to the difference between the casing velocity and that of the liquid discharged from the impeller. As such a compromise design is often used in which the casing is gradually enlarged so that the velocity is gradually reduced, from the velocity of the liquid leaving the impeller to that in the delivery pipe.

However, in actual practice it has been found that in volute type of casing there is only a slight increase in the efficiency of the pump, because a considerable loss of energy takes place in eddies developed in the casing.

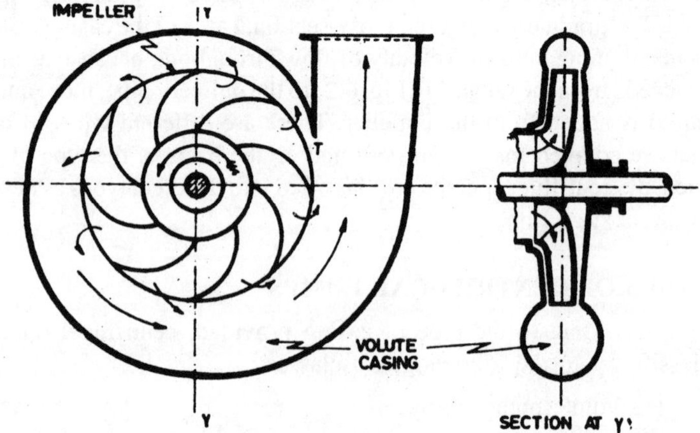

Fig. 6.2 : Volute pump.

A subsequent improvement over the simple volute casing was made by J. Thomson by prodiving a circular chamber between the impeller and the volute chamber as shown in Fig. 6.3. The circular chamber is known as *vortex* or *whirlpool chamber* and such a pump is known as *volute pump* with vortex chamber.

The vortex chamber is usually formed as a part of the casing with its side walls parallel or nearly parallel as shown in Fig. 3. It acts as a diffuser wherein the conversion of kinetic energy into pressure energy takes place as explained below. The liquid after leaving the impeller enters the vortex chamber with a whirling motion, that is the liquid particles move radially away from the centre following a rotary path while passing through this chamber. Since no work is done on the liquid as it passes through this chamber, its energy remains constant (except for the slight loss by friction).

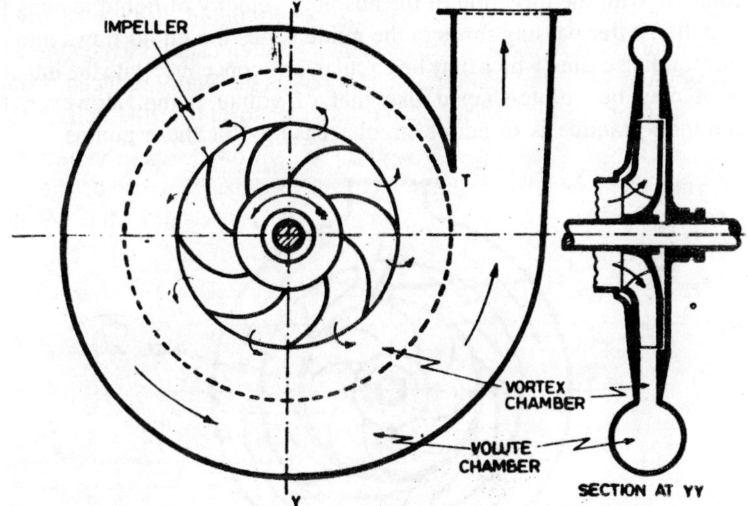

Fig. 6.3 : Volute pump with vortex chamber.

Therefore the torque produced for the liquid does not change and hence a free vortex is formed as the liquid passes through the vortex chamber. Since for a free vortex the velocity of whirl varies inversely as its radial distance from the centre, there is a reduction in velocity of flow of liquid as it passes through the vortex chamber. The reduction in velocity, is accompanied by an increase in pressure. As such a vortex chamber serves a dual purpose of reducing the velocity and increasing the efficiency of the pump by converting a large amount of kinetic energy into pressure energy. The liquid after leaving the vortex chamber passes through the vòlute chamber surrounding it, which further increases the efficiency of the pump.

Diffuser or Turbine Pump

In the diffuser pump, the impeller is surrounded by a series of guide vanes mounted on a ring called diffuser ring as shown in Fig. 6.4. The diffuser ring and the guide vanes are fixed in position. The adjacent guide vanes provide gradually enlarged passages for the flow of liquid. The liquid after leaving the impeller passes through these passages of increasing area, wherein the velocity of flow decreases and the pressure increases. The guide vanes are so designed that the liquid emerging from the impeller enters these passages without shock. This condition may however be achieved by making the tangent to the guide vane at the inlet tip to coincide with the direction of the absolute velocity of liquid leaving the impeller. After passing through the guide vanes the liquid flows into the surrounding casing which may be circular, and concentric with the impeller or it may be volute-shaped like that of volute pump. However, the common practice is to adopt circular casings for these pumps.

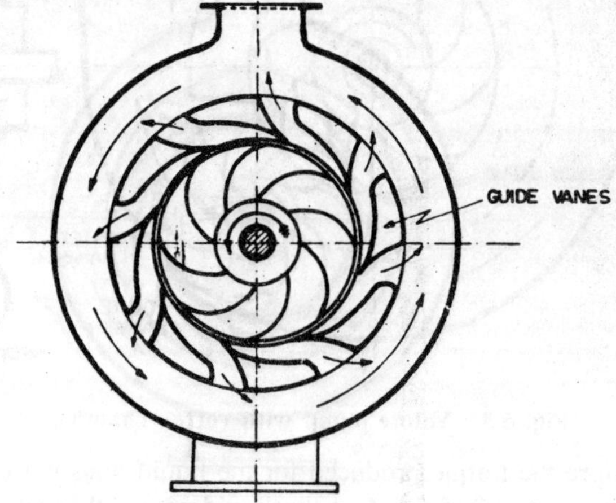

Fig. 6.4 : Diffuser (or turbine) pump.

These pumps which are provided with diffuser ring and guide vanes very much resemble a reversed turbine and hence they are also known as *turbine pumps*. It has been found from tests that a well designed diffuser pump is capable of converting as much as 75 percent of the kinetic energy of the liquid discharged from the impeller into pressure energy. However these pumps will work with maximum efficiency only for one rate of discharge at given impeller speed. This is so because the guide vanes will be correctly set or shaped for one rate of discharge only and for other discharges a loss of energy by shock or turbulence will occur at-the entrance to the guide vanes, thereby resulting in a low efficiency. Moreover turbine pumps are more costly than the simple volute pumps. As such the arrangement of diffuser ring is usually employed only in multistage pumps.

The centrifugal pumps may also .be classified on the basis of certain other factors as indicated below:

(a) Number of impellers per shaft

(b) Relative direction of flow through impeller.

(c) Number of entrances to the impeller.

(d) Disposition of shaft, and

(e) Working head.

According to the number of impellers provided the pumps may be classified as single-stage and multi-stage. A *single stage centrifugal pump* has only one impeller mounted on the shaft. A *multi-stage centrifugal pump* has two or more impellers connected in series, which are mounted on the same shaft and are enclosed in the same casing.

On the basis of the direction of flow of the liquid through the impeller the pump may be classified as radial flow pump, mixed flow pump and axial flow pump.

A *radial flow pump* is that in which the liquid flows through the impeller in the radial direction only. Ordinarily all the centrifugal pumps are provided with radial flow impellers. In mixed/low pumps the liquid flows through the impeller axially as well as radially, that is there is a combination of radial and axial flows. A mixed flow impeller is just a modification of radial flow type in this respect that the former is capable of discharging a large quantity of liquid. As such mixed flow pumps are generally used where a large quantity of liquid is to be discharged to low heights. In axial *flow pumps* the flow through the impeller is in the axial direction only. Axial flow pumps are usually designed to deliver very large quantities of liquid at relatively low heads. However, it is not justified to call axial flow pumps as centrifugal pumps, because there is hardly any centrifugal action in their operation.

Depending on the number of entrances to the impeller the centrifugal pumps may be classified as single suction pump and double suction pump. In a *single suction* (*or entry*) *pump* liquid is admitted from a suction pipe on one side of the impeller. In a double suction (or entry) pump liquid enters from both sides of the impeller. A double suction pump has an advantage that by this arrangement the axial thrust on the impeller is neutralised. Further it is suitable for pumping large quantities of liquid since it provides a large inlet area.

The centrifugal pumps may be designed with either horizontal or vertical disposition of shafts. Generally the pumps are provided with horizontal shafts. However, for deep wells and mines the pumps with vertical shafts are more suitable because the pumps with vertically disposed shafts occupy less space.

According to the head developed, the centrifugal pumps may be classified as *low head, median head* and *high head* pumps. A low head pump is the one which is capable of working against a total head upto 15 m. A medium head pump is that which is capable of working against a total head more than 15m but upto 40 m. A high head pump is the one which is capable of working against a total head above 40 m. Generally high head pumps are multi-stage pumps.

WORK DONE BY THE IMPELLER

The expression for the work done (or the energy supplied) by the impeller of a centrifugal pump on the liquid flowing through it may be derived in the same way as for a turbine. The liquid enters the impeller at its centre and leaves at its outer periphery. Fig. 5 shows a portion of the impeller of a centrifugal pump with one vane and the velocity triangles at the inlet and the outlet tips of the vans. For the sake of convenience the same system of notation is employed as that for turbines. Thus, V is absolute velocity of liquid, u is peripheral (or tangential) velocity of the impeller V_r is relative velocity of liquid, V_f is velocity of flow of liquid, and V_w is velocity of whirl of the liquid at the entrance to the impeller. Similarly V_1,u_1 V_{f1}, V_{w1} and V_{w1} represent their counterparts at the exit point of the impeller. Further θ represents the impeller vane angle at the entrance and ϕ represents the impeller vane angle at the outlet. Similarly α is the angle between the direction of the absolute velocity of entering liquid and the peripheral velocity of the impeller at the entrance, and β is the angle between the absolute velocity of leaving liquid and the peripheral velocity of the impeller at the exit point

At the entrance to the impeller since there are no guide vanes (as in the case of turbines) the direction of the absolute velocity of liquid at this point of the impeller is not directly known. However, for best efficiency of the pump it is commonly assumed that the liquid enters the impeller radially that is the absolute velocity of the liquid at the entrance to the impeller (or at the inlet tip of the impeller vane) is radial in direction. Thus, in this case a = 90° and the velocity of whirl V_w at inlet is equal to Zero. Further it is desired that the-liquid enters and leaves the vane without shock. This can be ensured if the inlet and outlet tips of the vane are parallel to the direction of the relative velocities at the two tips. As such it is assumed that the relative velocities V_r and V_{r1} are parallel to the tangents to the vane at the inlet and outlet tips respectively as shown in Fig. 6.5.

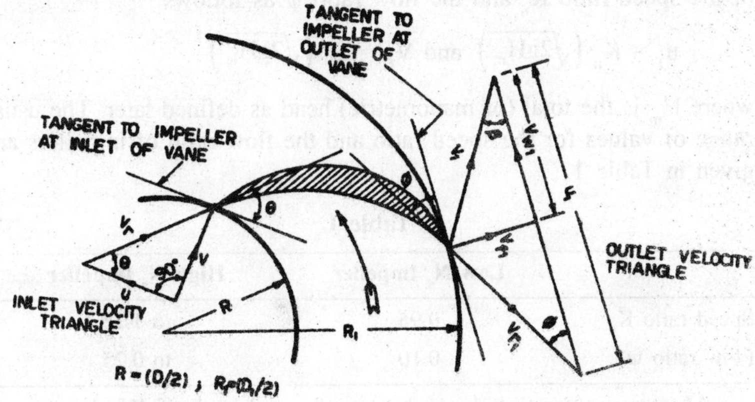

Fig. 5 : Velocity triangles for an Impeller vane.

In the case of a radially inward flow reaction turbine the work done per second by the liquid on the runner = (W/g) (V_{w1}, ± V_{w1}, u_1); where W is weight of liquid striking the runner per second. Since a centrifugal pump is just the reverse of a radially inward flow reaction turbine, the same analysis as used for turbines may be applied to pumps. Thus, the work done per second by the impeller on the liquid may be written as

$$\text{Work done} = \frac{W}{g}\left(V_{w1}u_1 - V_w u\right) \quad \text{...(1)}$$

where W is weight of liquid per second that passes through the impeller. Since in this case as stated earlier the liquid, enters the impeller radially, $\alpha = 90°$ and hence $V_w = 0$.

Thus equation 1 becomes

$$\text{Workdone} = \frac{W}{g}\left(V_{w1}u_1\right) \quad \text{...(2)}$$

and work done per unit weight of liquid

$$= \frac{1}{g}(V_{w1}u_1) \qquad ...(3)$$

Thus equation 3 represents the head imparted by the impeller to the liquid.

Further from the outlet velocity triangle of Fig. 5, $V_{w1} = (u_1 - V_{f1} \cot \phi)$. As in the case of turbines u_1 and V_{f1} can be expressed in terms of the speed ratio K_u and the flow ratio ψ as follows:

$$u_1 = K_u\left(\sqrt{2gH_m}\right) \text{ and } V_{f1} = \psi\left(\sqrt{2gH_m}\right)$$

where H_m is the total (or manometric) head as defined later. The usual range of values for the speed ratio and the flow ratio of impellers are given in Table 1.

Table 1

	Low N_s Impeller	**High N_s Impeller**
Speed ratio K_u	0.95	to 1.25
Flow ratio ψ	0.10	to 0.25

Moreover, equation 1 can be transformed in the following form:

Work done per kg per second

$$= \left(\frac{V_1^2 - V^2}{2g} + \frac{u_1^2 - u^2}{2g} + \frac{v_r^2 - V_{r1}^2}{2g}\right) \qquad ...(4)$$

Equation 4 shows that work done on the liquid consists of three parts. The first part represents the change in kinetic energy of the liquid, the second part represents effect of centrifugal head and the last part indicates the change in static pressure energy of the liquid if the losses in the impeller and the effect of difference in elevations of the inlet and the outlet points of the impeller are neglected. Usually equation 4 is known as *fundamental equation of centrifugal* pump.

HEAD OF PUMP

The head of a centrifugal pump may be expressed in the following two ways:

(a) Static head,

(b) Manometric head (or total head or gross head or effective head).

(a) *Static Head* : As shown in Fig. 6.6 the static head is the vertical distance between the liquid surfaces in the sump and the tank to which the liquid is delivered by the pump. Thus, if h_s is the vertical height of the centre line of the pump shaft above the liquid surface in the sump from which the liquid is being raised; and h_d is the vertical height of the liquid surface in the tank to which the liquid is delivered above the centre line of the pump shaft, then the static head (or lift) H_s may be expressed as

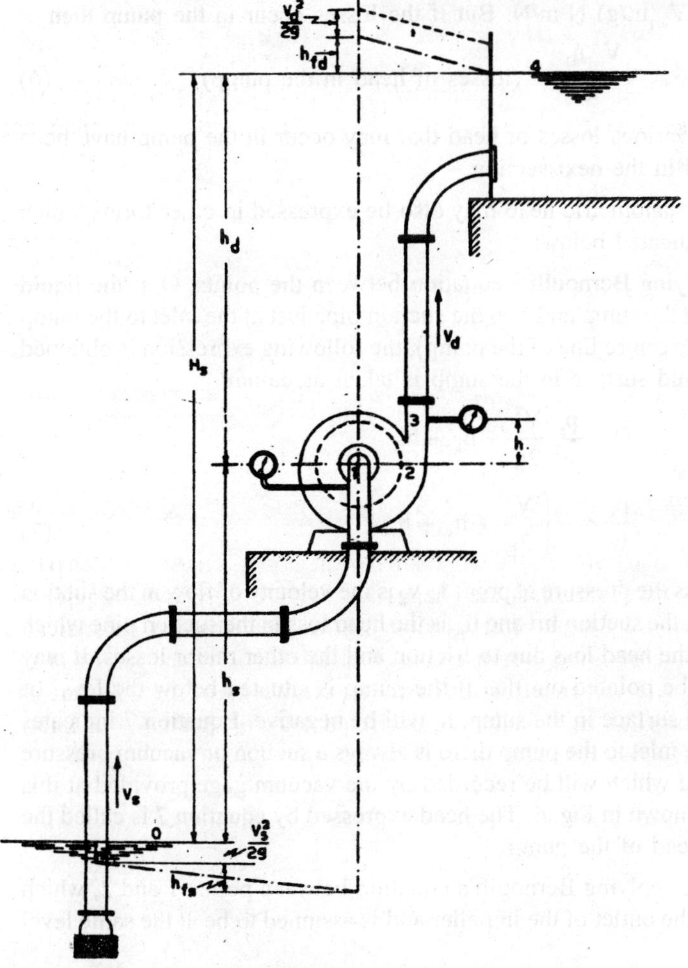

Fig. 6.6 : Head on a centrifugal pump.

$$H_s = h_s + h_d \quad ...(5)$$

The term h_s is known as static suction lift, and h_d is known as static delivery lift. Thus static head (or lift) is the net total vertical height through which the liquid is lifted by the pump.

(b) *Manometric head :* It is the total head that must be produced by the pump to satisfy the external requirements. If there are no energy losses in the impeller and the casing of the pump, then the manometric head H_m will be equal to the energy given to the liquid by the impeller, which for radial entry to the impeller = $(V_{w1}u_1/g)$ N-m/N. But if the losses occur in the pump then

$$H_m = \frac{V_{w1}u_1}{g} - \text{(losses of head in the pump)} \quad ...(6)$$

The various losses of head that may occur in the pump have been described in the next section.

The manometric head may also be expressed in other forms which are as indicated below:

Applying Bernoulli's equation between the points, O at the liquid surface in the sump and 1 in the suction pipe just at the inlet to the pump (*i.e.*, at the centre line of the pump), the following expression is obtained if the liquid surface in the sump is taken as datum.

$$O = \frac{p_s}{w}\frac{V_s^2}{2g} + h_s + h_{fs}$$

or

$$\frac{p_s}{w} = -\left(\frac{V_s^2}{2g} + h_s + h_{fs}\right) \quad ...(7)$$

where p_s is the pressure at point 1; V, is the velocity of flow in the suction pipe; h_s is the suction lift and h_{fs} is the head loss in the suction pipe which includes the head loss due to friction and the other minor losses. It may however be pointed out that if the pump is situated below the level of the liquid surface in the sump, h_s will be negative. Equation 7 indicates that at the inlet to the pump there is always a suction or vacuum pressure developed which will be recorded by the vacuum gage provided at this point as shown in Fig. 6. The head expressed by equation 7 is called the suction head of the pump.

Also, applying Bernoulli's equation between points 1 and 2, which is just at the outlet of the impeller and is assumed to be at the same level as point 1 then since the impeller imparts a head equal to ($V_{w1}u_1/g$) to the liquid the following expression is obtained.

$$\frac{p_s}{w} + \frac{V_s^2}{2g} + \frac{V_{w1}u_1}{g} = \frac{p_2}{w} + \frac{V_1^2}{2g} + h_{Li} \qquad ...(8)$$

where p_2 is the pressure and V_1 is the absolute velocity of the liquid leaving the impeller and h_{Li} is the loss of head in the impeller.

Further applying Bernoulli's equation between points 2 and 3, which is [as shown in Fig. 6] lying in the delivery pipe just at the exit from the pump, the following expression is obtained.

$$\frac{p_s}{w} + \frac{V_1^2}{2g} = \frac{p_d}{w} + \frac{V_d^2}{2g} + h + h_{Lc} \qquad ...(9)$$

where p_d is the pressure in the delivery pipe at point 3; V_d is the velocity of flow in the delivery pipe; h is the height of the point 3 above the centre line of the pump and h_{Lc} is the loss of head in the casing. The pressure in the delivery pipe can be recorded by the pressure gage provided at point 3, as shown in Fig. 6.

Adding equations 8 and 9 the following expression is obtained

$$\frac{p_s}{w} + \frac{V_s^2}{2g} + \frac{V_{w1}u_1}{2g} = \frac{p_d}{w} + \frac{V_d^2}{2g} + h + h_{Li} + h_{Lc} \qquad ...(10)$$

Equation 10 can however be obtained directly by applying Bernoulli's equation between the points 1 and 3. Rearranging the terms of equation 10 it may be written as

$$\frac{V_{w1}u_1}{g} - (h_{Li} + h_{Lc}) = \left(\frac{p_d}{w} + \frac{V_d^2}{2g} + h\right) - \left(\frac{p_s}{w} + \frac{V_s^2}{2g}\right)$$

or $\frac{V_{w1}u_1}{g}$ (losses of head in the pump)

$$= \left(\frac{p_d}{w} + \frac{V_d^2}{2g} + h\right) - \left(\frac{p_s}{w} + \frac{V_s^2}{2g}\right)$$

From equation 6, left hand side of the above equation is equal to H_m. Thus

$$H_m = \left(\frac{p_d}{w} + \frac{V_d^2}{2g} + h\right) - \left(\frac{p_s}{w} + \frac{V_s^2}{2g}\right) \qquad ...(11)$$

Equation 11 thus indicates that 'the menometric head is equal to the difference between the total energies of the liquid at inlet and exit from the pump.

If the gages are at the same level, h = 0, then equation 11 becomes

$$H_m = \left(\frac{p_d}{w} + \frac{V_d^2}{2g}\right) - \left(\frac{p_s}{w} + \frac{V_s^2}{2g}\right) \quad ...(12)$$

Furthermore if the suction abd delivery pipes are of the same diameter, $V_s = V_d$, then manometric head is equal to difference of the pressure heads at the inlet and the exit from the pump, that is

$$H_m = \left(\frac{p_d}{w} - \frac{p_s}{w}\right) \quad ...(13)$$

The suction pipe is usually slightly larger than the delivery pipe, but as the velocities are not excessive in either of the pipes, the difference in velocity heads is small and is usually neglected, in that case also equation 13 will hold good. Moreover the suction head (p_s/w) being negative, it is indicated by equation 13 that if suction and delivery pipes are of the same size then the manometric head is equal to the arithmetical sum of the readings of the vacuum and pressure gages placed at the same level on the suction and the delivery pipes just at the inlet and the exit from the pump respectively.

Equation 9 may also be written as

$$\frac{p_d}{w} - \frac{p_2}{w} = \frac{V_1^2}{2g} - \frac{V_d^2}{2g} - h - h_{Lc}$$

The left hand side of the above equation represents the increase in the pressure head between the point 2 at the outlet of the impeller and the point 3 at the outlet of the pump. This increase in pressure head is due to partial conversion of velocity head of liquid into pressure head and is equal to $\left(\frac{V_1^2}{2g} - \frac{V_d^2}{2g}\right)$ where V_1 is the velocity with which the liquid leaves the impeller and V_d is the velocity at point 3 in the delivery pipe. Generally this gain in pressure head is expressed as $k\left(\frac{V_1^2}{2g}\right)$, where k is a coefficient, the value of which varies from 0.4 for volute casing to 0.7 for a turbine pump. Thus,

$$\frac{V_1^2}{2g}-\frac{V_d^2}{2g}=k\frac{V_1^2}{2g} \quad ...(14)$$

Equation 8 may be written as

$$\frac{V_{w1}u_1}{g}-h_{L1}-h_{Lc}=\left(\frac{p_2}{w}-\frac{p_s}{w}\right)=\frac{V_1^2}{2g}-\frac{V_d^2}{2g}-h_{Lc}$$

or
$$H_m=\left(\frac{p_2}{w}-\frac{p_s}{w}\right)=\left(\frac{V_1^2}{2g}-\frac{V_d^2}{2g}\right)+\frac{V_d^2}{2g}-\frac{V_s^2}{2g}-h_{Lc}$$

Now if it is assumed that $V_d = V_s$ and equation 14 is introduced in the above expression then it becomes

$$H_m=\left(\frac{p_2}{w}-\frac{p_s}{w}\right)+\frac{V_1^2}{2g}-h_{Lc} \quad ...(15)$$

The equation 15 also provides another expression for the manometric head according to which

H_m = the actual pressure rise in the impeller + the pressure rise in the volute chamber or the guide blade ring (*i.e.*, between the points 2 and 3) – the loss of head in the casing.

Finally the manometric head may also be expressed by another expression as indicated below.

Applying Bernoulli's equation between points 3 and 4 (point 4, as shown in Fig. 6, is at the liquid surface in the delivery tank), the following expression is obtained if the datum is considered at point 3,

$$\frac{p_d}{w}-\frac{V_d^2}{2g}-h_d+h_{fd}+\frac{V_d^2}{2g}$$

or
$$\frac{p_d}{w}=h_d+h_{fd} \quad ...(16)$$

where h_d is the delivery lift and h_{jd} is the head loss in the delivery pipe which includes the head loss due to friction and the other minor losses.

The head $\left(h_d+h_{fd}+\frac{V_u^2}{2g}\right)$ is called the delivery head of the pump.

Now introducing equations 7 and 16 in equation 12 it becomes

$$H_m=h_s+h_d+h_{fs}+h_{fd}+\frac{V_d^2}{2g} \quad ...(17)$$

Equation 17 represents another expression for the manometric head. It is observed from equation 17 that all the manometric head is not used to lift the liquid against the state head, since some of it is used to overcome the various losses in the pipes. Hence

H_m = static head + friction and minor head loss in suction and delivery pipes + the velocity head in the delivery pipe.

If the velocity head in the delivery pipe $\left(\frac{V_d^2}{2g}\right)$ is relatively small it may be neglected and then equation 17 becomes

$$H_m = h_s + h_d + h_{fs} + h_{fd} \qquad ...(17a)$$

LOSSES AND EFFICIENCIES

(A) **Losses :** The various losses occurring during the operation of a centrifugal pump may be classified as follows:

(1) Hydraulic losses.

(2) Mechanical losses.

(3) Leakage loss.

(1) *Hydraulic losses :* The hydraulic losses that may occur in a centrifugal pump installation may be grouped as

(a) Hydraulic losses in the pump.

(b) Other hydraulic losses.

The hydraulic losses that may occur within the pump consist "f the following:

(i) Shock or eddy losses at the entrance to and the exit from the impeller.

(ii) Friction losses in the impeller.

(iii) Friction and eddy losses in the guide vanes (or diffuser) and casing.

It can be seen from Fig. 6.5 that for the given values of the blade angles θ and ϕ), and the speed of rotation, there can be only one rate of discharge that will ensure tangential entry to the impeller and tangential exit from the impeller. But often the pump is required to operate under varying conditions which results in the variation in the rate of discharge. As such at the entrance and the exit of the impeller the shock losses generally occur. Furthermore at the exit from the impeller there occurs a loss of energy due to the more or less abrupt change in the direction of the velocity of liquid as it enters the casing.

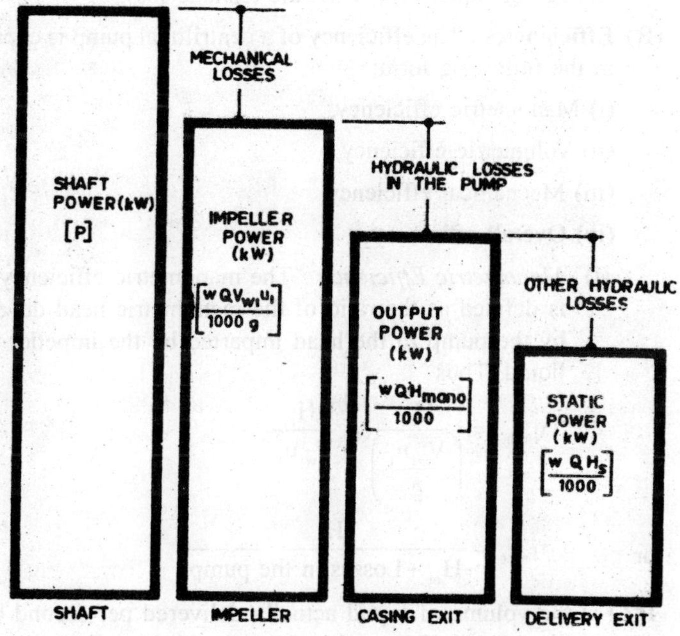

Fig. 6.7 : Head losses In centrifugal pump.

The other hydraulic losses consist of the following:

(i) Friction and other minor losses in the suction pipe.

(ii) Friction and other minor losses in the delivery pipe.

(2) **Mechanical losses:** The mechanical losses occur in the centrifugal pump on account of the following :

(i) Disc friction between the impeller and the liquid which fills the clearance spaces between the impeller and the casing.

(ii) Mechanical friction of the main bearings and glands.

(3) **Leakage loss :** In centrifugal pumps as ordinarily built, it is not possible to provide a completely water tight seal between the delivery and suction spaces. As such there is always a certain amount of liquid which slips or leaks from the high pressure to the low pressure points in the pump and it never passes through the delivery pipe. The liquid which escapes or leaks from a high

pressure zone to a low pressure zone carries with it energy which is subsequently wasted in eddies. This loss of energy due to leakage of liquid represents the leakage loss.

(B) **Efficiencies :** The efficiency of a centrifugal pump is expressed in the following form:

(i) Manometric efficiency,

(ii) Volumetric efficiency,

(iii) Mechanical efficiency,

(iv) Overall efficiency.

(i) *Manometric Efficiency :* The manometric efficiency η_{mano} is defined as the ratio of the manometric head developed by the pump to the head imparted by the impeller to the liquid. Thus

$$\eta_{mano} = \frac{H_m}{\left(\frac{V_{w1}u_1}{g}\right)} = \frac{gH_m}{V_{w1}u_1} \quad ...(18)$$

or
$$\eta_{mano} = \frac{H_m}{H_m + \text{Losses in the pump}}$$

If Q is the volume of liquid actually delivered per second by the pump and w is the specific weight of the liquid then

$$\eta_{mano} = \frac{wQH_m}{wQ\left(\frac{V_{w1}u_1}{g}\right)} \quad ...(19)$$

$$\eta_{mano} = \frac{\text{Power actually delivered by the pump}}{\text{Power imparted by the impeller}}$$

or
$$\eta_{mano} = \frac{\text{Output of the pump}}{\text{Power imparted by the impeller}} \quad ...(20)$$

[Power imparted by the impeller = Power delivered by the motor to the shaft (or shaft power) – Power lost in mechanical losses]

(ii) *Volumetric Efficiency* : The volumetric efficiency η_v is defined as the ratio of die quantity of liquid discharged per second from the pump to the quantity passing per second through the impeller. As stated earlier these two quantities differ by the rate ΔQ at which the liquid from the impeller leaks through the clearances between the impeller and the casing and finds its way back to the eye of the impeller. Thus

$$\eta_v = \frac{Q}{(Q+\Delta Q)} \quad \text{...(21)}$$

where Q is the quantity of liquid actually discharged per second from the pump.

(ii) *Mechanical Efficiency* : The mechanical efficiency η_{mech} is defined as the ratio of the power actually delivered by the impeller to the power supplied to the shaft by the prime mover or motor. Thus

$$\eta_{mech} = \frac{w(Q+\Delta Q)\left(\frac{V_{w1}u_1}{g}\right)}{\text{Power given to the shaft}} \quad \text{...(22)}$$

or

$$\eta_{mech} = \frac{\left(\frac{V_{w1}u_1}{g}\right)}{\text{Energy head given to the shaft}}$$

or

$$\eta_{mech} = \frac{\left(\frac{V_{w1}u_1}{g}\right)}{\left(\frac{V_{w1}u_1}{g}\right) + \text{(mechanical head losses in bearing)}} \quad \text{...(22a)}$$

(iv) *Overall Efficiency* : The overall efficiency η_o of the pump is defined as the ratio of the power output from the pump to the power input from the prime mover driving the pump. Thus

$$\eta_o = \frac{wQH_m}{\text{Power given to the shaft}} \quad \text{...(23)}$$

The overall efficiency is also equal to the product of all the three efficiencies described above. That is

$$\eta_o = (\eta_{mano}) \times (\eta_v) \times (\eta_{mech})$$

$$= \frac{H_m}{\left(\frac{V_{w1}u_1}{g}\right)} \times \frac{Q}{(Q+\Delta Q)} \times \frac{w(Q+\Delta Q)\left(\frac{V_{w1}u_1}{g}\right)}{\text{Power given to the shaft}}$$

$$= \frac{wQH_m}{\text{Power given to the shaft}}$$

which is same as equation 23.

(C) **Effect of Vane Angle ϕ on Manometric Efficiency.:** If the loss of head in the pump is neglected then the manometric head may be expressed as

$$H_m = \frac{V_{w1}u_1}{g} - \frac{V_1^2}{2g} \qquad ...(24)$$

From the outlet velocity triangle shown in Fig. 5

$$V_1^2 = (V_{w1}^2 + V_{f1}^2)$$

and $\quad V_{w1} = (u_1 - V_{f1} \cot \phi)$

Then $\quad H_m = \dfrac{u_1^2 + V_{f1}^2 \operatorname{cosec}^2\phi}{2g} \qquad ...(25)$

Under the ideal condition assumed above the manometric efficiency of the pump will become

$$h_{mano} = \frac{gH_m}{V_{w1}u_1} = \frac{(u_1^2 + V_{f1}^2 \operatorname{cosec}^2\phi)}{2u_1(u_1 + V_{f1} \cot\phi)} \qquad ...(26)$$

Now assuming a constant value for the flow ratio $\Psi \left[= \dfrac{V_{f1}}{\sqrt{2gH_m}} \right]$ = 0.25 and computing the value of u_1 in terms of H_m from equation 25 for different values of ϕ, it will be observed that as the value of ϕ varies from 90° to 20°, the value of η_{mano} increases from 0.47 to 0.73. It is obvious from equation 26 that a further decrease in the value of ϕ will result in the increase in the efficiency, but it is impracticable to have the values of ϕ less than 20°, since it will result in the long and narrow passages causing high frictional losses. As such in practice the value of ϕ is not reduced below 20°, but some other devices in the form of guide vanes etc., are adopted which will convert a part of the velocity energy into useful pressure energy, thereby increasing the efficiency of the pump.

(D) **Effect of Finite Number of Vanes of the Impeller on Head and Efficiency :** Equation 3 or 4 represents the head imparted by the impeller to the liquid only if there are infinite number of vanes (or blades) in the impeller and it is usually known as Eider's head. Similarly the velocity triangles shown in Fig. 5 will be developed only if there are infinite number of vanes in the impeller and these velocity triangles are also known as Euler's velocity triangles. However in practice the impeller has only a finite number of vanes. Therefore although the vanes are designed in accordance with the consideration of Euler's velocity triangles, but the actual velocity triangles developed in the case of impellers with finite number of vanes are different from the Euler's velocity triangles. On account of secondary or circulatory flow effects developed in the impeller the actual velocity of whirl V_{w1} developed in the case of impeller with finite number of vanes is always less than that corresponding to the Euler's velocity triangles. Consequently the actual head H_i imparted by the pump impeller (with finite number of vanes) to the liquid is always less than the Euler's head H_e (given by equation 3 or 4). The difference between the actual head H_i imparted by the impeller with finite number of vanes to the liquid and the Euler's head H, is not considered as loss of head. Further the ratio (H_i/H_e) is usually termed as vane efficiency or *vane effectiveness* $\in$ (Greek 'epsilon') *i.e.*,

$$\in = \frac{H_i}{H_e}$$

Experiments made on impellers of identical size, speed and vane angle but with different number of vanes have indicated that as the number of vanes is increased the value of $\in$ increases and approaches unity. In other words, it means that the actual head imparted by the impeller (with finite number of vanes) to the liquid, approaches the Euler's head as the number of vanes is increased. In addition to the number of vanes, the value of $\in$ also depends on the shape of the vane and the outlet vane angle. In general for radial flow pumps the value of $\in$ varies from 0.6 to 0.8 as the number of vanes is increased from 4 to 12. However, for impellers with vanes more than 24 the value of $\in$ may be taken as unity.

A portion of the head H_i actually imparted by the impeller (with finite number of vanes) to the liquid will be lost in the pump and hence the net manometric head H_m available from the pump may be expressed as

$H_m = H_i$ –losses of head in the pump.

Further the ratio of the manometric head H_m available from a pump and the head H_i actually imparted by the impeller to the liquid is termed as *hydraulic efficiency* η_k of the pump *i.e.*,

$$\eta_k = \frac{H_i - \text{losses of head in the pump}}{H_i}$$

or $$\eta_k = \frac{H_m}{H_i}$$

However, the manometric efficiency η_{mano} of a pump has been defined earlier as (equation 18)

$$\eta_{mano} = \frac{H_m}{\left(\frac{V_{w1} u_1}{g}\right)} = \frac{H_m}{H_e}$$

Thus combining the above noted expressions the manometric efficiency becomes

$$\eta_{mano} = \frac{H_m}{H_i} \times \frac{H_i}{H_e} = (\eta_k \times \in)$$

Again for $\in = 1$; $H_i = H_e$ and $\eta_{mano} = \eta_k$

As stated earlier since the actual shape of the velocity triangles for impellers with finite numbers of vanes being not known, the impeller vanes are designed in accordance with the Euler's velocity triangles. As such in the various illustrative examples the value of $\in$ has been assumed as 1 and accordingly H_i has been considered to be same as H_e.

MINIMUM STARTING SPEED

When the pump is started, there will be no flow of water until the pressure difference in the impeller is large enough to overcome the gross or manometric head. If the impeller is rotating, but there is no flow, then the water is rotating in a forced vortex. Therefore a centrifugal head or pressure head caused by the centrifugal force on the rotating water will be $\left(\frac{u_1^2 - u^2}{2g}\right)$. The flow will commence only if

$$\left(\frac{u_1^2 - u^2}{2g}\right) \geq H_m$$

or $$\left[\frac{\left(\frac{\pi D_1}{60}\right)^2}{2g} - \frac{\left(\frac{\pi DN}{60}\right)^2}{2g}\right] \geq H_m$$

or $$\left[\left(\frac{\pi N}{60}\right) - (D_1^2 - D^2)\right] \geq (2gH_m)$$

For determining the minimum speed required for the pump to commence the flow, the above expression may be written as

$$\left[\left(\frac{\pi N}{60}\right)^2 - (D_1^2 - D^2)\right] \geq (2gH_m) \qquad ...(27)$$

from which the requited value of N may be computed.

LOSS OF HEAD DUE TO REDUCED OR INCREASED FLOW

The efficiency of a pump will be maximum only when it is running and discharging at its designed speed. But if the discharge is either reduced or increased, then there will be a loss of head due to shock at the entrance to the impeller, which will result in lowering the efficiency of the pump.

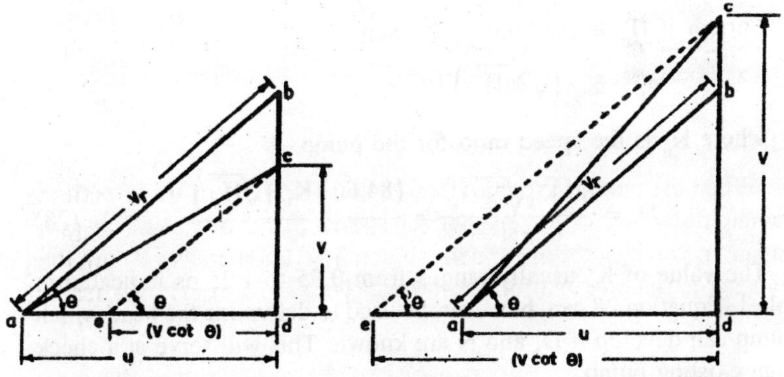

Fig. 6.8 : Inlet velocity triangle with reduced and increased flow

As shown in Fig. 6.8, let abd be the inlet velocity triangle for the pump when running under normal conditions. The vanes at inlet tip will be parallel to ab. Now if the radial flow through the pump is reduced (or increased) from bd to cd, while the speed of rotation remains the same, the velocity triangle will be represented by acd, ac being the new relative velocity. But the angle of the vane at inlet will be the same as before, the relative velocity, therefore, wilt no longer be parallel to the vane and the shock will occur.

Now as the new velocity of flow cd is fixed and also the water must pass along the vane, it follows that the velocity diagram will be triangle ecd, ec being parallel to ab. Therefore a tangential change of velocity ae will suddenly take place resulting in the shock causing the head loss. The loss of head due to sudden change in velocity

$$= \left[\frac{(\text{change in velocity})^2}{2g}\right]$$

∴ Loss of head at the entrance

$$= \left[\frac{(ae)^2}{2g}\right] = \left[\frac{(u - V \cot\theta)^2}{2g}\right]$$

DIAMETERS OF IMPELLER AND PIPES

(a) Outside Diameter of Impeller

If the outside diameter of the impeller is D_1 and the speed of the pump shaft is N r.p.m., then

$$u_1 = (\pi D_1 N/60)$$

Further if H_m is the total head, then

$$u_1 = K_u \left(\sqrt{2gH_m}\right)$$

where K_u is the speed ratio for the pump.

$$D_1 = \frac{60\, K_u \sqrt{2gH_m}}{\pi N} = \frac{(84.60)\, K_u \left(\sqrt{H_m}\right)}{N} \quad ...(28)$$

The value of K_u usually ranges from 0.95 to 1.25 as indicated in Table 1. Equation 28 can, however, be used to determine the head which a pump can develop if D_1 and N are known. This will serve at a check for an existing pump.

(b) Inlet Diameter of Impeller

The inlet diameter D is $\frac{2}{3}$ to $\frac{1}{3}$ of D_1 depending upon specific speed N_s or total head H_m. However in most of the cases, $D = (0.5)\, D_1$.

(c) Least Diameter of Impeller

An expression for the minimum diameter of an impeller can be derived on the basis of the fact that for the pump to start pumping, the centrifugal head must be equal to the total head H_m. Thus

$$\left[\frac{u_1^2 - u^2}{2g}\right] = H_m$$

or $$\left[\left(\frac{\pi D_1 N}{60}\right)^2 - \left(\frac{\pi D N}{60}\right)^2\right] = 2gH_m$$

or $$\left(D_1^2 - D^2\right) = \left(\frac{60}{\pi N}\right)^2 \times (2gH_m)$$

Taking $D = 0.5\ D_1$; and solving for D_1

$$D_1 = \frac{2 \times 60 \times \left(\sqrt{2gH_m}\right)}{\left(\sqrt{3}\right)\pi N} = \frac{(97.68)\left(\sqrt{H_m}\right)}{N} \qquad \text{...(29)}$$

A comparison of equation 28 and 29 indicates that for speed ratio K_u equal to $\left(\frac{2}{\sqrt{3}}\right)$, both the equations become same.

(d) Diameter of Suction Pipe

If d_s is the diameter of the suction pipe and V_s is the velocity of flow in suction pipe, then the amount of water to be pumped is given by

$$Q = \left[\frac{(\pi d_s^2)\ V_s}{4}\right]$$

or $$d_s = \left[\sqrt{\frac{(4Q)}{(\pi V_s)}}\right] \qquad \text{...(30)}$$

Usually V_s is 1.5 to 3 m/s.

(e) Diameter of Delivery Pipe

If d_d is the diameter of the delivery pipe and V_s is velocity of flow in delivery pipe, the

$$Q = \left[\left(\frac{\pi}{4}\right) d_d^2 V_d\right]$$

or $$d_d = \left[\sqrt{\frac{(4Q)}{\pi V_d}}\right] \qquad ...(31)$$

Usually V_d is 1.5 to 3.5 m/s. The value of V_d is generally equal to or slightly higher than that of V_s.

SPECIFIC SPEED

In order to compare the performance of different pumps, it is necessary to have some term which will be common to all the centrifugal pumps. The term ordinarily used for this purpose is the 'specific speed'. This is a type characteristic and can be used to predict the behaviour of one pump based on tests of similar, but different sized pumps.

The specific speed of a centrifugal pump may be defined as *the speed in revolutions per minute of a geometrically similar pump of such a size that under corresponding conditions it would deliver 1 litre of liquid per second against a head of 1 metre.* It is represented by N_s. and it may be expressed by a relation as derived below:

Discharge $\quad Q = (k\pi B_1 D_1)\, V_{f1}$

or $\quad Q \propto B_1 D_1 V_{f1}$

or $\quad Q \propto D_1^2 V_{f1}$; since $B_1 \propto D_1$

Also $\quad V_{f1} = \psi\left(\sqrt{2gH_m}\right) = \text{or } V_{f1} \propto \left(\sqrt{H_m}\right)$

Thus $\quad Q \propto D_1^2\left(\sqrt{H_m}\right)$

or $$\frac{Q}{D_1^2\left(\sqrt{H_m}\right)} = \text{constant} \qquad ...(32)$$

Further $\quad u_1 = (\pi D_1 N/60)$

or $\quad D_1 \propto (u_1/N)$

Also $\quad u_1 = K_u\left(\sqrt{2gH_m}\right);\ u_1 \propto \left(\sqrt{2gH_m}\right);$

Thus $\quad D_1 \propto \left(\sqrt{H_m}/N\right); \left(\sqrt{H_m}/N\right)$

or $$\frac{\sqrt{H_m}}{D_1 N} = \text{constant} \qquad ...(33)$$

Substituting the value of D_1 from equation 33 in equation 32 it becomes

$$Q \propto (H_m^{3/2}/N^2) \qquad ...(34)$$

or $$N^2 \propto (H_m^{3/2}/Q)$$

or $$N \propto \left(\frac{H_m^{3/4}}{\sqrt{Q}}\right)$$

or $$N = C\left(\frac{H_m^{3/4}}{\sqrt{Q}}\right); \text{ where C is a constant}$$

or $$\frac{N\sqrt{Q}}{H_m^{3/4}} = C \qquad ...(35)$$

Now according to the definition of the specific speed, putting Q = 1 litre/second and H_m = 1 m, then C = N = N_s. Thus

$$N_s = \frac{N\sqrt{Q}}{H_m^{3/4}} \qquad ...(36)$$

Equation 36 represents the, expression for the specific speed of a centrifugal pump, which is based on the unit discharge. It is the most commonly adopted expression for the specific speed of the pumps. The value of N_s usually varies from about 300 to 15,000 for a single impeller. Further the value of N_s is same is SI and metric system of units and it is equal to 0.67 times N_s in F.P.S system, where Q is in gallons per minute and H_m is in feet.

The values of Q and H_m to be substituted in equation 36 are those corresponding to the maximum efficiency of the pump at its normal working speed.

For a multi-stage pump the value of H_m to be used in equation 36 is obtained by dividing the total head developed by the number of stages. Similarly for a double suction pump half the actual discharge delivered by the pump is taken as Q.

Sometimes another definition of the specific speed of a centrifugal pump is used, which is based on unit power, though it is not very common. According to this basis the specific speed is defined as 'the speed in revolutions per minute of a geometrically similar pump of such a size that under corresponding conditions it would absorb 1 kW power when working against a head of 1 metre'. If the power required to drive the pump is P, then

$$P = [(wQH_m)/(\eta_0)]$$

Assuming η_0 to be constant and substituting for Q from equation (34), we have

$$P \propto (H_m^{5/2}/N^2);\ \text{or}\ P = \text{constant}\ (H_m^{5/2}/N^2) \qquad ...(37)$$

Hence $N = [C_1 H_m^{5/4}/P^{1/2})$ where C_1 is a constant.

Now according to the definition of the specific speed, putting $H_m = 1$ and $P = 1$, then $C_1 = N = N_s$. Thus

$$N_s = \frac{N\sqrt{Q}}{H_m^{5/4}} \qquad ...(38)$$

Equation 38 is, however, not very common with the pumps. As such if nothing is mentioned, then the specific speed of a pump generally refers to that obtained on the basis of the unit discharge criterion, as represented by equation 36.

Centrifugal pumps may also be classified on the basis of the specific speed. Table 2 gives the different types of centrifugal pumps with the corresponding ranges of specific speeds. The values of N, given in this table are as given by equation 36.

Table 2

S.No	Type of pump	Specific speed N_s
1.	Slow speed radial flow	300 to 900
2.	Medium speed radial flow	900 to 1500
3.	High speed radial flow	1500 to 2400
4.	Mixed flow or screw type	2400 to 5000
5	.Axial flow or propeller type	3400 to 15000

MODEL TESTING OF PUMPS

The large-sized pumps are usually manufactured only after testing their small scale models, because from the test results of the models it would be possible to predict the performance of the prototype in advance and also it would be possible to make necessary alterations in the design of the prototype pump. The performance of the prototype pump will be correctly predicted by its model tests only if the model bears a complete similitude with its prototype. The necessary conditions for the complete similarity to exist between the model of a pump and its prototype may be obtained as indicated below.

The variables involved in this case are discharge Q, manometric or total head H, speed of rotation of impeller N, impeller diameter D, power P, mass density and viscosity p of the fluid pumped. Again as in the case of turbines considering gH as the variable in place of H, these variables may be grouped into the dimensionless parametres viz., discharge number $\left(\frac{Q}{ND^3}\right)$, head number $\left(\frac{gh}{N^2D^2}\right)$, power number $\left(\frac{p}{\rho gHND^3}\right)$ and Reynolds number $\left(\frac{\mu}{\rho ND^2}\right)$ By dividing the square root of discharge number by head number raised to power (3/4) the following dimensionless parameter n_s is obtained

$$n_s = \frac{N\sqrt{Q}}{(gN)^{3/4}}$$

The parameter n_s is known as *dimensionless specific speed or shape number* of the pump.

For complete similarity to exist between the model and the prototype pumps the above noted parameters must have the same value for the model and the prototype pumps. However, the flow in the prototype pump is turbulent, if the flow in the model pump is also turbulent then even if the Reynolds numbers are not equal for the model and the prototype pumps, the similarity between them can be ensured. Thus for complete similarity to exist between the model and the prototype pumps the necessary conditions are as given below:

$$\left(\frac{Q}{NF^3}\right)_m = \left(\frac{Q}{ND^3}\right)_p \qquad ...(39)$$

$$\left(\frac{gH}{N^2D^2}\right)_m = \left(\frac{gH}{N^2D^2}\right)_p \qquad ...(40)$$

$$\left(\frac{p}{\rho gHND^3}\right)_m = \left(\frac{p}{\rho gHND^3}\right)_p \qquad ...(41)$$

$$\left[\frac{N\sqrt{Q}}{(gH)^{3/4}}\right]_m = \left[\frac{N\sqrt{Q}}{(gH)^{3/4}}\right]_p \qquad ...(42)$$

Since in most of the cases $g_m = g_p$ and $\rho_m = \rho_p$ the above conditions may be simplified.

However, the efficiencies of the model and the prototype pumps are not same and it may therefore be a source of error in predicting the performance of the actual pump on the basis of the model test results. On account of the losses in the model and the prototype being disproportionate, the model efficiency is usually lower than that of the prototype. However, by applying suitable correction factor, the efficiency of the actual pump can be predicted from the efficiency obtained for its model. Based on the test efficiencies of single stage centrifugal pumps, G.F. Wislicenus has suggested the following expression relating the efficiencies of the model and the prototype pumps.

$$\frac{0.95-\eta_m}{0.95-\eta_p}=\left[\frac{0.658+\log_{10} Q_p}{0.658+\log_{10} Q_m}\right]^2 \qquad ...(43)$$

in which Q is the discharge in litres per minutes.

MULTI-STAGE PUMPS

The head produced by a centrifugal pump depends on the rim speed of the impeller. To increase the rim speed, either the rotative speed or the diameter of the impeller or both must be increased. Increasing either of these has the effect of increasing the stress in the impeller material. For this reason it is usually not possible to produce very high head with ones impeller. Normally a pump with a single impeller can be used to deliver the required discharge against a maximum head of about 100 m. But if the liquid is required to be delivered against a still larger head then it can be done by using two or more pumps in series. However, this method of producing higher heads has been completely replaced by the modern multi-stage pumps.

A multi-stage pump consists of two or more identical impellers mounted on the same shaft, and enclosed in the same casing. All the impellers are connected in series, so that liquid discharged with increased pressure from one impeller passes through the connecting passages to the inlet of the next impeller and so on, till the discharge from the last impeller passes into the delivery pipe. The impellers are surrounded by guide vanes which are generally provided within the connecting passages, and are meant for the recuperation of the velocity energy of the liquid leaving the impeller into pressure energy. According to the number of impellers fitted in the casing a multi-stage pump is designated as two stage, three-stage, etc.; Fig. 9 shows a three-stage centrifugal pump.

Fig. 6.10 indicates the variations of the velocity and the pressure head of the liquid as it passes through the pump. As the liquid passes through each impeller the absolute velocity of the liquid increases to Vi and in each connecting passage the absolute velocity decreases again to Vo; but the pressure head continuously increases.

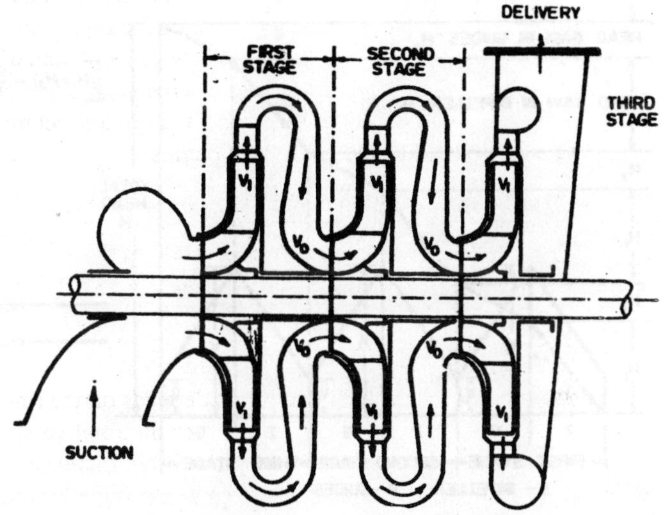

Fig. 6.9 : Three-stage centrifugal pump.

If H_1 and H_2 are the pressure heads gained by the liquid in each impeller and the surrounding guide vanes respectively, then the pressure head impressed on the liquid at each stage is $H_m = (H_1 + H_2$. Now, if there are n impellers then since at each stage the pressure head will be raised by the same amount, the total head H developed by this multi-stage pump will be

$$H = n(H_m) = n(H_1 + H_2) \qquad ...(44)$$

Since the same liquid flows through each impeller, the discharge of a multi-stage pump is same as the discharge passing through each impeller of the series.

The number of stages to be adopted depends on the limitations of speed of the driving motor and also upon the discharge and the total head. If these factors are known, the head per stage can be fixed on the basis of the fact that specific speed N_s per impeller should not be less than about 700, with a' provisional maximum limit of head per stage of 160 m.

PUMPS IN PARALLEL

The multi-stage pumps or the pumps in series as described earlier are employed for delivering a relatively small quantity of liquid against very high heads. However, when a large quantity of liquid is required to be pumped against a relatively small head, then it may not be possible for a single pump to deliver the required discharge.

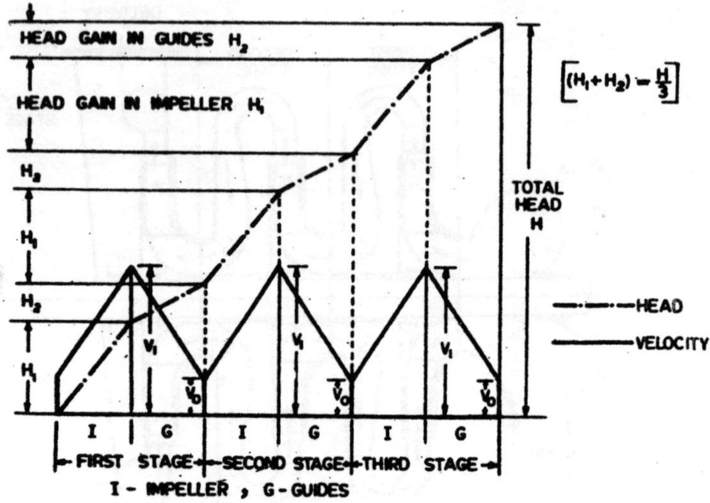

Fig. 6.10 : Variations of velocity and pressure head in a three-stage pump.

In such cases two or more pumps are used which are so arranged that each of these pumps working separately lifts the liquid from a common sump and delivers it to a common collecting pipe through which it is carried to the required height (Fig. 6.11). Since in this case each of the pumps delivers the liquid against the same head, the arrangement is known as *pumps in parallel*. If Q_1, Q_2 Q_3 .... Q_n are the discharging capacities of n pumps arranged in parallel then the total discharged delivered by these pumps will be

$$Q_t = (Q_1 + Q_2 + Q_3 \text{ .... } Q_n) \qquad ...(45)$$

If the discharging capacity of each of the n pumps is same, equal to Q then the total discharge delivered by these pumps will be

$$Q_t = nQ \qquad ...(46)$$

PERFORMANCE OF PUMPS-CHARACTERISTIC CURVES

A pump is usually designed for one speed, flow rate and head, but in actual practice the operation may be at some other condition of head or flow rate, and for the changed conditions the behaviour of the pump may be quite different. As indicated earlier, if the flow through the pump is less than the designed quantity, the value of the velocity of flow of liquid through the impeller will be changed, and as a result, the values of V_w and V_{w1} will be altered, thereby changing the head developed by the pump, and at the same time the losses will increase so that the efficiency of the pump will be lowered.

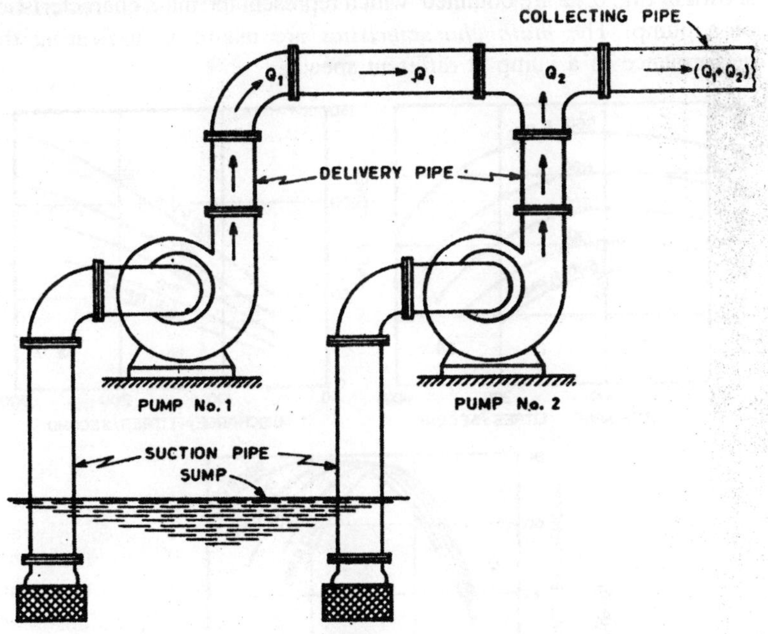

Fig. 6.11 : Two centrifugal pumps arranged in parallel.

Therefore, in order to predict the behaviour and performance of a pump under varying conditions, tests are performed, and the results of the tests are plotted. The curves thus obtained are known as the characteristic curves of the pump. The following three types of characteristic curves are usually prepared for the centrifugal pumps :

(a) Main and operating characteristics.

(b) Constant efficiency or Muschel curves.

(c) Constant head and constant discharge curves.

(a) Main and Operating Characteristics

In order to obtain the main Characteristic curves of a pump it is operated at different speeds. For each speed the rate of flow Q is varied by means of a delivery valve and for the different values of Q the corresponding values of manometric head H_m, shaft power P, and overall efficiency η_0 are measured or calculated. The same operation is repeated for different speeds of the pump. Then H_m v/s Q; P v/s Q and η_0 v/s Q curves for different speeds are plotted, so that three sets of curves, as shown in Fig. 6.12 are obtained, which represent the main characteristics of a pump. The *main characteristics* are useful in indicating the performance of a pump at different speeds.

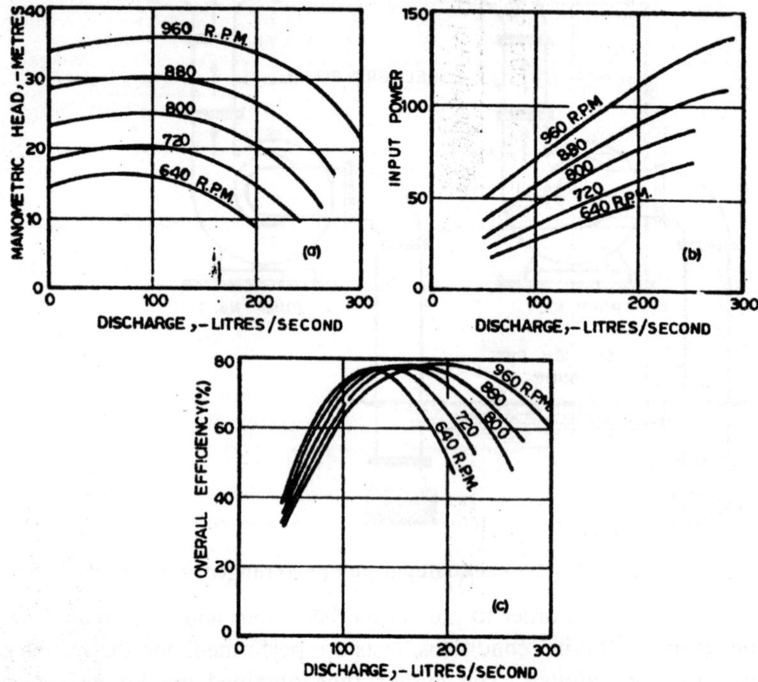

Fig. 6.12 : Main characteristics of a centrifugal pump.

During the operation a pump is normally required to run at a constant speed, which is its designed speed (same as the speed of the driving motor). As such that particular set of main characteristics which corresponds to the designed speed is mostly used in the operations of a pump and is, therefore, known as the *operating characteristics*. A typical set of such characteristics of a pump is shown in Fig. 6.13. The values of the head and the discharge corresponding to the maximum efficiency are known as the normal (or designed) head and the normal (or designed) discharge of a pump. The head corresponding to zero or no discharge is known as the *'shut-off head'* of the pump. From these characteristics it is possible to determine whether the pump will handle the necessary quantity of liquid against the desired head and what will happen if the head is increased or decreased. The P v/s Q curve will show what size motor will be required to operate the pump at the required conditions and whether or not the motor will be overloaded under any other operating conditions.

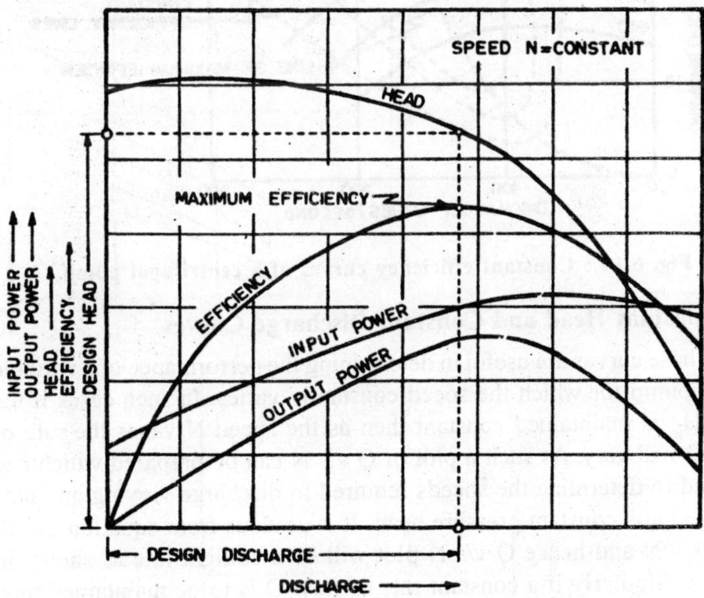

Fig. 6.13 : Operating characteristic curves of a centrifugal pump.

(b) Constant Efficiency Curves

As shown in Fig. 14 the constant or iso-efficiency curves may be obtained from H_m, v/s Q and η_0 v/s Q curves of Fig. 6.12. In order to plot the iso-efficiency curves, horizontal lines representing constant efficiencies are drawn on the η_0 v/s Q curves. The points, at which these lines cut the efficiency curves at various speeds, are transferred to the corresponding H_m v/s Q curves. The points corresponding to the same efficiency are then joined by smooth curves which represent the iso-efficiency curves. From these curves the line of maximum efficiency as shown in Fig. 14 may be obtained. The iso-efficiency curves facilitate the direct determination of the range of operation of a pump with a particular efficiency.

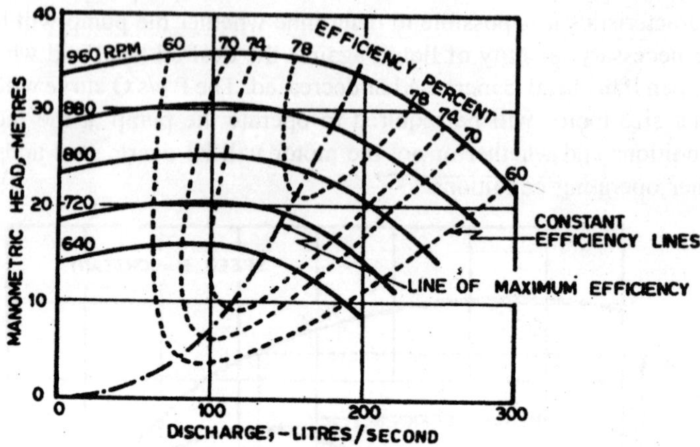

Fig. 6.14 : Constant efficiency curves of a centrifugal pump.

(c) Constant Head and Constant Discharge Curves

These curves are useful in determining the performance of a variable speed pump for which the speed constantly varies. In such cases if the head H_m is maintained constant then as the speed N vanes the rate of flow Q will vary. As such a plot of Q vis N can be prepared which can be used to determine the speeds required to discharge varying amounts of liquid at a constant pressure head. It is obvious from equation 24' 39 that Q ~ N and hence Q v/s N plot will be a straight line as shown in Fig. 15. Similarly if a constant rate of flow Q is to be maintained then as N varies, H_s will vary. Thus a plot of H_m v/s N can be prepared which can be used to determine the speeds required to discharge a certain quantity of liquid at different pressure heads. From equation 40 it is evident that $H_m \sim N^2$ and hence H_m vis N plot is a parabolic curve as shown in Fig. 15. Similarly it is obvious from equation 41 that P - N^3, as shown in Fig. 6.15.

LIMITATION OF SUCTION LIFT

From equation 47 the absolute pressure head at the inlet to the pump may be expressed as

$$\frac{p_a}{w} + \frac{p_s}{w} = \frac{p_a}{w} - \left[\frac{V_s^2}{2g} + h_s + h_{fs}\right] \qquad ...(47)$$

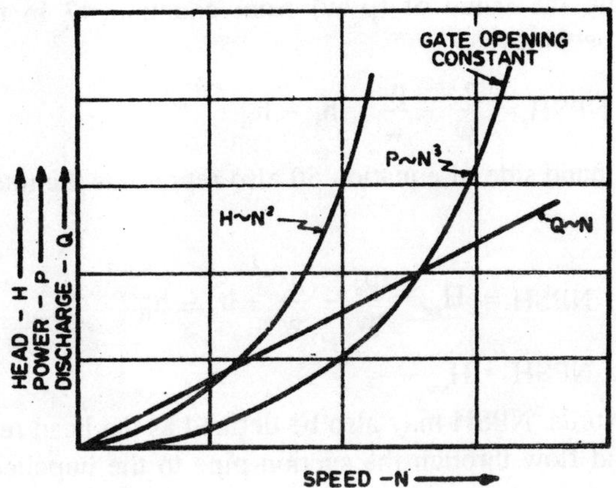

Fig. 6.15 : Q v/s N; H_m v/s N; and P v/s N curves of a centrifugal pump.

where p_a is the atmospheric pressure acting on the free liquid surface in the sump, p_s in the gage pressure at the inlet to the pump, V_s is the velocity of flow in the suction pipe, h_s is the suction lift and h_{fs} is the head lost in the foot valve, strainer and the suction pipe. It is not possible to create at the pump inlet, an absolute pressure lower than the vapour pressure of the liquid. Thus, if p_v is the vapour pressure of the liquid in absolute units, then in the limiting case $(p_a + P_s) = P_v$i and hence from equation 47 the limiting value of the suction lift h_s is obtained as

$$h_s = \left(\frac{p_a - p_v}{w}\right) - \frac{V_s^2}{2g} - h_{fs} \qquad ...(48)$$

The suction lift should in no case be more than that given by equation 48, because a greater h, may result in a rapid vaporization of the liquid due to the reduction of the pressure, which may ultimately lead to cavitation.

NET POSITIVE SUCTION HEAD (NPSH)

The *net positive suction head* (or NPSH) is defined as the absolute pressure head at the inlet to the pump, minus the vapour pressure head (in absolute units) corresponding to the temperature of the liquid pumped, plus the velocity head at this point. Thus

$$NPSH = \left(\frac{p_a}{w} + \frac{p_s}{w}\right) - \frac{p_v}{w} + \frac{V_s^2}{2g} \qquad ...(49)$$

Introducing the value of (p_s/w) from equation 47 in the above expression it becomes

$$\text{NPSH} = \frac{p_a}{w} - \frac{p_s}{w} - h_s - h_{fs} \qquad ...(50)$$

The right hand side of equation 50 also represents the total suction head H_{sv}, *i.e.*,

$$\text{NPSH} = H_{sv} = \frac{p_a}{w} - \frac{p_v}{w} - h_s - h_{fs} \qquad ...(51)$$

and hence $\text{NPSH} = H_{sw}$ ...(51a)

In other words, NPSH may also be defined as the head required to make the liquid flow through the suction pipe to the impeller.

The term NPSH is very commonly used in the pump industry. For any pump installation a distinction is usually made between the required NPSH and the available NPSH. The required NPSH varies with the pump design, the speed of the pump and its capacity. The required NPSH for a pump may be determined experimentally as indicated below and its magnitude must be given by the manufacturer of the pump. When the pump is installed the available NPSH can be determined from the above noted expression. Both the available and the required NPSH vary with the discharge in such a manner that with the increase in the discharge the available NPSH decreases but the required NPSH increases. In order to have cavitation free operation of a centrifugal pump the available NPSH should be greater than the required NPSH. It may also be mentioned that the NPSH requirements for a given pump as determined experimentally, by definition are tied to the size of the suction pipe. Therefore, the available NPSH of a given installation should be based on the same size of the suction pipe.

For determining the required NPSH, the pump is operated at the designed speed with different suction lifts and for each setting of the pump, the head v/s discharge and efficiency v/s discharge characteristics are obtained. In each case it will be observed that with the decrease in the head the discharge increases upto a certain value of the head. With further reduction in the head there is only slight increase in the discharge till a certain value of discharge is reached, when further reduction in head causes no increase in the discharge. In other words, at a certain value of discharge there is an abrupt reduction in the head developed by the pump. Further corresponding to this discharge an abrupt reduction in the efficiency of the pump also occurs. The abrupt reduction in the head as well as the efficiency of the pump indicates the inception of the cavitation. As such the value of the discharge at which there is an abrupt reduction in the head as well as the efficiency of the pump is the maximum discharge which the pump is able to provide without cavitation. For the different settings of the pump the values of such maximum discharges are determined and corresponding to each of these discharges the value of NPSH is also computed by equation 49, which thus represents the required NPSH for each setting of the pump. Usually the values of the required NPSH are plotted against the corresponding values of the discharge to obtain a curve of required NPSH for the pump. Further if a pump is required to be operated at different speeds then by adopting the same procedure, for each of the different speeds a curve for required NPSH may be obtained.

CAVITATION IN CENTRIFUGAL PUMPS

If the pressure at the suction side of the pump drops below the vapour pressure of the liquid men the cavitation may occur. The cavitation in a pump can be noted by a sudden drop in efficiency and head.

Thoma's Cavitation Factor

As in the case of turbines, for pumps also, Thoma's cavitation factor is used to indicated whether cavitation will occur. For pumps, Thoma's cavitation factor is defined as

$$\sigma = \frac{[(p_a - p_v)/w] - (h_s + h_{fs})}{H_m} \qquad ...(52)$$

By comparing equation 52 with 50 and 51, cavitation factor may also be expressed as

$$\sigma = \frac{NPSH}{H_m} = \frac{H_{sv}}{H_m}$$

The cavitation will occur if the value of σ is less than the critical value, σ_c at which the cavitation just begins. The critical value σ_c is that at which a drop in the efficiency or head or some other property begins. The value of σ_c any specified operating conditions depends upon the design of the particular pump, and in any important installation it should be determined experimentally from the model tests. However, σ_c depends on the specific speed of a pump and for the pumps of normal design the following empirical relation may be used to determine σ_c.

$$\sigma_c = 0.103\ (N_s/1000)^{4/3} \qquad ...(54)$$

where N_s is the specific speed of the pump defined by equation 36.

Equation 52 indicates that the value of <y can be increased by reducing the suction lift h_s. In some cases, especially when hot liquids are to be handled, the pump may have to be initialled either at the liquid surface or even below the liquid surface in the sump. In the first case, $h_s = 0$, but in the second case, $h_s < 0$, which indicates that there is positive pressure at the inlet of the pump. A higher value of a also ensures that there is sufficient NPSH available.

Suction Specific Speed

In addition to Thoma's cavitation factor, suction specific speed is another cavitation parameter, which is commonly used in the case of pumps to indicate whether the cavitation will occur. The *suction specific speed S* may be defined as the speed in r.p.m. of a geometrically similar pump which under corresponding conditions would deliver 1 litre of liquid per second when the total suction head is 1 m (in absolute units). According to this definition the expression for suction specific speed may be obtained by replacing the manometric head H_m in equation 36 for the specific speed by the total suction head H_{sv}. Thus

$$S = \frac{N\sqrt{Q}}{H_{sv}^{3/4}} \qquad ...(55)$$

Further by combining equations 36, 53 and 55, the following relationship between σ, N_s and S may be obtained

$$\sigma = \left(\frac{N_s}{S}\right)^{4/3} \qquad ...(56)$$

For most of the centrifugal and propeller pumps for cavitation free flow the limiting value of the suction specific speed have been found to lie between 4700 and 6700 when S is expressed in [r.p.m./litres/second)$^{1/2}$/m$^{3/4}$] units.

By having the same value of the Thoma's cavitation factor <s or the suction specific speed S for the model and the prototype pumps the similarity in respect of cavitation can be established. However for the pumps the suction specific speed is a more pertinent cavitation parameter for ensuring similarity in the model and the prototype pumps in respect of cavitation. This is so because although both a and S depend on the inlet flow conditions, σ also depends on the specific speed N_s of the

pump. As such a constant value of σ can be obtained only if in addition to the inlet flow conditions, the N_s is also held constant and only then it will ensure similar cavitation conditions. On the other hand suction specific speed is independent of the specific speed of the pump and therefore a constant value of S alone will ensure similar cavitation conditions in the model and the prototype pumps. For instance, if the specific speed of pump is changed by changing the outside diameter of the impeller, leaving the inlet flow conditions unchanged, it can be assumed 'that the cavitation conditions remain the same. The suction specific speed S will not change since the inlet flow conditions remain the same, but σ will change since its value depends on the head H and hence on the outer diameter of the impeller. As such even if the value of σ is not same, as long as the value of S is same for the model and prototype pumps, similar cavitation conditions can be ensured.

PRIMING DEVICES

As stated earlier, before starting a centrifugal pump, it must be primed. Small pumps are usually primed by pouring the liquid through a funnel into the casing from some external source. The air vent provided in the casing is opened to facilitate the exit of the air. When all the air has been removed from the suction pipe and the pump casing, the air vent is closed and the pump is primed.

Large pumps are usually primed by evacuating the casing and the suction pipe with the aid of an air pump or a stream ejector, the liquid is thus sucked into the suction pipe from the sump.

In some pumps, their internal construction is such that special arrangements containing a supply of liquid are provided in the suction pipe, which-facilitate automatic priming of the pump. Such pumps are known as 'self priming pumps'.

CENTRIFUGAL-PUMP TROUBLES AND REMEDIES

Some of the troubles commonly experienced during the operation of the centrifugal pumps and the remedial measures to be taken for the same are as listed below:

(1) Pump Fails to Start Pumping

(i) Pump is probably not properly primed. Reprime the pump, opening the air vent until a steady and unbroken stream of liquid is obtained.

(ii) Total static head is probably much higher than that for which the pump is designed. Check the same with accurate vacuum and pressure gages, or may be determined by actual measurement the difference in elevation between pump and suction liquid level, and also between pump and point of discharge Add to this static head the loss of head due to friction in pipes and that in the other fittings used in the installation.

(iii) Wrong direction of rotation of the impeller. Arrow on the pump case shows the proper direction of rotation.

(iv) Impeller may be clogged. Examine carefully for solids or foreign matter lodged in the impeller.

(v) Suction lift too high. Check with vacuum gage or by actual measurement where possible, adding to the static suction lift the loss of head in the pipe and fittings.

(vi) Strainer or suction line may be clogged. This would cause an excessive suction lift.

(vii) Speed may be too low. Check the speed with a techeometer and compare it with the one given on the name plate of the pump. When the pump is being driven by electric motor, check up to see whether the voltage may not be too low.

(2) Pump Working but not upto Capacity and Pressure

(i) Air may be leaking into pump through suction line or through stuffing boxes.

(ii) Speed may be to low.

(iii) Discharge head may be higher than anticipated.

(iv) Suction lift may be too high.

(v) Foot valve or end of suction pipe may not have sufficient submergence, or it may be partly clogged or entirely too small.

(vi) Impeller may be partly clogged or too small in diameter.

(vii) Rotation may be in the wrong direction.

(viii) Wearing rings may be worn, impeller may be damaged, shaft may be loose, stuffing box packing may be defective.

(xi) If hot or volatile liquids are being pumped, there may not be sufficient positive head on the suction.

(3) Pump Starts and then Stops Pumping

(i) Improperly primed or leaky suction line.

(ii) Air pockets in suction line.

(iii) Air may be entering suction pipe because of liquid being, delivered in suction tank or sump too near the pump suction line.

(iv) Suction lift may be too high.

(4) Pumps Take too Much Power

(i) Speed may be too high.

(ii) Head may be too low and pump delivers too much liquid.

(iii) Liquid may have too high a specific gravity.

(iv) Pump may be operating in wrong direction.

(v) Shaft may be bent, impeller may be rubbing on casing, stuffing boxes may be too light wearing rings may be worn.

SOLVED EXAMPLES

Example 1:

A centrifugal pump has an impeller 0.5 m outer diameter and when running at 600 r.p.m. discharges water at the rate of 8000 litres/minute against a head of 8.5 m. The water enters the impeller without whirl and shock. The inner diameter is 0.25 m, and the vanes are set back at outlet at an angle of 45° and the area of flow which is constant from inlet to outlet of the impeller is 0.06 m². Determine (a) the manometric efficiency of the pump, (b) the vane angle at inlet, and (c) the least speed at which the pump commences to work:

Solution:

$$V_f = V_{f1}$$

$$= \frac{8000 \times 10^{-3}}{60 \times 0.6} = 2.22\,\text{m/s}$$

$$u = \frac{\pi DN}{60}$$

$$= \frac{\pi \times 0.25 \times 600}{60} = 7.85\ \text{m/s}$$

$$u_1 = \frac{\pi D_1 N}{60}$$

$$= \frac{\pi \times 0.50 \times 600}{60} \, 15.71 \text{m/s}$$

From the inlet velocity triangle

$$\tan \theta = \left(\frac{V_f}{u}\right)$$

$$= \left(\frac{2.22}{7.85}\right) = 0.2828$$

$$\theta = 15° \, 47'.$$

From the outlet velocity triangle

$$V_w 1 = \left[u_1 - \left(\frac{V_{f1}}{\tan 45°}\right)\right]$$

$$= (15.71\text{-}2.22)$$

$$= 13.49 \text{ m/s}$$

From equation 18

$$\eta_{mano} = \frac{gH}{V_{w1} u_1}$$

$$= \frac{9.81 \times 8.5}{13.49 \times 15.71} = 0.3935 \text{ or } 39.35\%$$

For the least speed, from equation 27

$$\left[\left(\frac{\pi N}{64}\right)^2 \left(D_1^2 - D^2\right)\right] = 2g H_m$$

or $$\left(\frac{\pi N}{60}\right)^2 [(0.50)^2 - (0.25)^2] = (2 \times 9.81 \times 8.5)$$

$\therefore$ $$N = 570 \text{ r.p.m.}$$

Example 2:

A centrifugal pump has to discharge 225 litres of water per second against a head of 25 m when the impeller rotates at a speed of 1 500 r.p.m. Determine (a) the impeller diameter, and (b) the vane angle at the outlet edge of the impeller. Assume that η_{mano} = 0.75; the loss of head in pump in metres due to fluid resistance is 0.03 V_l^2, where Y_l m is the absolute velocity of water leaving the impeller, the area of the impeller *outlet surface is (1.2 D_1^2) m^2, where D_1 is the impeller diameter in m, and water enters the impeller without whirl.*

Solution:

$$\eta_{mano} = (gH/V_{w1}u_1);$$

or $$0.75 = [25 \times g)/V_{w1}u_1]$$

$\therefore$ $$[(V_{w1}u_1)/g] = 33.33 \text{ m} \qquad ...(i)$$

From equation 6, die manometric head is $H_m = (V_{w1}u_1/g)$ – losses of head in the pump.

$\therefore$ Losses in the pump

$$= (33.33–25) = 8.33 \text{ m}$$

Thus $$0.03\, V_1^2 = 8.33$$

$\therefore$ $$V_1 = 16.66 \text{ m/s}$$

$$V_{f1} = \frac{Q}{A} = \frac{225 \times 10^{-3}}{1.2 \times D_1^2} = \frac{0.188}{D_1^2} \text{ m/s}$$

$$u_1 = [(\pi DiN)/60] = 78.54\, D_1 \text{ m/s}$$

Substituting the above value of u_1 in (i)

$$V_{w1} = \frac{33.33 \times 9.81}{78.54 \times D_1} = \frac{4.16}{D_1} \text{ m/s}$$

Now from the outlet velocity triangle of Fig. 5

$$V_{f1} = \left(V_1^2 - V_{w1}^2\right)^{1/2}$$

or $$(0.188/D_1^2) = [(16.66)^2 - (4.16/D_1)^2]^{1/2}$$

or $$278D_1^4 - 17.31\, D_1^2 - 0.035 = 0$$

$\therefore$ $$D_1 = 0.253\,4 \text{ m or } 253.4 \text{ mm.}$$

Thus $$V_{f1} = 2.93 \text{ m/s};$$

$$u_1 = 19.90 \text{ m/s}$$

and $$V_{w1} = 16.42 \text{ m/s}$$

$$\tan\phi = [V_{f1}/(u_1 - V_{w1})]$$

$$= [2.93/(19.90 - 16.42)]$$

$$= 0.8420$$

$\therefore$ $$\phi = 40° 6'.$$

Example 3:

Show that the pressure rise in the impeller of a centrifugal pump is given by $[V_f^2 + u_1^2 - V_{f1}^2 \text{cosec}^2 \phi/2g\}$ *provided the frictional and other losses in-the impeller are neglected.*

Solution:

Applying Bernoulli's equation between inlet and outlet tips of the impeller

$$\frac{p}{w} + \frac{V^2}{2g} = \frac{p_1}{w} + \frac{V_1^2}{2g} - \frac{V_{w1}u_1}{g}$$

For radial entry $V = V_f$, thus

$$\frac{p_1 - p}{w} = \frac{V_f^2}{2g} + \frac{V_{w1}u_1}{g} - \frac{V_1^2}{2g}$$

From the outlet velocity triangle

$$V_{w1} = (u_1 - V_{f1} \text{ Cot } \phi)$$

$$V_1^2 = V_{f1}^2 + V_{w1}^2 = [V_{f1}^2 + (u_1 - V_{f1} \cot \phi)^2]$$

since $(1 + \cot^2 \phi) = \text{cosec}^2 \phi$

$$V_1^2 = V_{f1}^2 \text{ cosec}^2 \phi + u_1^2 - 2u_1 V_{f1} \cot \phi$$

Hence
$$\frac{p_1 - p}{w} = \frac{V_f^2}{2g} + \frac{u_1(u_1 - V_{f1} \cot \phi)}{g} - \frac{V_{f1}^2 \text{ cosec}^2\phi + u_1^2 - 2u_1 V_{f1} \cot \phi}{2g}$$

$$\therefore \quad \frac{p_1 - p}{w} = \left[\frac{V_f^2 + u_1^2 - V_{f1}^2 \text{ cosec}^2 \phi}{2g}\right]$$

Example 4:

Prove that, in general, for a centrifugal pump running at speed N and giving a discharge Q, the manometric head is expressible in the form, $H_m = [AN^2 + BNQ + CQ^2]$ *where A, B and C are constants.*

Solution:

If the loss of head in the pump is neglected, the manometric head is given by the pressure rise through the impeller, together with a certain percentage of the kinetic head at the exit of the impeller which is recovered in the volute chamber or the diffuser ring.

As derived in illustrative example 3,

$$\text{Pressure rise} = \frac{1}{2g}\left[V_f^2 + u_1^2 - V_{f1}^2 \operatorname{cosec}^2 \phi\right]$$

Let the percentage of kinetic head recovered in diffuser ring = (kV_1^2)m. Then it can be shown that

$$kV_1^2 = \frac{c}{2g}\left[V_{f1}^2 + u_1^2 - 2u_1 V_{f1} \cot\phi + V_{f1}^2 \cot^2\phi\right]$$

where c is a constant

$$H_m = \frac{1}{2g}[u_1^2(1+c) - 2cu_1 V_{f1} \cot\phi + V_{f1}^2$$

$$\times c + c\cot^2\phi - \operatorname{cosec}^2\phi) + V_f^2]$$

Assuming $V_f = V_{f1}$,

$$H_m = \frac{1}{2g}[Au_1 + Bu_1 V_{f1} + CV_{f1}^2]$$

Since $MI = (\pi D_1 N/60)$. $\therefore u_1 \sim N$

and $V_{f1} = (Q/A_1)$; $\therefore V_{f1}^2 \sim 2$

$$\therefore \quad H_m = \frac{1}{2g}[AN^2 + BNQ + CQ^2]$$

Example 5:

A centrifugal pump has an impeller 0.29 m diameter running at 960 r.p.m. with an effective outlet vane angle of 28°. The velocity of flow (assumed uniform throughout the system) is 2 mis. The static suction lift is 2.8 m. The energy losses in metres of water are: in suction pipe, 0.61 m; in impeller, 0.49 m; in volute casing, 0.88 m. From these particulars calculate the readings of vacuum or pressure gages placed (i) at inlet to the pump, (ii) at impeller outlet (in clearance space between impeller and volute), (iii) at pump outlet or delivery flange, 0.2 m above the centerline of the pump.

Solution:

From equation 47 the pressure head at the inlet to the pump is given by

$$\frac{p_s}{w} = -\left[\frac{V_s^2}{2g} + h_s + h_{fs}\right]$$

or $$\frac{p_s}{w} = -\left[\left(\frac{2\times2}{2\times9.81}\right) + 2.8 + 0.61\right]$$

$$= -3.614\text{m}$$

From equation 48

$$\frac{p_s}{w} + \frac{V_s^2}{2g} + \frac{V_{w1}u_1}{g} = \frac{p_2}{w} + \frac{V_1^2}{2g} + h_{Li}$$

Now $u_1 = [(\pi D_1 N)/60] = [(\pi \times 0.29 \times 960/60]$

$= 14.58$ m/s

$V_{w1} = (u_1 - V_{fl} \cot \phi) = (14.58 - 2 \cot 28°)$

$= 10.82$ m/s

$V_1 = (V_{w1}{}^2 + V_{f1}{}^2) = [(10.82)^2 + (2)^2]^{1/2}$

$= 11.00$ m/s

Thus by substitution in the above equation

$$-3.614 + 0.204 + \left(\frac{10.82\times14.58}{9.81}\right) = \frac{p_2}{w} + \left(\frac{11\times11}{2\times9.81}\right) + 0.49$$

∴ Pressure head at impeller outlet is

$(p_2/w) = 6.01$m

From equation 49

$$\frac{p_2}{w} + \frac{V_1^2}{2g} = \frac{p_d}{w} + \frac{V_d^2}{2g} + h + h_{Lc}$$

or $$6.01 + \left(\frac{11\times11}{2\times9.81}\right) = \frac{p_d}{w} + \left(\frac{2\times2}{2\times9.81}\right) + 0.20 + 0.88$$

or $$\frac{p_d}{w} = 10.89 \text{ m}$$

i.e., the pressure head at pump outlet is 10.89 m.

Example 6:

A centrifugal pump lifts water against a static head of 40 m. of which 4 m is suction lift. The suction and delivery pipes are both 150 mm diameter; the head loss in the suction pipe is 2.3 m and in the delivery pipe, 7.4 m. The impeller is 420 mm diameter and 25 mm wide at the mouth; it revolves at 1200 r.p.m. and its effective vane angle at exit is 35°. If η_{mano} = 52% and η_0 = 72%, determine the discharge relayed by *the pump and power required to drive the pump. Also find the pressure head indicated at the suction and delivery branches of the pump.*

Solution:

If V_p = Velocity in pipes,

then discharge $Q = [\pi(0.15)^{2/4}]\ V_p = 0.0177\ V_p$

Also $Q = \pi D_1 B_1 \times V_{f1}$ (neglecting the vane thickness)

or $Q = (\pi \times 0.42 \times 0.025)\ V_{f1} = 0.0177\ V_p$

or $V_{f1} = 0.537\ V_p \approx 0.54\ V_p$

$u_1 = (\pi D_1 N/60) = (\pi \times 0.42 \times 1200/60)$

$= 26.39$ m/s

From equation 17, since $V_d = V_p$

$$H_m = (h_s + h_d) + h_{fs} + h_{fd} + \frac{V_p^2}{2g}$$

or $$H_m = 40 + 2.3 + 7.4 + \frac{V_p^2}{2g} = 49.7 + \frac{V_p^2}{2g}$$

But $$\eta_{mano} = \frac{gH_m}{V_{w1}u_1};$$

or $$0.82 = \frac{9.81 \times H_m}{u_1(u_1 - V_{f1}\cos\phi)}$$

or $$H_m = \frac{(0.82 \times 26.39)(26.39 - 0.54\ V_p \cot 35°)}{9.81}$$

Equating the two values of H_m, we have

$$49.7 + \left(\frac{V_p^2}{2 \times 9.81}\right) = \frac{(0.82 \times 26.39)(26.39 - 0.54\ V_p\ 1.428)}{9.81}$$

Solving this equation, we get

$V_p = 4.42$m/s

Hence discharge

$$Q = (0.0177 \times 4.42) = 0.078\ m^3/s$$

$$H_m = \left(49.7 + \frac{(4.42)^2}{2 \times 9.81}\right) = 50.70\text{m}$$

$$\text{Shaft Power} = \frac{wQH_m}{\eta_0}$$

$$= \frac{9810 \times 0.078 \times 50.70}{0.72}$$

$$= 53.881 \text{ W} = 53.881 \text{ kW}$$

Since velocity head

$$= \frac{(4.42)^2}{2 \times 9.81} = 0.996\text{m} \approx 1\text{m}$$

Thus at pump suction the pressure head is given by equation 7 as

$$\frac{p_s}{w} = -\left(\frac{V_s^2}{2g} + h_s + h_{fs}\right)$$

Similarly at pump delivery the pressure head is given by equation 13 as

$$\frac{p_d}{w} = \left(H_m + \frac{P_s}{w}\right)$$

Example 7:

Water is pumped from a low level reservoir through a main pipe pipeline of 0.45 m diameter and 1400 m length. The pump is located at the low level reservoir. At a point along the main line at a distance of 450 m from the high level reservoir, a branch line of 0.3 m diameter and 360 m length takes off to discharge 180 Us in the atmosphere.

Level of water surface in high level reservoir = + 30.00 m

Level of water in the open end of the 0.3 m diameter branch line = 25.50 m

Level of water in the low level reservoir = 75.00 m

Darcy's frictional coefficient for both pipes = 0.032

Determine the flow rate into the high level reservoir and the theoretical power of the pump assuming the delivery valve of the pump to be at + 20.00 m.

Solution:

Let Q be the total discharge delivered by the pump; Q_1 be the discharge through the branch pipe delivered in the atmosphere; and Q_2 be the discharge delivered to the high level reservoir.

Thus by continuity equation, we have

$$Q = Q_1 + Q_2$$

But $$Q_1 = 1801/s = 0.18m^3/s$$

$$Q = 0.18 + 62 \qquad ...(i)$$

From equations 7, 8 and 9, we have with usual notation

$$O = \frac{p_s}{w} + \frac{V_s^2}{2g} + h_s + h_{fs} \qquad ...(ii)$$

$$\frac{p_s}{w} + \frac{V_s^2}{2g} + \frac{V_{w1}u_1}{g} = \frac{p_s}{w} + \frac{V_1^2}{2g} + h_{Li} \qquad ...(iii)$$

$$\frac{p_2}{w} + \frac{V_1^2}{2g} = \frac{p_d}{w} + (20 - 18 - h_s) + h_{Lc} + \frac{V_d^2}{2g} \qquad ...(iv)$$

Adding equations (ii), (iii) and (iv) we get

$$\frac{V_{w1}u_1}{g} - h_{Li} - h_{Lc} = \frac{p_d}{w} + \frac{V_d^2}{2g} + 2 + h_{fs}$$

or $$H_m = \frac{p_d}{w} + \frac{V_d^2}{2g} + 2 + h_{fs} \qquad ...(v)$$

Since the pump is located close to the low level reservoir, h_{fs} may be neglected. Thus from equation (v) we have

$$\frac{p_d}{w} + \frac{V_d^2}{2g} = H_m - 2$$

Let h_f be the head loss due to friction in the main pipeline between the delivery point of the pump and the point where the branch line takes off; h_{f1} be the head loss due to friction in the branch line; and h_{f2} be the head loss due to friction in the main pipeline between the point where the branch line takes off and the high level reservoir.

Thus applying Bernoulli s equation between the delivery point of the pump and the high level reservoir, neglecting the minor losses we get

$$\frac{p_d}{w} + \frac{V_d^2}{2g} + 20 = 30 + h_f + h_{f2}$$

or $$\frac{p_d}{w} + \frac{V_d^2}{2g} = 10 + h_f + h_{f2} \qquad ...(vii)$$

Similarly applying Bernoulli's equation between the delivery point of the pump and the open end of the branch line, neglecting the minor losses, we get

$$\frac{p_d}{w} + \frac{V_d^2}{2g} = 20.5 + h_f + h_{f1}$$

or $$\frac{p_d}{w} + \frac{V_d^2}{2g} = 5.5 + h_f + h_{f1} \qquad ...(viii)$$

From equations (vii) and (viii), we have

$$h_{fl} - h_{fl} = 4.5 \qquad ...(ix)$$

The velocity of flow in the branch line is

$$V_1 = \frac{0.18}{(\pi/4)\,(0.3)^2} = 2.546 \text{ m/s}$$

$\therefore$ $$h_{f1} = \frac{0.032 \times 360 \times (2.546)^2}{2 \times 9.81 \times 0.3} = 12.687 \text{ m}$$

Thus from equation (ix), we get

$$h_{fl} = (12.687 - 4.5) = 8.187\text{m}$$

If V_2 is the velocity of flow in the main pipeline between the branch line and the high level reservoir, then

$$h_{f2} = \frac{0.032 \times 450 \times V_2^2}{2 \times 9.81 \times 0.45}$$

or $$8.187 = \frac{0.032 \times 450 \times V_2^2}{2 \times 9.81 \times 0.45}$$

$\therefore$ $$V_2 = 2.240 \text{ m/s}$$

Thus $$Q_2 = \frac{\pi}{4} \times (0.45)^2 \times 2.240 = 0.356\text{m}^3/\text{s}$$

$\therefore$ $$Q = (0.18 + 0.356) = 0.536 \text{ m/s}$$

If V is the velocity of flow in the main pipeline between the delivery point of the pump and the point where the branch line takes off, then

$$V = \frac{0.536}{\left(\frac{\pi}{4}\right)(0.45)^2} = 3.370 \text{ m/s}$$

and $$h_f = \frac{0.032 \times (1400 - 450) \times (3.370)^2}{2 \times 9.81 \times 0.45}$$

$$= 39.104\text{m}$$

Thus from equation (vii), we have

$$\frac{p_d}{w}+\frac{V_d^2}{2g}=(10+39.104+8.187)=57.291$$

Thus from equation (vi), we have

$$H_m=(57.291+2)=59.291 \text{ m}$$

Theoretical power of the pump is given by

$$P=wQH_m$$

in SI units

$$P=\frac{9.810\times0.536\times59.291}{1000}=311.762\text{kW}$$

in metric units

$$P=\frac{1000\times0.536\times59.291}{75}=423.733\text{h.p.}$$

Example 8:

A centrifugal pump running at 1400 r.p.m., has the characteristics as indicated below :

Discharge, Q litres/second	*12.5*	*18.7*	*25.0*	*31.3*	*37.5*	*43.8*	*50.0*
Head H_m metres	*28.3*	*27.4*	*26.4*	*25.0*	*23.4*	*20.7*	*18.0*
η_0 (percent)	*65*	*70*	*73*	*74*	*72*	*69*	*63*

Draw the operating characteristics of the pump and determine its specific speed. If the pump is driven by a motor, determine the power of the motor when operating at maximum efficiency.

Solution:

The curves of head and efficiency are plotted on a base of discharge shown in the accompanying figure.

From the plot, maximum η_0 = 74.1% and the corresponding

$$H_m=25.3 \text{ m and } Q=30.4 \text{ l/s.}$$

$$\therefore \quad N_s=\frac{N\sqrt{Q}}{H_m^{\;3/4}}$$

$$=\frac{1400\times(30.4)^{1/2}}{(25.3)^{3/4}}=684.3$$

$$\eta_0=\frac{(wQH_m)}{P}$$

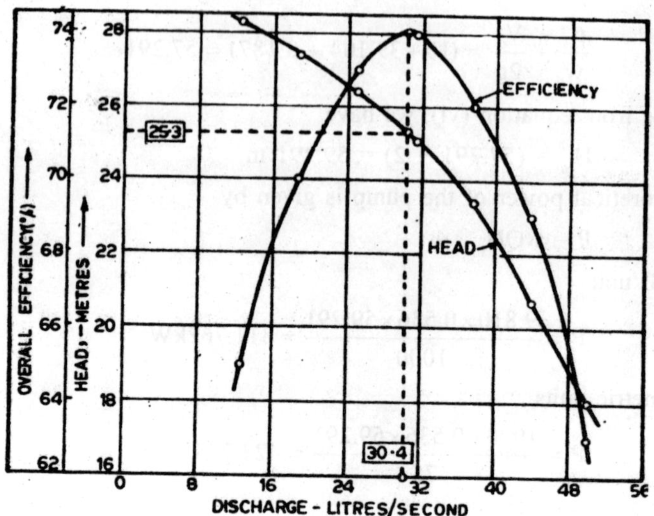

Fig. 6.16

$$\therefore \text{ Power } P = \frac{(wQH_m)}{\eta_0}$$

$$= \frac{9810 \times (30.4 \times 10^{-3}) \times 25.3}{0.741}$$

$$= 10\ 182\text{W} = 10.182\text{kW}$$

Example 9:

Determine the number of the impellers required for a multistage pump to lift 4 200 litres/minute against a total head of 185 m. at a speed of 750 r.p.m. The specific speed is not to exceed 700.

Solution:

The limiting N_s for each impeller = 700

Thus $$N_s = \frac{N\sqrt{Q}}{H_m^{3/4}};$$

or $$700 = \frac{750\left(\frac{4200}{60}\right)^{1/2}}{H_m^{3/4}}$$

$\therefore$ Head per stage $= H_m = 18.62$ m

$$\text{Number of stages} = \frac{185}{18.62} = 10.$$

Example 10:

Water is to be pumped out of a deep well under a total head of 95 m. A number of identical pumps of design speed 1000 r.p.m. and specific speed 900 r.p.m., with a rated capacity of 150 litres/second are available. How many pumps will be needed and how should they be connected ?

Solution:

If//m is the manometric head developed by each pump then since

$$N_s = \frac{N\sqrt{Q}}{H_m^{3/4}};$$

$$900 = \frac{1000(150)^{1/2}}{H_m^{3/4}}$$

$$\therefore \quad H_m = 32.5\text{m}$$

$$\therefore \text{ Number of pumps needed} = \frac{95}{32.5}$$

Since the total head required to be developed is more than the head developed by each pump, the pumps should be connected in series.

Example 11:

Two centrifugal pumps have the head and discharge character sties as follows :

Discharge Q (litres/second)	*0*	*4*	*8*	*12*	*16*	*20*	*24*
Pump 1 Head H_{m1} (metres)	*50.0*	*51.8*	*50.8*	*48.0*	*42.5*	*32.5*	*18.3*
Pump 2 Head H_{m2} (metres)	*46.7*	*45.9*	*44.2*	*40.3*	*34.3*	*26.0*	*17.0*

Both pumps are installed together and are required to pump water through a pipe 150 mm diameter having $f = 0.02$. Calculate the heads under which pumps are working and discharge in litres per second pumped by them if :

(a) The pumps are connected in series : static lift is 65 m and the suction and the delivery pipes are 800 m long.

(b) *The pumps are connected in parallel; static lift is 15 m and the suction and the delivery pipes are 360 m long.*

Solution:

(a) When the pumps are connected in series then neglecting the velocity head in the delivery pipe, the total head required to be developed by both the pumps is

$$H_m = (h_s + h_d) + (h_{fs} + h_{fd})$$

or $$H_m = 65 + \frac{0.02 \times 800 \times (Q \times 10^{-3})^2}{[\pi(0.15)^2/4]^2 \times 2 \times 9.81 \times 0.15}$$

or $$H_m = 65 + \frac{Q^2}{57.44} \qquad ...(i)$$

where Q is the discharge in litres/second. The computed values of H_m for different values of Q in respect of equation (i) are as given below:

Q	0	4	8	12	16	20	24
H_m	65.0	65.3	66.1	67.5	69.5	72.0	75.0

Curves of H_{m1}, H_{m2} and $(H_{m1} + H_{m2})$ are plotted on a base of Q as shown in the accompanying figure (a). On the same graph H_m, v/s Q curve in respect of the above noted values is also plotted. It may be seen that the curves of H_m and $(H_{m1} + H_{m2})$ intersect at a point corresponding to which Q = 17.6 litres/second; H_{m1} = 39.0 m and H_{m2} = 31.5 m.

(b) Let Q_1 and Q_2 the discharges in litres/second, flowing through the pumps 1 and 2 respectively which are connected in parallel. Then the total discharge flowing through the suction and delivery pipes = $(Q_1 + Q_2)$.

The total head required to be developed is

$$H_m = (h_s + h_d) + (h_{fs} + h_{fd})$$

or $$H_m = 15 + \frac{0.02 \times 360 \times [(Q_1 + Q_2) \times 10^{-3})^2}{[\pi(0.15)^2/4]^2 \times 2 \times 9.81 \times 0.15}$$

or $$H_m = 15 + \frac{(Q_1 + Q_2)}{127.64} \qquad ...(ii)$$

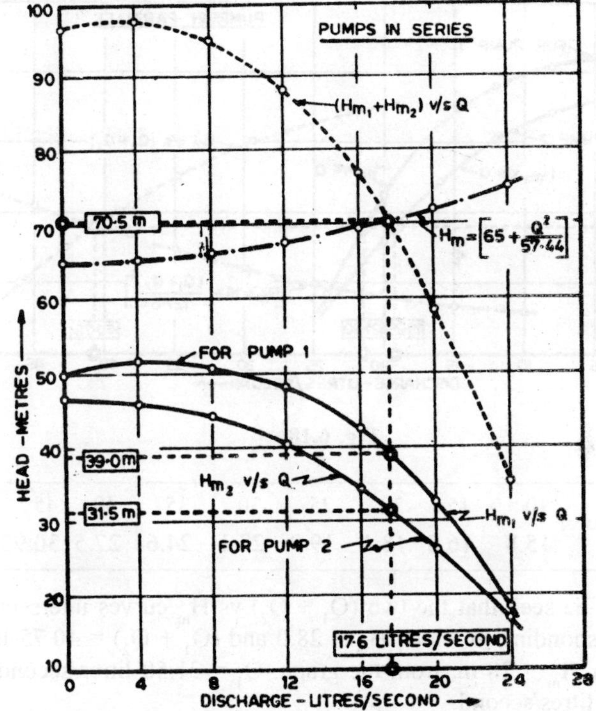

Fig. 6.17

Curves of H_{m1} and H_{m2} are plotted on the base of Q as shown in the accompanying figure (b). From these curves for the same values of H_m, the values of Q_1, Q_2, and $(Q_1 + Q_2)$ are obtained which are as given below. Curve of $(Q_1 + Q_2)$ v/s H_m is also plotted on the same graph.

H_m	20	25	30	35	40	45
Q_1	23.4	22.3	20.8	19.2	17.2	14.5
Q_2	22.7	20.5	18.2	15.6	12.2	6.0
$(Q_1 + Q_2)$	46.1	42.8	39.0	34.8	29.4	20.5

Similarly for different values of $(Q_1 + Q_2)$ the corresponding value of H_m in respect of equation (ii) are computed as indicated below; and these are also plotted on the same graph.

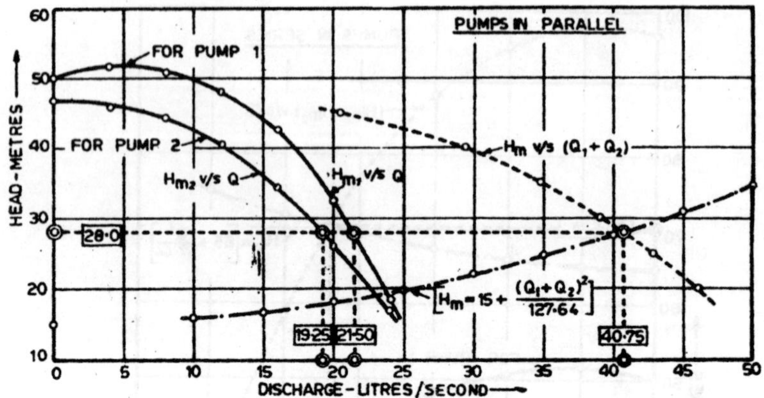

Fig. 6.18

$(Q_1 + Q_2)$	10	15	20	25	30	35	40	45	50
H_m	15.8	16.8	18.1	19.9	22.1	24.6	27.5	30.9	34-6

It may be seen that the two $(Q_1 + Q_2)$ vs H_m curves intersect at a point corresponding to which $H_m = 28.0$ and $(Q_1 + Q_2) = 40.75$ litres/second. For $H_m = 28$ m, from the graph, $Q_1 = 21.50$ litres/second and $Q_2 = 19.25$ litres/second.

Example 12:

A three-stage centrifugal pump has impellers 375 mm diameter and 18 mm wide at outlet. The outlet vane angle is 45° and the vanes occupy 8% of the outlet area. The manometric efficiency is 84% and the overall efficiency is 75%. What head the pump will generate when running at 900 r.p.m., discharging 60 litres/second ? What is the input power?

Solution:

$$V_{f1} = \frac{60+10^{-3}}{\pi \times 0.375 \times 0.018 \times 0.92}$$

$$= 3.08 \text{ m/s}$$

$$u_1 = \left[\frac{(\pi D_1 N)}{60}\right]$$

$$= \frac{\pi \times 0.375 \times 900}{60}$$

$$= 17.67 \text{ m/s}$$

Now from the outlet velocity triangle of Fig. 5,

$$V_{w1} = (u_1 - Vf_1 \cot 45^o) = (17.67 - 3.08)$$

$$= 14.59 \text{ m/s}$$

$$\eta_{mano} = \frac{gH_m}{V_{w1}u_1}$$

or $$0.84 = \frac{9.81 \times H_m}{14.59 \times 17.66}$$

$$H_m = 22.06 \text{ m}$$

$\therefore$ Total head $H = 3H_m = 3(22.06) = 66.18\text{m}$

$$\eta_0 = (wQH)/P$$

or $$P = \frac{9810 \times 60 \times 10^{-3} \times 66.18}{.75}$$

$$= 51\ 938, \text{ W} = 51.938 \text{ kW}.$$

Example 13:

A pump impeller is 375 mm diameter and it discharges water with component velocities of 2 and 12 mis in the radial and tangential directions respectively. The impeller is surrounded by a concentric cylindrical chamber with parallel sides the outer diameter being 450 mm. If the flow in this chamber is a free spiral vortex, find the component velocities of water on leaving and increase in pressure if there is no loss.

Solution:

For free cylindrical vortex from equation 7.57

or $$12 \times (187.5) = V_2 \times (225)$$

$$V_2 = \frac{12 \times 187.5}{225} = 10 \text{ m/s}$$

For radial flow from equation 7.65

$$V_1R_1 = V_2R_2$$

$$2 \times (187.5) = V_2 \times (225)$$

$$V_2 = \frac{2 \times 187.5}{225} = 1.67 \text{ m/s}$$

Hence the component velocities of water on leaving are 1.67 m/s and 10 m/s in the radial and tangential directions respectively.

For spiral vortex motion the increase in pressure is given by equation 7.71.

$$\frac{p_2}{w_2} - \frac{p_1}{w} = \left[\frac{(12)^2 - (10)^2}{2\times 9.81}\right] + \left[\frac{(2)^2 - (1.67)^2}{2\times 9.81}\right] = 2.304\text{m}$$

EXERCISES

1. Explain with neat sketches the working of a single-stage centrifugal pump.
2. Under what headings the centrifugal pumps are classified?
3. State the difference between a closed, semi-closed and open impeller.
4. Discuss the various methods adopted to increase the efficiency of a centrifugal pump by altering the shape of the casings of chamber surrounding the impeller.
5. Why is the efficiency of a volute casing of a centrifugal pump as an energy conversion device low? How is a whirlpool or vortex chamber superior in performance?
6. If a centrifugal pump does not deliver any water when started, what may be the probable causes and how can they be remedied?
7. Define static and manometric head of a centrifugal pump. State the different types of head losses which may occur in a pump installation.
8. What are the different efficiencies of a centrifugal pump?
9. What is meant by 'Priming' of a pump? What are the different priming arrangements employed for small and big pumping units?
10. Why can the suction lift of a pump not exceed a certain limit?
11. A centrifugal pump of the radial type delivers 5 000 litres per minute against a total head of 38 m when running at a speed of 1450 r.p.m. If the outer diameter of Ac impeller is 300 mm and its width at the outer periphery is 13 mm, find the vane angle at exit, Assume manometric efficiency as 80%.
12. The impeller of a centrifugal pump has 1.2 m outside diameter. It is used to lift 1800 litres of water per second against a head of 6 m. Its vanes make an angle of 150° with the direction of

motion at outlet and runs at 200 r.p.m If the radial velocity of flow at outlet is 2.5 m/s, find the manometric efficiency. Also find the lowest speed to start the pump, if the diameter of the impeller at inlet is equal to half the diameter at exit.

13. A centrifugal pump is required to deliver 280 litres of water per second against a head of 16.m. If the vanes of the impeller are radial at outlet and the velocity of flow is constant equal to 2 m/s, find the proportions of the pump. Assume η_{mano} = 80% and the ratio of breadth to diameter at outlet as 0.1.

14. A centrifugal pump is required to discharge 600 litres of water per second against a head of 12 m, when running at a speed of 750 r.p.m. The manometric efficiency is to be 80%, the loss of head in the pump being assumed to be 0.025 V_1^2 of water, where V_1 is the absolute velocity of water leaving the impeller. Water enters the impeller without whirl and the velocity of flow is 3 m/s. Determine (a) the impeller diameter and the outlet area; and (b) the vane angle at the outlet edge of the impeller.

15. A centrifugal pump running at 1 450 r.p.m. has the characteristics as given below:

Discharge (Litres/second)	11.3	16.9	22.6	28.3	34.0	39.6	45.2
Head (metres)	25-8	25.0	24.1	23.2	21.4	18.9	15.8
Efficiency (%)	65	70	73	74	72	69	62

Draw the operating characteristics of the pump and determine its specific speed.

The pump lifts water against a static head of 12 m through a long pipeline in which the loss of head in metres, due to friction is given by the expression, hr = 0.012 Q^2, where Q is the discharge in litres/sec. The minor losses in the pipe may be neglected. Determine the power required to drive the pump.

16. The axis of a centrifugal pump is 2.5 m above water level in the sump and the static lift from the pump centre is 33.5 m. The friction losses in the suction and the delivery pipes are 1 m and 7.5 m respectively. The suction and delivery pipes are each 75mm diameter. The impeller is 300 mm diameter and 18 mm wide at outlet and its speed is 1700 r.p.m. The water at inlet has

radial flow and the vane angle at outlet is 32° to the tangent to the periphery. Compute the discharge of the pump and the power required to drive the pump. Assume $\eta_{mano} = 77\%$ and $\eta_0 = 72\%$.

17. A four-stage centrifugal pump has impellers 380 mm diameter and 19 mm wide at outlet. The outlet vane angle is 45° and the vanes occupy 8% of the outlet area. The manometric efficiency is 84% and the overall efficiency 72%. Determine the head generated by the pump when running at 900 r.p.m., and discharging 59 litres/second. Also determine the power required to drive the pump.

18. A two-stage centrifugal pump is required for a fire engine for a duty of 3 660 litres per minute at a head of 75 m. If the overall efficiency of the pump is 75% and the specific speed per stage is about 1 300, find (a) the running speed in r.p.m. and (b) the power of the driving engine. If the actual manometric head developed is 65% of the theoretical head, there is no slip, the outlet vane angle is 30°, and the radial flow velocity at exit is 0-15 times the tip speed at exit, find the diameter of the impellers.

19. A multi-stage centrifugal pump is to be designed to deliver 700 litres of water per second against a manometric head of 60 m. There are to be four equal impellers keyed to the same shaft, which has speed of 350 r.p.m. The vanes are curved back so that the direction of the relative velocity at exit makes an angle of 120° with the direction of the corresponding peripheral velocity and the impeller is surrounded by guides. Assume that the water enters radially and the velocity of flow through the impeller is 0.27 of the peripheral velocity and that the losses in the pump amount to one-third of the velocity head at discharge from the impeller. Find : (a) the outer dimensions of the impeller; (b) the manometric efficiency; (c) the angle of the guides.

20. Two centrifugal pumps operating in parallel are employed for lifting water against a static head of 18 m, the delivery pipe being 0.5 m in diameter and 15000 m long having f = 0.032. Determine the head at the pipe inlet and the discharge; given the following data for the two pumps :

Q Litres/second	0	15	30	45	60
Pump 1 H_{m1} metres	45.0	43.5	40.0	32.7	23.3
Pump 2 H_{m2} metres	40.7	39.0	33.7	27.3	20.7

Explain (i) double suction impeller, (ii) shut-off head, (iii) manometric efficiency, (iv) net positive suction head as applied to a suction pump.

21. Assuming that the radial component of flow of fluid through a centrifugal pump remains constant and that the fluid enters radially, prove that the ratio of pressure head H_p to velocity head H_v created by an impeller neglecting the losses is given by an equation

$$\frac{H_p}{H_v} = \left(\frac{u_1 - V_{f1} \cot \beta}{u_1 - V_{f1} \cot \beta}\right)$$

where u_1 is rim speed of impeller at outlet, V_{f1} is velocity of flow and p is the exit angle of impeller blades. Hence show that ideal efficiency of the impeller is given by

$$\eta_0 = \frac{u_1 - V_{f1} \cot \beta}{2u_1}$$